Advanced Optical Techniques in Biosciences

Advanced Optical Techniques in Biosciences is an interdisciplinary resource that deals with the in-depth analysis of several industrially and economically important physicochemical properties that have a wide range of applications for the pharmaceutical sector, food sciences, the textile industry, drug delivery, and tissue engineering. This comprehensive volume explores cutting-edge optical technologies revolutionizing biomedical research and diagnostics. From molecular-level investigations using carbon nanotubes to advanced imaging techniques like fluorescence lifetime imaging microscopy (FLIM) and quantitative phase imaging (QPI), readers will discover how optical science is transforming our understanding of biological systems. The text covers innovative approaches, including photoacoustic imaging for regenerative medicine, quantum dot applications in high-resolution microscopy, and terahertz spectroscopy for biomedical research. Practical applications range from analyzing millet starch digestibility and detecting adulterated vegetable oils to quantifying exosomes and adapting expansion microscopy for virus research. The final chapter examines how deep learning (DL) models are enhancing computer-assisted diagnosis of Alzheimer's disease, highlighting the intersection of optical technologies with artificial intelligence. This essential resource provides graduate students, researchers, and clinicians with a thorough understanding of modern optical diagnostic tools and their transformative potential in addressing critical biomedical challenges.

Key Features:

- Provides comprehensive coverage of cutting-edge optical technologies.
- Discusses interdisciplinary applications across multiple industries.
- Highlights the integration of artificial intelligence into optical technologies.

Dr. Nirmal Mazumder is an Associate Professor at the Department of Biophysics, Manipal School of Life Sciences, Manipal Academy of Higher Education (MAHE), Manipal, India. His research interests include the development of microscopy-spectroscopy techniques, biomedical devices, biophotonics, smart materials, and microfluidics and its biomedical applications. He has published more than 125 research articles in the peer-reviewed international journals, 20 books, 45 book chapters, and several poster/oral/invited presentations in National and International conferences.

Dr. Yury V. Kistenev is a Professor at the Tomsk State University (TSU). He received his PhD degree in optics in 1987 and a doctoral degree in physics and mathematics in 1997. He is the author of more than 150 journal papers and has written four books as author and editor. His current research interests include laser molecular imaging, laser spectroscopy, and machine learning.

Dr. Igor K. Lednev is a Distinguished Professor and Williams-Raycheff Endowed Chair in Chemistry at the University at Albany, State University of New York. He earned his PhD from the Moscow Institute of Physics and Technology. Dr. Lednev is the founding director of the Center for Biophotonics Technology and Artificial Intelligence and co-founder of two startup companies. He served as an advisory member on the White House Subcommittee for Forensic Science and has co-authored over 300 publications and 12 patents, achieving an h-index of 80. His honors include the Charles Mann Award for Applied Raman Spectroscopy and the Gold Medal Award from the Society for Applied Spectroscopy.

Series in Optics and Optoelectronics

Handbook of Optoelectronics, Second Edition (Three-Volume Set)
John P. Dakin, Robert G. W. Brown

Optical MEMS, Nanophotonics, and Their Applications
Guangya Zhou, Chengkuo Lee

Thin-Film Optical Filters, Fifth Edition
H. Angus Macleod

Laser Spectroscopy and Laser Imaging: An Introduction
Helmut H. Telle, Ángel González Ureña

Fourier Optics in Image Processing
Neil Collings

Holography
Principles and Applications
Raymond K. Kostuk

An Introduction to Quantum Optics, Second Edition
Photon and Biphoton Physics
Yanhua Shih

Polarized Light and the Mueller Matrix Approach, Second Edition
José J. Gil, Razvigor Ossikovski

Introduction to Holography, Second Edition
Vincent Toal

Optical Coherence Tomography in Dentistry
Scientific Developments to Clinical Applications
Anderson S. L. Gomes, Denise M. Zezell, Cláudia C. B. O. Mota, John M. Girkin

Holography: Principles and Applications, Second Edition
Raymond K. Kostuk

CAD-Based Optical Design with Quadoa
Rafael G. González-Acuña

Anisotropy of Metamaterials: Beyond Conventional Paradigms
Vladimir Mityushev, Tatjana Gric, Radoslaw Kycia and Natalia Rylko

Optical Science and Engineering
Elias Glytsis

Advanced Optical Techniques in Biosciences
Nirmal Mazumder, Yury V. Kistenev and Igor K. Lednev

For more information about this series, please visit: https://www.crcpress.com/Series-in-Optics-and-Optoelectronics/book-series/TFOPTICSOPT

Advanced Optical Techniques in Biosciences

Edited by
Nirmal Mazumder, Yury V. Kistenev
and Igor K. Lednev

CRC Press
Taylor & Francis Group
Boca Raton London New York

CRC Press is an imprint of the
Taylor & Francis Group, an **informa** business

Designed cover image: Nirmal Mazumder, Yury V. Kistenev and Igor K. Lednev

First edition published 2026
by CRC Press
2385 NW Executive Center Drive, Suite 320, Boca Raton FL 33431

and by CRC Press
4 Park Square, Milton Park, Abingdon, Oxon, OX14 4RN

CRC Press is an imprint of Taylor & Francis Group, LLC

ISBN: 9781032937687 (hbk)
ISBN: 9781032943008 (pbk)
ISBN: 9781003570011 (ebk)

DOI: 10.1201/9781003570011

Typeset in Times
by codeMantra

Contents

Contributors

Maryam Abdul Ajees
Department of Biophysics, Manipal School of Life Sciences
Manipal Academy of Higher Education
Manipal, India

M. M. Balamurali
Chemistry Division, School of Advanced Sciences
Vellore Institute of Technology
Chennai, India

Shyju Bhaskar
Department of Biophysics, Manipal School of Life Sciences
Manipal Academy of Higher Education
Manipal, India

Vyasraj G. Bhat
Department of Biophysics, Manipal School of Life Sciences
Manipal Academy of Higher Education
Manipal, India

Arghya Bhattacharya
Department of Pharmacology
Bengal School of Technology
Sugandha, India

Rajib Biswas
Department of Physics
Tezpur University
Tezpur, India

Alexey V. Borisov
Laboratory of Laser Molecular Imaging and Machine Learning
National Research Tomsk State University
Tomsk, Russia

Ishita Chakraborty
Department of Biophysics, Manipal School of Life Sciences
Manipal Academy of Higher Education
Manipal, India

Kaushik Chanda
Department of Chemistry
Rabindranath Tagore University
Bhopal, Assam, India

Olga P. Cherkasova
Institute of Automation and Electrometry, Siberian Branch
Russian Academy of Sciences
Novosibirsk, Russia
and
National Research Centre "Kurchatov Institute"
Moscow, Russia

Baishali Das
Department of Biophysics, Manipal School of Life Sciences
Manipal Academy of Higher Education
Manipal, India

Subir Das
Department of Chemistry
University of Zurich
Zurich, Switzerland

Budheswar Dehury
Department of Bioinformatics, Manipal School of Life Sciences
Manipal Academy of Higher Education
Manipal, India

Gitanjal Deka
Prospexis Research PVT. Ltd
Bhangagarh, Dispur, Assam, India

Sruti Bagchi Ghosh
Department of Pharmaceutical Chemistry
Calcutta Institute of Pharmaceutical Technology & AHS
Howrah, India

Ankur Gogoi
Department of Physics
Jagannath Barooah University
Jorhat, India

Vibha Kamath
Department of Biophysics
Manipal School of Life Sciences, Manipal Academy of Higher Education
Manipal, India

Manikanth Karnati
Department of Biophysics, Manipal School of Life Sciences
Manipal Academy of Higher Education
Manipal, India

Talitha K. Kurian
Department of Biophysics, Manipal School of Life Sciences
Manipal Academy of Higher Education
Manipal, India

Tatiana B. Lepekhina
Laboratory of Laser Molecular Imaging and Machine Learning
National Research Tomsk State University
Tomsk, Russia

Nandana Bijulal
Department of Biophysics, Manipal School of Life Sciences
Manipal Academy of Higher Education
Manipal, India

Viktor V. Nikolaev
Laboratory of Laser Molecular Imaging and Machine Learning
National Research Tomsk State University
Tomsk, Russia

Hemanth Noothalapati
Faculty of Life and Environmental Sciences
Shimane University
Nishikawatsu-Cho, Japan
and
Department of Chemical Engineering
Indian Institute of Technology Hyderabad
Hyderabad, India
and
Department of Biomedical Engineering
Chennai Institute of Technology
Chennai, India

Ajeetkumar Patil
Manipal Institute of Applied Physics
Manipal Academy of Higher Education (MAHE)
Manipal, India

Luca Pesce
CISUP
University of Pisa
Pisa, Italy

Varad Mahesh Phadke
Manipal Institute of Applied Physics
Manipal Academy of Higher Education (MAHE)
Manipal, India

Md. Hafizur Rahman
Department of Quality Control and Safety Management
Faculty of Food Sciences and Safety
Khulna Agricultural University
Khulna, Bangladesh

Pietro Ricci
Department of Physics and Astronomy
University of Florence
Florence, Italy
and
Lab. Europeo di Spettroscopie Non-Lineari
LENS
Sesto Fiorentino, Italy

Gagan Raju
Department of Biophysics, Manipal School of Life Sciences
Manipal Academy of Higher Education
Manipal, India
and
Manipal School of Information and Science
Manipal Academy of Higher Education (MAHE)
Manipul, India

Santanu Roy
Department of Computer Science (Data Science), CSDS
Dayananda Sagar College of Engineering
Bangalore, India

Norma Sammán
Centro Interdisciplinario de Investigaciones en Tecnologías y Desarrollo Social para el NOA
Facultad de Ingeniería
Universidad Nacional de Jujuy
San Salvador de Jujuy, Argentina

Giuseppe Sancataldo
Department of Physics and Chemistry
University of Palermo
Palermo, Italy

Smritimala Sarmah
Department of Physics
B. Barooah College
Assam, India

Kazuo Umemura
Tokyo University of Science
Tokyo, Japan

Valeria Vetri
Department of Physics and Chemistry
University of Palermo
Palermo, Italy

Olga A. Zakharova
Laboratory of Laser Molecular Imaging and Machine Learning
National Research Tomsk State University
Tomsk, Russia

Guan-Yu Zhuo
Institute of Biophotonics
National Yang Ming Chiao Tung University
Taipei, Taiwan

Hala Zuhayri
Laboratory of Laser Molecular Imaging and Machine Learning
National Research Tomsk State University
Tomsk, Russia

1 Biosciences at Molecular Levels Illuminated by Near-Infrared Optical Responses of Single-Walled Carbon Nanotubes

Kazuo Umemura
Tokyo University of Science

1.1 UNIQUE OPTICAL RESPONSES OF NANO CARBONS

Some specific nano carbons such as single-walled carbon nanotubes (SWNTs) have unique optical responses (Segawa et al. 2016; Panwar et al. 2019). In particular, photoluminescence (PL) and absorbance of SWNTs in near-infrared (NIR) areas are beneficial for biological applications (Ackermann et al. 2022; Krasley et al. 2024; Wang et al. 2019). Because most of the optical responses of biomolecules are not within the NIR areas, it is easy to distinguish the optical responses of SWNTs and biomolecules. Various biosensing devices using SWNTs have been reported.

Although many researchers of SWNTs have focused on the development of practical biodevices such as biosensors, the unique optical properties of nano carbons are also applicable for biosciences. In this chapter, SWNT applications for biosciences using specific optical responses will be introduced, as well as those in biotechnologies.

1.2 CHIRALITY OF SWNTS

NTs have extraordinary mechanical structures, electrical properties, and optical properties (Segawa et al. 2016; Sanchez-Valencia et al. 2014). To understand the specific features of SWNTs, graphene sheets can be imagined as the starting point. There are many articles that have described the origins of the specific features of SWNTs. For example, Segawa et al. provided sophisticated illustrations (Segawa et al. 2016). A graphene sheet is the starting point to understand SWNT structures (Figure 1.1). It is a well-known hexagonal carbon structure. Then, you can imagine rolling up the graphene sheet to prepare cylindrical structures. There are various

DOI: 10.1201/9781003570011-1

possible directions to roll up and various possible diameters. It means that various chiralities of SWNTs are available.

Chirality is one of the important concepts to consider for the physicochemical properties of SWNTs (Nugraha et al. 2011; Avery et al. 2016; Venkataraman et al. 2019). Chiralities of SWNTs are defined with chiral vectors:

$$\boldsymbol{C}_h = n\boldsymbol{a}_1 + m\boldsymbol{a}_2$$

Basic lattice vectors $\boldsymbol{a}_1$ and $\boldsymbol{a}_2$ are defined as connecting the two vertices of the hexagonal lattice of graphene structures (see the left corner of Figure 1.1). Chiral indexes m and n are integers to define one period of SWNT cylinders. Geometric structures of SWNTs are determined with the chiral vectors. For example, circumference lengths of SWNT cylinders are magnitudes of chiral vectors:

$$C_h = \sqrt{3}a_{c-c}\sqrt{n^2 + nm + m^2}$$

$\boldsymbol{a}_{c\text{-}c}$ is the carbon atom distance (~0.144 nm). Diameters of SWNTs can be calculated as follows:

$$d_l = \frac{\sqrt{3}a_{c-c}}{\pi}\sqrt{n^2 + nm + m^2}$$

Physicochemical properties of SWNTs are also determined with chiralities. For example, it is known that SWNTs become metallic if $m = n$ (Moon and Kim 2019;

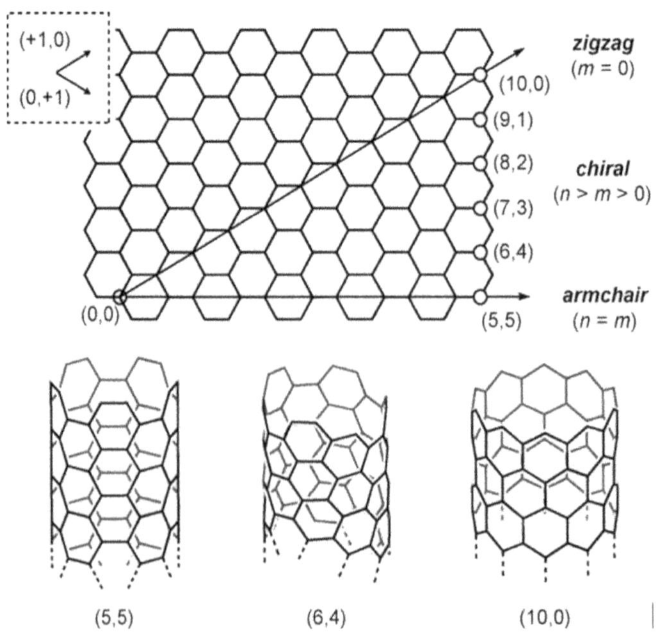

FIGURE 1.1 Chiral index (n,m) for CNTs. (Reprinted with permission from Segawa et al. 2016.)

Anantram and Leonard 2006). Other SWNTs should be zero band gap semiconductors.

From the viewpoints of actual experiments, synthesized carbon nanotube powders include multiple chirality SWNTs. Amorphous carbons and multi-walled carbon nanotubes (MWNTs) are also included in many cases. In addition, defects of SWNT structures have effects on the mechanical, electrical, and optical properties of SWNTs. Preparation of highly purified defect-free SWNTs has been intensively studied, but the use of perfect SWNT powders is not usual for various biological applications (Sanchez-Valencia et al. 2014; Zhang et al. 2017; Yang et al. 2014; Yang et al. 2020; Rao, Voggu, and Govindaraj 2009). As commercial products, (6,5) rich SWNT powders and (7,6) rich SWNT powders, obtained from Sigma Aldrich Co. Ltd., are convenient for practical uses at this moment. Although they contain several other chirality SWNTs, optical responses of (6,5) SWNTs or (7,6) SWNTs stand out compared to other chirality SWNTs.

1.3 PREPARATION OF HYBRIDS OF NTS WITH ORGANIC MOLECULES

Bare SWNT molecules are not water-soluble. Although organic solvents such as 1,2-dichloroethane (DCE), are available to prepare mono-dispersed SWNT suspensions, aggregates are not negligible for long-term storage (Hayashida and Umemura 2016; Sun et al. 2016). For biological applications, preparation of stable aqueous suspensions of SWNTs is one of the important experimental procedures.

There are two major approaches to solubilizing SWNTs. One is wrapping bare SWNTs with water-soluble organic molecules by physisorption (Karachevtsev et al. 2011; Angelikopoulos et al. 2010; Barisci et al. 2004). In this way, resonance structures of SWNT carbons are present, thus, extraordinary optical properties of SWNTs should also be present. In this chapter, this physisorption method will be described. Another method is surface modification of SWNTs by synthetic chemical techniques (Lucente-Schultz et al. 2009; Zhao, Song, and Pehrsson 2002; Pan et al. 2003; Wang et al. 2007). For example, OH- or COOH- groups are attached by covalent bonding. With this approach, stable water-soluble SWNTs can be fabricated. Instead, the optical properties of SWNTs should be affected by the modification of resonant structures of SWNTs.

For preparing aqueous SWNT suspensions, the use of surfactants such as sodium dodecyl sulfate (SDS) and sodium cholate (SC) is a widely known method (Singh et al. 2008; Flavel et al. 2014; Sa and Kornev 2011). As a typical protocol, SWNT powders are added to a surfactant aqueous solution. Then, the mixture of SWNT bundles is isolated by sonication with a probe-type equipment (Blanch, Lenehan, and Quinton 2010; Yamamoto et al. 2008; Fernandes et al. 2015). Surfactant molecules wrap the surfaces of the isolated SWNT molecules. Finally, aggregates and insoluble impurities are removed by centrifugation. The fabricated conjugates are described, such as SDS-SWNT hybrids or SC-SWNT hybrids. The surfactant molecules attach to the SWNT surfaces with physisorption (Xu, Yang, and Yang 2010).

Although preparation methods of mono-dispersed SWNT suspensions using surfactants are convenient, there are several major problems in using the conjugates for biological applications. One of the major problems is the existence of the surfactants themselves. SDS is a typical denaturation agent for protein molecules (Xu and Keiderling 2004; Gudiksen, Gitlin, and Whitesides 2006). Excess surfactant molecules are included in the SWNT suspensions, not just the ones attached to SWNT surfaces. The use of SWNT suspensions with surfactants may have some effects on biomolecules. Stability of the prepared suspensions is another problem. Although the SWNT suspensions with surfactants can be stored, aggregates or precipitation occasionally appear. The adsorption of surfactant molecules is not irreversible, so that detachment of surfactant molecules from SWNT surfaces should constantly occur. If excess surfactant molecules are removed by dialysis or other methods, aggregate formation of SWNTs is expected.

Preparation of mono-dispersed SWNT suspensions has been reported with various organic molecules such as peptides, proteins, fluorescent dyes, artificial polymers, and DNA for various biological applications (Brown, Jespersen, and Nygård 2008; Hirano et al. 2010; Bomboi et al. 2011; Zhang et al. 2013; Kruss et al. 2013; Antonucci, Kupis-Rozmyslowicz, and Boghossian 2017). For example, Koh et al. (2011) demonstrated conjugating several fluorescent dyes with SWNTs. Optical responses and thermal stabilities of the conjugates were affected by the types of dye molecules. Carboxymethyl cellulose (CMC) is one of the popular compounds to solubilize SWNTs (Umemura, Izumi, and Oura 2016; Chia et al. 2021; Chia et al. 2023; Jang et al. 2025; Ito, Yajima, and Homma 2016). Stable suspensions could be prepared, although CMC is a rather cheap compound.

DNA molecules are widely used to solubilize SWNTs as well as surfactants. Both single-stranded DNA (ssDNA) and double-stranded DNA (dsDNA) realize stable DNA-SWNT suspensions (Zheng et al. 2003; Nakashima et al. 2003). Numbers and sequences of DNA bases can be variously designed (Tu et al. 2009). The affinity between nucleotides and carbon atoms of SWNTs has been intensively studied by molecular dynamics (MD) simulations (Johnson, Johnson, and Klein 2008). Molecular models of DNA-SWNT conjugates have been proposed based on the MD, in particular, for the use of ssDNA (Johnson et al. 2006; Karachevtsev et al. 2008; Zhao and Johnson 2007). Figure 1.2 shows a geometry model of DNA on SWNT surfaces proposed by Zhao and Johnson (2007).

Different configurations were expected due to the charges on SWNT surfaces. One of the advantages of the use of ssDNA is chirality-specific adsorption of DNA bases. Tu et al. (2009) reported that several single chirality SWNTs could be purified by attaching DNA molecules that had specific sequences. They suggested, for example, that affinity between (9,1) SWNTs and $(TCC)_{10}$ DNA oligomers should be high. By attaching the specific DNA oligomers, purification of (9,1) SWNTs using ion exchange chromatography (IEX) was realized. Many other combinations were similarly proposed, and actually, 12 different chiralities of SWNTs were well purified by attaching DNA molecules with specific base sequences (Figure 1.3). Absorbance spectra of the purified single-chirality SWNTs revealed beautifully ordered peaks specifically originated from different chiralities.

Morphologies and physicochemical properties of DNA-SWNT hybrids have been characterized by various microscopes and spectroscopies. For an intuitive understanding of the morphology of DNA-SWNT hybrids, here I refer to several reports

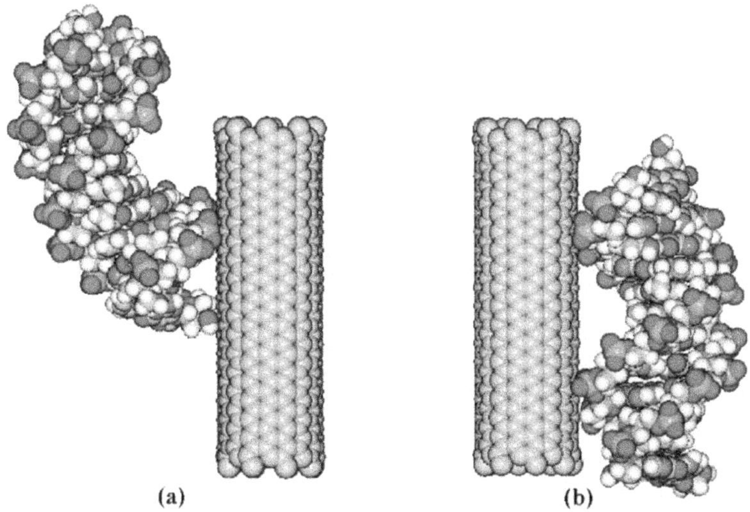

FIGURE 1.2 Adsorption geometries of DNA on a SWNT at $t = 3$ ns, for initial configurations with axes of DNA and SWNT parallel to each other. (a) DNA on an uncharged SWNT; (b) DNA on a positively charged SWNT ($q = +0.05e$ per carbon atom). (Reprinted with permission from Zhao et al. 2007.)

of atomic force microscopy (AFM). If SWNTs were dispersed in volatile organic solvents without surfactants or other organic molecules, it is enough to simply deposit the SWNT suspension on mica or glass substrates. Although bundles might be formed on the substrate surfaces during drying, individual SWNT molecules can be observed by AFM in air. In the case of SWNT suspensions with surfactants, excess surfactant molecules should be removed somehow. In the case of DNA-SWNT hybrids, usually some buffer solutions are used to stabilize pH values. If a DNA-SWNT suspension was dropped on a substrate surface and just dried, salt aggregates/crystals should not be neglected. There are well-known DNA molecules that have a negative charge, which are easy to adsorb on positively charged surfaces. The addition of positive divalent ions to DNA solutions or the functionalization of substrate surfaces with 3-aminopropyltrimethoxysilane (APTES) are effective to stably attach DNA molecules to the substrate surfaces. The scheme is available for DNA-SWNT hybrids. After depositing DNA-SWNT suspension on a positively charged substrate surface, the sample can be rinsed with pure water to remove buffer solutions.

Although many AFM observations of SWNT conjugates were demonstrated in air with dried samples, a few reports have revealed AFM images in liquid (Hayashida and Umemura 2016; Umemura et al. 2016). Hayashida and Umemura (2016) compared the same SWNT molecules in air and in a buffer solution. With dried samples in air, the same SWNT molecules were observed with both attractive and repulsive forces. Figure 1.4 shows typical AFM images of DNA-SWNT hybrids and SWNT molecules. Basically, SWNT molecules were observed as rod-like structures on a mica surface. SWNTs are straight, but flexible structures were expected from the gentle curve features. In the case of DNA-SWNT hybrids, apparent diameters in a

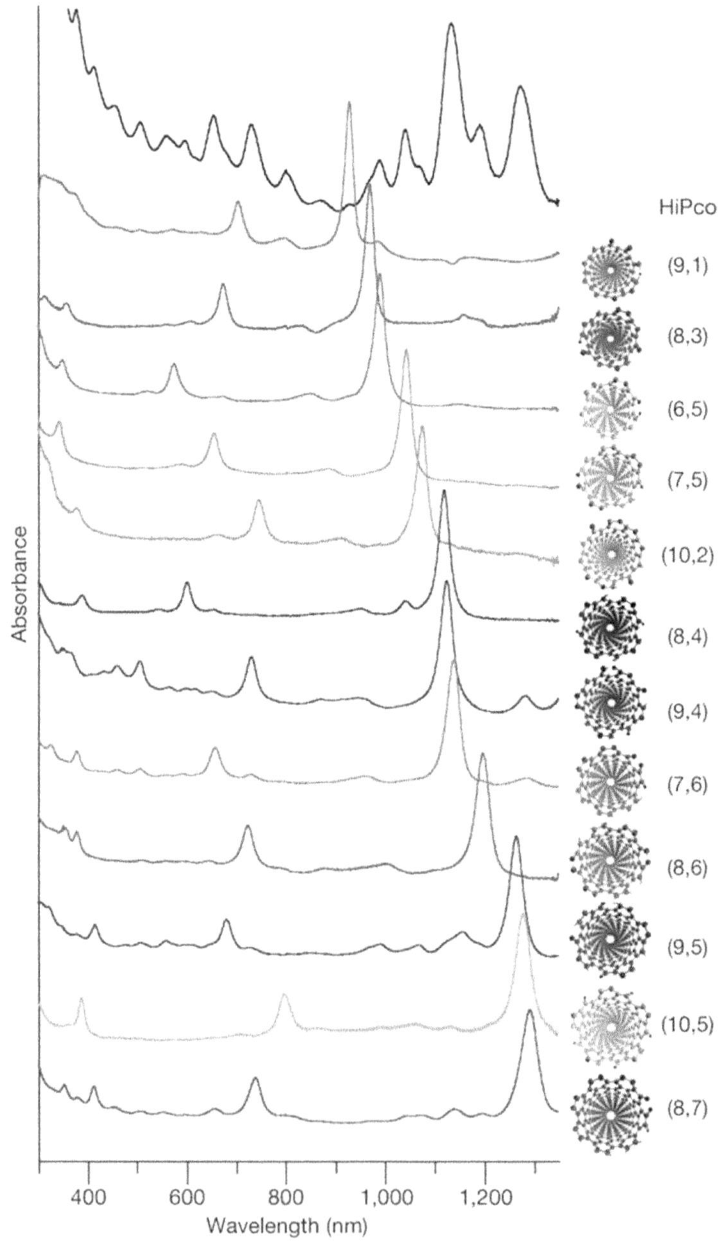

FIGURE 1.3 Ultraviolet–visible–near-infrared absorption spectra of 12 purified semiconducting SWNTs (ranked according to the measured E_{11} absorption wavelength) and the starting HiPco mixture. The structure of each purified SWNT species (viewed along the tube axis) and its (*n*,*m*) notation are given at the right side of the corresponding spectrum. Each purified SWNT was obtained according to procedures described in the Supplementary Information. The elution buffer was replaced with de-ionized water before spectral measurement. The baseline level of each spectrum was offset to facilitate visual comparison. (Reprinted with permission from Tu et al. 2009.)

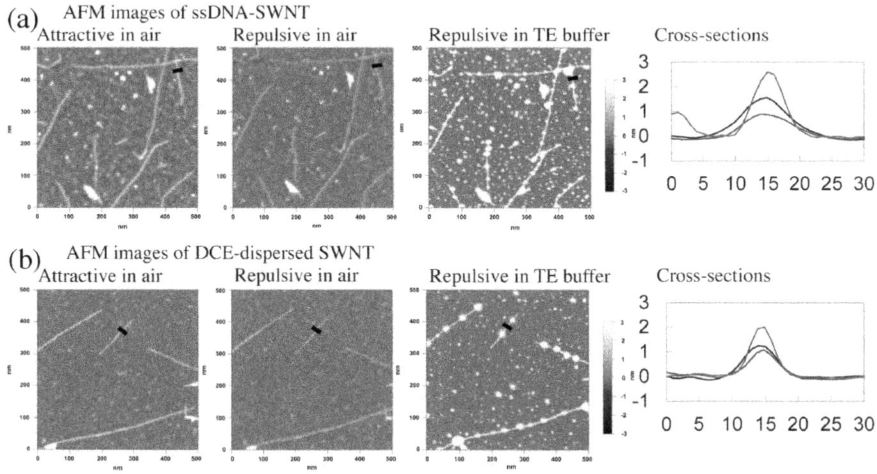

FIGURE 1.4 AFM images and cross-sections of DNA-SWNT hybrids and DCE-dispersed bare SWNTs. (a) ssDNA-SWNT hybrids. (b) DCE-dispersed bare SWNTs. From left to right, LP mode in air, HM mode in air, LM mode in TE buffer solution, and cross-sections of a DNA-SWNT or SWNT from AFM images. In cross-sections, red lines: LM mode in TE buffer solution, black lines: LP mode in air, blue lines: HM mode in air. Scan size was 500 nm × 500 nm. (Reprinted with permission from Hayashida and Umemura 2016.)

buffer solution were more than twice those in air. For establishing various biological applications of SWNTs, more accumulations of microscopic evaluations of SWNT hybrids in aqueous solutions are expected.

1.4 DETECTION OF BIOCHEMICAL REACTIONS BY NEAR-INFRARED SPECTROSCOPY OF NANO CARBONS

Among numerous biological applications of mono-dispersed SWNTs, this chapter introduces research studies focusing on specific optical responses of SWNTs. There are several characteristic optical responses of SWNTs.

Firstly, SWNTs have broad absorbance properties from ultraviolet areas to infrared areas (Krupke et al. 2003; Umemura et al. 2020; Yamazaki and Umemura 2021; Palwai et al. 2007; Gong, Peng, and Liu 2013). Yang et al. (2008) described a low-density nanotube array as an extremely dark material. Secondly, mono-dispersed SWNTs reveal PL in the NIR area (He et al. 2017; Ackermann et al. 2022; Hemmer et al. 2016). Thirdly, quenching of fluorescent dyes adsorbed on SWNT surfaces is a unique phenomenon (Horii, Ohura, and Umemura 2020; Umemura et al. 2017; Hedderman et al. 2004; Koh et al. 2011; Satishkumar et al. 2007; Chiu, Dementev, and Borguet 2011).

One of the pioneer studies, Zhao, Ergul, and Zhao (2015), demonstrated quantitative evaluation of caffeine antioxidant properties using NIR absorption spectra of dsDNA-SWNT suspension. In their work, dsDNA of salmon testes was employed to wrap SWNTs. Commercially available SWNTs from Carbon Nanotechnologies Inc. were used. Figure 1.5 shows examples of absorbance spectra of the dsDNA-SWNT suspension reacting with 0.59 mM hydroxyl radicals in the absence and presence of 50 mM caffeine. Probably several different chiralities of SWNTs were included in the suspension because of the multiple NIR absorbance peaks. It was obvious that absorbance peaks became smaller over time (Figure 1.5a). It could be understood that SWNTs were oxidized with the hydroxyl radicals (Nakayama et al. 2018; Petit et al. 1999; Fagan et al. 2006). The decrease of absorbance peaks was suppressed in the presence of caffeine. It suggests the antioxidant potency of caffeine. They succeeded in quantitatively comparing the antioxidant ability of caffeine, regular coffee, decaffeinated coffee, and vitamin C. Ishibashi et al. (2018) demonstrated similar experiments for Japanese tea and pure catechin, with both absorbance spectra and PL spectra.

There are numerous biological applications using PL of SWNTs (Cohen and Williams 2024; Acharya et al. 2024; Chen et al. 2011; Ackermann et al. 2022; Liu et al. 2009; Xu et al. 2019). For example, Nissler et al. (2021) demonstrated biosensing using PL of SWNTs. They purified (6,5), (7,5), (9,4), and (7,6) SWNTs, which had ~990, 1,040, 1,115, and 1,130 nm PL wavelengths, respectively, based on the aqueous two-phase extraction (ATPE) protocol proposed by Li et al. (2019). They also examined several different DNA sequences to wrap the SWNT surfaces and then found that $(GT)_{40}$-(6,5) SWNT hybrids revealed the maximum fluorescence response to the neurotransmitter dopamine and long-term stability. Figure 1.6 indicates a part of their data. PL intensities of $(GT)_{40}$-(6,5) SWNT hybrids drastically increased when dopamine (10 µM, Figure 1.6d) was added. Furthermore, different PL responses were observed with the addition of riboflavin and ascorbic acid and changes in pH values.

In general, PL responses of SWNTs are more drastic than those of absorbance responses. On the other hand, the duration of increases in SWNT absorbance induced by adding catechin was longer than that of increases in PL intensities (Umemura et al. 2019).

Fluorescent dyes such as fluorescein (FAM) and rhodamine are popular compounds to illuminate biomolecules (Lavis 2017; Berezin and Achilefu 2010; Lavis and Raines 2008). For DNA engineering, functionalization of ends of DNA fragments with fluorescent dyes is a popular technique to detect biochemical reactions of

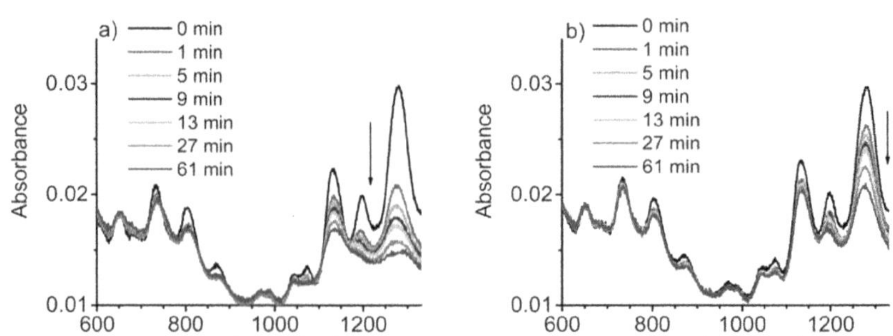

FIGURE 1.5 Time-dependent optical absorption spectra of dsDNA-SWNT solutions reacting with 0.59 mM hydroxyl radicals in the (a) absence and (b) presence of 50 mM caffeine. (Reprinted with permission from Zhao et al. 2015.)

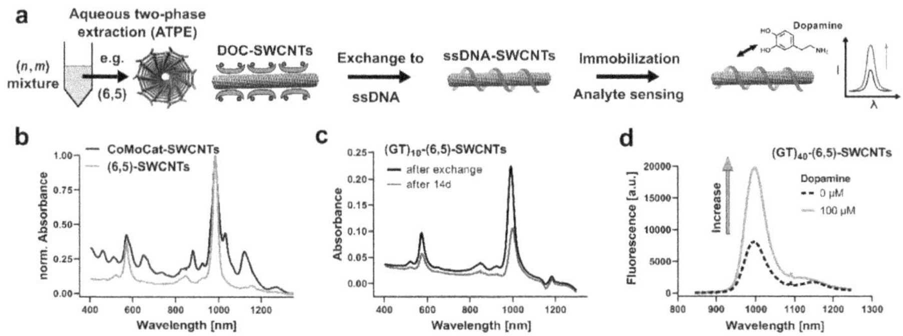

FIGURE 1.6 Assembly of chirality-pure near-infrared fluorescent carbon nanotube-based sensors. (a) Schematic of the workflow. Aqueous two-phase extraction (ATPE), yielding chirality-pure SWNTs (e.g., (6,5)-SWNTs) from a parental mixture. Exchange of the organic corona from the surfactant sodium deoxycholate (DOC) to single-stranded (ss) DNA or other (bio)polymers leads to monochiral functionalized SWNTs that serve as sensors for small molecules such as the neurotransmitter dopamine. (b) Exemplary normalized absorbance spectra of parental CoMoCat-SWNTs and purified (6,5)-SWNTs in 1% DOC. (c) Nine different ssDNA-(6,5)-SWNTs were assessed. Here, exemplary absorbance spectra of $(GT)_{10}$-(6,5)-SWNTs directly after surface exchange and after 14 days are shown. (d) ssDNA-SWNTs immobilized on a glass surface and exposed to analytes that change their NIR fluorescence in a DNA sequence-dependent way. Here, the exemplary monochiral fluorescence spectrum of a sensor for the neurotransmitter dopamine is shown. (Reprinted with permission from Nissler et al. 2021.)

the DNA molecules (Bourdoncle et al. 2006; Chenoweth, Viger, and Dervan 2007). However, when the fluorescent dyes are attached to SWNT surfaces, the photoluminescence of the dyes is quenched. The quenching phenomenon of dyes caused by SWNTs has been applied in biological applications (Satishkumar et al. 2007; Cavuslar and Unal 2015; Wang et al. 2014; Heller et al. 2011). The use of the quenching phenomenon of graphene-related materials has also been intensively studied (Lin et al. 2016; Zhou et al. 2012; Liu et al. 2011).

As an impressive work, Yang et al. (2008) reported the detection of biomolecular interactions using SWNTs. As shown in Figure 1.7, they attached FAM at one end of oligo DNA molecules. When the FAM labeled DNA (FAM-DNA) is adsorbed on SWNT surfaces, PL from FAM is quenched, even though the FAM-DNA-SWNT hybrids are excited by optimal light sources. Then, the FAM-DNA molecules are hybridized with another oligo DNA molecules, which have complementary sequences. Because distances between FMA molecules and SWNT surfaces are increased as a result of the hybridization, PL from FAM is recovered. Therefore, DNA hybridization could be detected by changes in PL intensity from FAM. It means that SWNTs were used as a quencher.

The adsorption manner of ssDNA molecules on SWNT surfaces has been studied by using the quenching phenomenon (Umemura et al. 2017). In many reports, mixtures of SWNT powder and DNA solutions were sonicated for several tens of minutes by a probe-type sonicator to fabricate DNA-SWNT hybrids. The hybridization of DNA molecules and SWNTs was evaluated by the use of FAM-labeled DNA. As a result, PL from FAM decreased by more than 90% within 2 min of sonication when a probe-type sonicator was used. When a bath-type sonicator was employed, PL from FAM gradually decreased over 30 min. When the mixture of DNA and SWNTs was just stored without sonication, PL decreased up to 40% after 30 min of storage. In addition, compared to the use of a probe-type sonicator, the data variability was much larger for the use of a bath-type sonicator and the storage.

From the viewpoints of experimental protocols, it should be noted that excess molecules are included in the samples. Excess organic molecules that were used to solubilize SWNTs remained in the suspension in many cases. Even when focusing on SWNTs of a particular chirality, SWNTs of other chiralities are also contained, as well as amorphous carbon. In order to obtain valuable results, optimized experimental protocols are required to extract the desired optical information among the various possible reactions.

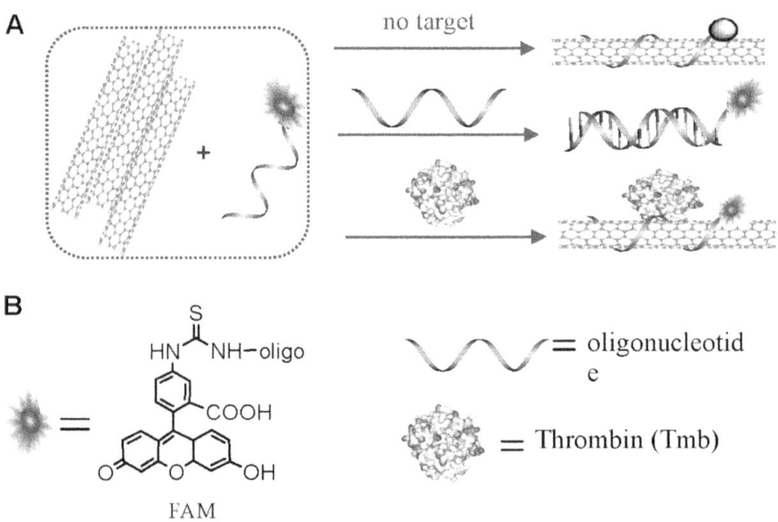

P1: 5'-CAGTTACATTCTC**C**CAGTTGATT-FAM-3'

P2: 5'-TCTCTCAGTCCGTGGTAGGGCAGGTTG GGGTGACT-FAM-3'

T1: 5'-AATCAACTG**G**GAGAATGTAACTG-3'

T2: 5'-AATCAACTG**T**GAGAATGTAACTG-3'

FIGURE 1.7 (a) Scheme for signaling biomolecular interactions by the assembly of SWNTs and dye-labeled ssDNA. The illustration of the dye-DNA-functionalized SWNT is only a graphic representation and does not accurately depict the precise manner in which DNA binds to SWNTs. (b) Structures of FAM, the FAM-labeled oligonucleotides, P_1 and P_2, the perfect cDNA (T_1) and one mismatched DNA (T_2) of P_1, and thrombin. Oligo represents the position of attachment of tethered oligonucleotides. (Reprinted with permission from Yang et al. 2008.)

1.5 MICROSCOPIC DETECTION OF NEAR-INFRARED RESPONSES OF SWNTS

Microscopic studies are one of the recent topics of biological applications of SWNTs, although there are numerous macroscopic reports based on NIR spectroscopy. The microscopic studies include molecular-level research, cellular-level research, and individual-level (in vivo) research (Welsher et al. 2009; Cherukuri et al. 2004; Yi et al. 2012; Liang et al. 2016; Hong, Gao, and Cai 2009; Tsyboulski, Bachilo, and Weisman 2005). In this section, molecular-level studies will be mainly introduced.

As one of the excellent pilot works, Card et al. (2021) observed PL from individual SWNT molecules using an optical microscope. SWNTs were dispersed with $(GT)_{15}$ oligo DNA molecules. The mono-dispersed DNA-SWNT hybrids were embedded in hydrogel films on a cover slip. It suggests that the sample was in "gel" states, not liquids and not dried solids. They succeeded in observing NIR PL from individual SWNTs in the hydrogel film using an NIR fluorescence microscope. To excite the SWNTs, a 730 nm diode laser (1.5 W) was employed. Figure 1.8 shows an example of detecting calcium with PL of individual SWNTs. PL spectra of three individual

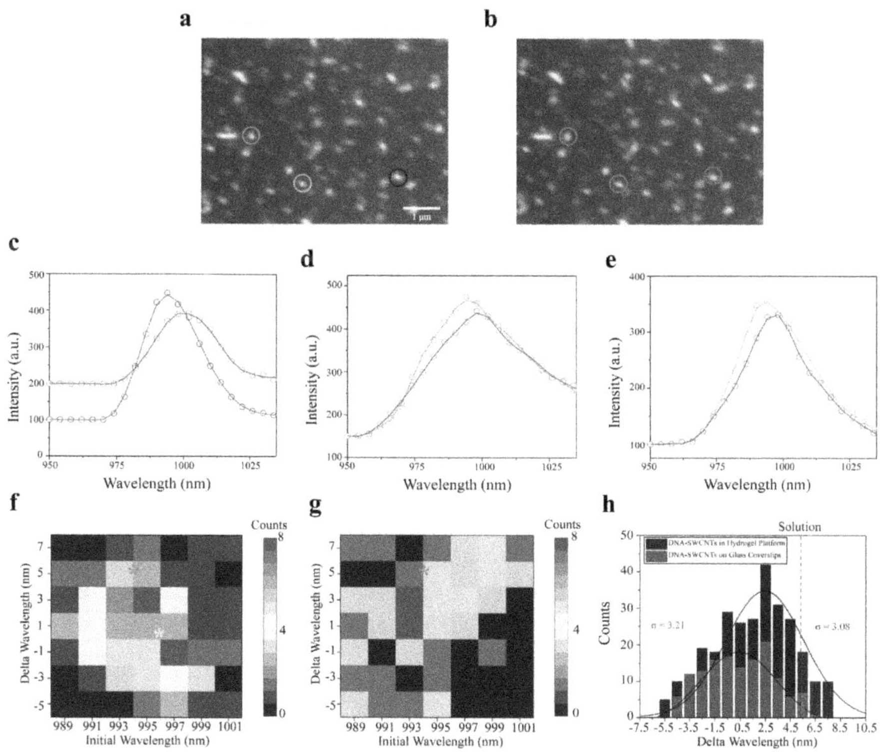

FIGURE 1.8 Single DNA-SWNT calcium detection. Magnified broadband NIR fluorescence images of individual SWNTs in the hydrogel platform (a) before and (b) after addition of 1 M calcium chloride. The green, blue, and cyan circles each designate a single ROI, presumably a single SWNT, which is again found in the image after calcium addition (red circles) (c–e).

SWNTs (see circles in Figure 1.8a and b) were changed before and after injection of calcium into the hydrogel, including DNA-SWNT hybrids. The intensity and peak wavelengths of each SWNT were slightly changed due to the injection of calcium (Figure 1.8c–e). They suggested that the detection of PL spectra was better with hydrogels than with bare coverslips.

Although the observation of individual SWNTs was excellent, it is hard for usual biologists to use them in their routine work. Recently, Tachikawa et al. (2024) demonstrated the detection of changes in PL intensities of (6,5) SWNTs induced by papain enzymes. In the work, CMC-SWNTs were covalently attached to silica microbead surfaces. Then, PL from a microbead was visualized. Because numerous SWNTs were attached to a bead surface, it is easier to use. Even in this case, the amount of SWNTs was much smaller than that for macroscopic spectroscopy.

1.6 BIOSCIENCES ILLUMINATED BY NEAR-INFRARED RESPONSES OF SWNTS

There are numerous reports of biological applications using specific optical responses of SWNTs, as introduced in previous sections. Although it is reasonable that many of the research studies focus on industrially valuable results and practical medical/environmental results, SWNTs have the potential to be applicable as a research tool for biological sciences. In this section, several thought-provoking research studies for SWNT applications for pure-/basic biosciences are presented.

Although it is not an application using optical responses of SWNTs, Balavoine et al. (1999) demonstrated helical crystallization of proteins along MWNT surfaces. Streptavidin molecules were assembled in a helical manner on MWNT surfaces. Although the purpose of the work is to develop new biosensors, the techniques might be applicable to investigate the assembly mechanism of biomolecules. For example, some of the protein molecules such as RecA and Rad51 associate with DNA surfaces in helical manners (Petalcorin et al. 2007). Formation of helical structures of amyloid beta peptides is also an attractive research subject (Castelletto et al. 2009).

Heller et al. (2011) used the modulation of SWNTs to detect the change in the secondary structures of peptides from the bombolitin family attached to the SWNT surface. Figure 1.9 shows examples of PL modulation profiles induced according to the peptide sequences. PL spectra from SWNTs (Figure 1.9a–c) revealed different profiles by replacing a few amino acids of the peptides (Figure 1.9g). It is also modulated by the addition of cyclotrimethylenetrinitramine (RDX). It was speculated that RDX molecules bound to the peptides on SWNT surfaces (Figure 1.9i). Based on those results, the authors proposed a chaperone sensor. The techniques might be applicable to investigate secondary structures of biomolecules from the viewpoints of pure biological sciences.

Chirality-specific responses of SWNT PL induced by proteins were reported by Hirayama et al. (2023). DNA-wrapped SWNTs were attached to aminated frustule surfaces. Frustules are shells of photosynthetic planktons, which are micron-sized nanoporous biosilica. Compared with artificially synthesized microbeads, the sizes and shapes of frustules are rather uniform. Therefore, SWNT-wrapped nanoporous micro silica were fabricated. Figure 1.10 shows an example of PL maps of the

SWNT-functionalized frustule suspension. *Y*-axis and *X*-axis indicate excitation and emission wavelengths of SWNT PL, respectively. Because several different chiralities of SWNTs were included, several PL spots appeared. Although actual PL wavelengths were in the near-infrared area (1,000–1,300 nm), spots were colored with visible colors. When an aliquot of potassium permanganate solution was added to the functionalized frustule suspension, PL from SWNTs was quenched once (Figure 1.10a and d). Next, dithiothreitol (DTT) or papain enzyme was further

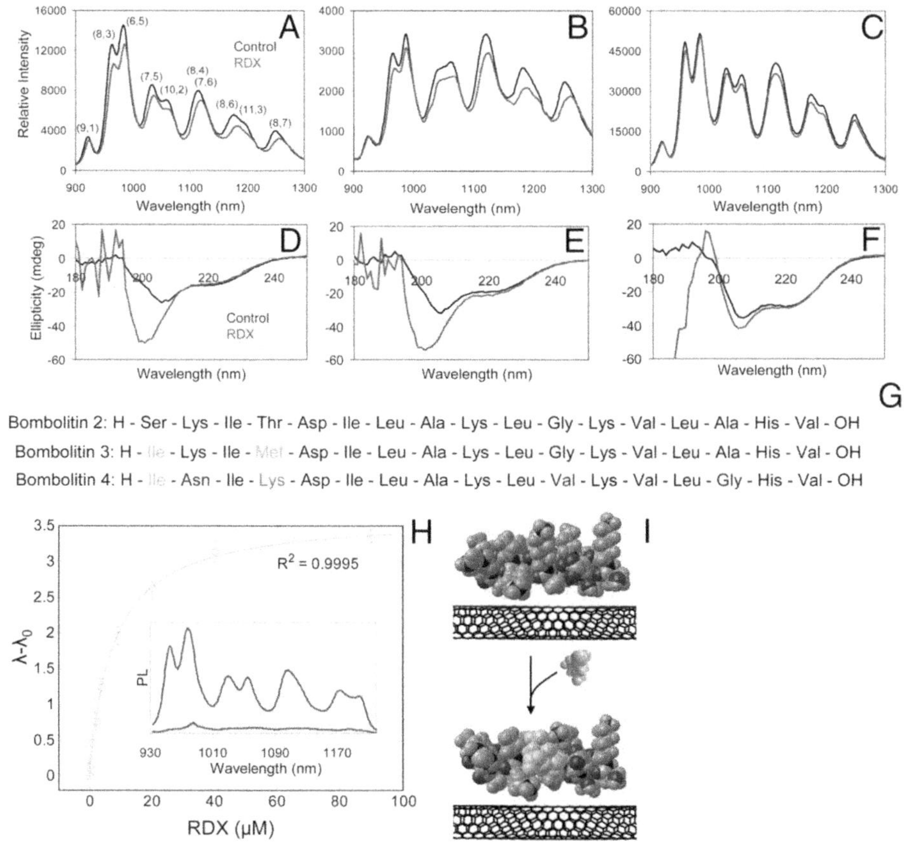

FIGURE 1.9 Peptide-SWNT chaperone sensor for RDX. Photoluminescence spectra of peptide-suspended nanotubes before (black) and after (grey) addition of RDX. PL spectra of nanotubes suspended by (a) bombolitin II, (b) bombolitin III, and (c) bombolitin IV. (d–f) CD spectra of the same peptides in the absence of nanotubes before (black) and after (grey) addition of RDX. (g) Sequences of the three bombolitin peptides highlighting differences between the variants. (h) Center wavelength of the (11,3) nanotube peak of bombolitin-II-solubilized SWNT plotted versus RDX concentration (grey circles). The data are fit to a first-order Langmuir adsorption isotherm (light grey curve). PL spectra of bombolitin-II-solubilized SWNT before (black peak with higher amplitude) and after (black peak with lower amplitude) dialysis using a 20,000 Da molecular weight cutoff membrane (inset). (i) Energy-minimized structures of bombolitin II in water and in the presence of both water and RDX; a conformational change with RDX is computed. (Reprinted with permission from Heller et al. 2011.)

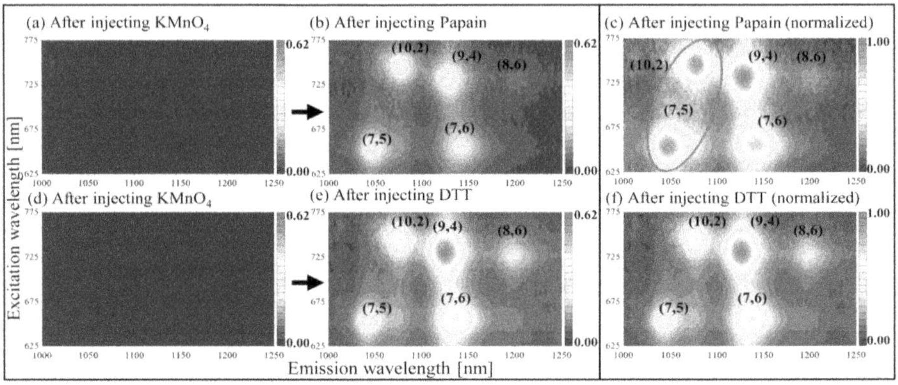

FIGURE 1.10 PL maps of DNA-SWNT hybrids on frustule surfaces. (a) Quenched with 5 μM $KMnO_4$. (b) After the addition of papain (final conc. 0.5 mg/mL) to the sample of Figure 1.1a. (c) Normalized image in Figure 1.1b. The PL intensity of the (9,4) SWNT was defined as 1. (d) Quenched with 5 μM $KMnO_4$. (e) After the addition of DTT (final conc. 0.5 mg/mL) to the sample of Figure 1.1d. (f) A normalized picture of Figure 1.1e. The PL intensity of the (9,4) SWNT was defined as 1. (Reprinted with permission from Hirayama et al. 2023.)

added to the frustule suspension. Those are recognized as reductant molecules. By the reduction of SWNTs, PL intensities were recovered (Figure 1.10b and e). DTT is a strong reductant, so intensities of recovered PL spots with DTT (Figure 1.10e) were much larger than those with papain (Figure 1.10b).

However, when intensities of each spot were normalized by setting the intensity of (9,4) SWNT to 1, a unique comparison was available. In the case with papain, PL recovery of (10,2) and (7,5) SWNTs was as strong as that of (9,4) SWNT. But the recovery of (10,2) and (7,5) was not moderate with DTT. The comparative results revealed that there were different chirality-specific PL responses between DTT and papain. It suggests biomolecules such as papain enzymes can be investigated by the chirality-specific optical responses of SWNTs.

Although NIR optical responses of SWNTs and other nano carbons have been mainly applied for bioengineering, there are also high expectations for applications to pure biosciences.

Fluorescence spectra of the chosen regions of interest (ROIs) before and after calcium addition were obtained. Delta wavelength (nm) vs initial wavelength for individual SWNTs (f) on glass coverslips or (g) in the hydrogel platform after addition of calcium was obtained. The green and yellow asterisks denote the locations of the average solution-phase values and average hydrogel platform or glass values, respectively. (h) Stacked histogram of DNA-SWNT delta wavelength on glass coverslips (red) or in the hydrogel platform (blue). The dotted line represents solution-phase data after the addition of 1 M calcium chloride. The solid black lines represent Gaussian fits with widths (σ) denoted on the graphs (reprinted with permission from Card et al. 2021).

REFERENCES

Acharya, R, TV Patil, SD Dutta et al. 2024. Single-walled carbon nanotube-based optical nano/biosensors for biomedical applications: role in bioimaging, disease diagnosis, and biomarkers detection. *Advanced Materials Technologies* 9 (20): 2400279.

Ackermann, J, JT Metternich, S Herbertz, and S Kruss. 2022. Biosensing with fluorescent carbon nanotubes. *Angewandte Chemie-International Edition* 61 (18): e202112372.

Anantram, MP, and F Leonard. 2006. Physics of carbon nanotube electronic devices. *Reports on Progress in Physics* 69 (3): 507–561.

Angelikopoulos, P, A Gromov, A Leen, O Nerushev, H Bock, and EEB Campbell. 2010. Dispersing individual single-wall carbon nanotubes in aqueous surfactant solutions below the CMC. *Journal of Physical Chemistry C* 114 (1): 2–9.

Antonucci, A, J Kupis-Rozmyslowicz, and AA Boghossian. 2017. Noncovalent protein and peptide functionalization of single-walled carbon nanotubes for biodelivery and optical sensing applications. *ACS Applied Materials & Interfaces* 9 (13): 11321–11331.

Avery, AD, BH Zhou, J Lee et al. 2016. Tailored semiconducting carbon nanotube networks with enhanced thermoelectric properties. *Nature Energy* 1 (4): 16033.

Balavoine, F, P Schultz, C Richard, V Mallouh, TW Ebbesen, and C Mioskowski. 1999. Helical crystallization of proteins on carbon nanotubes: a first step towards the development of new biosensors. *Angewandte Chemie-International Edition* 38 (13–14): 1912–1915.

Barisci, JN, M Tahhan, GG Wallace et al. 2004. Properties of carbon nanotube fibers spun from DNA-stabilized dispersions. *Advanced Functional Materials* 14 (2): 133–138.

Berezin, MY, and S Achilefu. 2010. Fluorescence lifetime measurements and biological imaging. *Chemical Reviews* 110 (5): 2641–2684.

Blanch, AJ, CE Lenehan, and JS Quinton. 2010. Optimizing surfactant concentrations for dispersion of single-walled carbon nanotubes in aqueous solution. *Journal of Physical Chemistry B* 114 (30): 9805–9811.

Bomboi, F, A Bonincontro, C La Mesa, and F Tardani. 2011. Interactions between single-walled carbon nanotubes and lysozyme. *Journal of Colloid and Interface Science* 355 (2): 342–347.

Bourdoncle, A, AE Torres, C Gosse et al. 2006. Quadruplex-based molecular beacons as tunable DNA probes. *Journal of the American Chemical Society* 128 (34): 11094–11105.

Brown, S, TS Jespersen, and J Nygård. 2008. A genetic analysis of carbon-nanotube-binding proteins. *Small* 4 (4): 416–420.

Card, M, M Gravely, SZM Madani, and D Roxbury. 2021. A spin-coated hydrogel platform enables accurate investigation of immobilized individual single-walled carbon nanotubes. *ACS Applied Materials & Interfaces* 13 (27): 31986–31995.

Castelletto, V, IW Hamley, RA Hule, and D Pochan. 2009. Helical–Ribbon formation by a β-amino acid modified amyloid β-peptide fragment. *Angewandte Chemie-International Edition* 48 (13): 2317–2320.

Cavuslar, O, and H Unal. 2015. Self-assembly of DNA wrapped carbon nanotubes and asymmetrical cyanine dyes into fluorescent nanohybrids. *RSC Advances* 5 (29): 22380–22389.

Chen, Z, XB Zhang, RH Yang, Z Zhu, Y Chen, and WH Tan. 2011. Single-walled carbon nanotubes as optical materials for biosensing. *Nanoscale* 3 (5): 1949–1956.

Chenoweth, DM, A Viger, and PB Dervan. 2007. Fluorescent sequence-specific dsDNA binding oligomers. *Journal of the American Chemical Society* 129 (8): 2216–2217.

Cherukuri, P, SM Bachilo, SH Litovsky, and RB Weisman. 2004. Near-infrared fluorescence microscopy of single-walled carbon nanotubes in phagocytic cells. *Journal of the American Chemical Society* 126 (48): 15638–15639.

Chia, L, Y Huang, P Lu, and AN Bezbaruah. 2021. Surface modification of carbon nanotubes using carboxymethyl cellulose for enhanced stress sensing in smart cementitious composites. *IEEE Sensors Journal* 21 (13): 15218–15229.

Chia, LA, Y Huang, WJ Xia, P Lu, and DW Zhang. 2023. Carbon nanotube (CNT) reinforced cementitious composites using carboxymethyl cellulose (CMC) treatment for enhanced dispersion, mechanical, and piezoresistive properties. *Construction and Building Materials* 377: 131104. doi:10.1016/j.conbuildmat.2023.131104.

Chiu, CF, N Dementev, and E Borguet. 2011. Fluorescence quenching of dyes covalently attached to single-walled carbon nanotubes. *Journal of Physical Chemistry A* 115 (34): 9579–9584.

Cohen, Z, and RM Williams. 2024. Single-walled carbon nanotubes as optical transducers for nanobiosensors in vivo. *ACS Nano* 18 (52): 35164–35181.

Fagan, JA, BJ Landi, I Mandelbaum et al. 2006. Comparative measures of single-wall carbon nanotube dispersion. *Journal of Physical Chemistry B* 110 (47): 23801–23805.

Fernandes, RMF, B Abreu, B Claro et al. 2015. Dispersing carbon nanotubes with ionic surfactants under controlled conditions: comparisons and insight. *Langmuir* 31 (40): 10955–10965.

Flavel, BS, KE Moore, M Pfohl, MM Kappes, and F Hennrich. 2014. Separation of single-walled carbon nanotubes with a gel permeation chromatography system. *ACS Nano* 8 (2): 1817–1826.

Gong, H, R Peng, and Z Liu. 2013. Carbon nanotubes for biomedical imaging: the recent advances. *Advanced Drug Delivery Reviews* 65 (15): 1951–1963.

Gudiksen, KL, I Gitlin, and GM Whitesides. 2006. Differentiation of proteins based on characteristic patterns of association and denaturation in solutions of SDS. *Proceedings of the National Academy of Sciences of the United States of America* 103 (21): 7968–7972.

Hayashida, T, and K Umemura. 2016. Atomic force microscopy of DNA-wrapped single-walled carbon nanotubes in aqueous solution. *Colloids and Surfaces B-Biointerfaces* 143: 526–531.

He, XW, NF Hartmann, XD Ma et al. 2017. Tunable room-temperature single-photon emission at telecom wavelengths from sp3 defects in carbon nanotubes. *Nature Photonics* 11 (9): 577–582.

Hedderman, TG, SM Keogh, G Chambers, and HJ Byrne. 2004. Solubilization of SWNTs with organic dye molecules. *Journal of Physical Chemistry B* 108 (49): 18860–18865.

Heller, DA, GW Pratt, JQ Zhang et al. 2011. Peptide secondary structure modulates single-walled carbon nanotube fluorescence as a chaperone sensor for nitroaromatics. *Proceedings of the National Academy of Sciences of the United States of America* 108 (21): 8544–8549.

Hemmer, E, A Benayas, F Légaré, and F Vetrone. 2016. Exploiting the biological windows: current perspectives on fluorescent bioprobes emitting above 1000 nm. *Nanoscale Horizons* 1 (3): 168–184.

Hirano, A, Y Maeda, XF Yuan et al. 2010. Controlled dispersion and purification of protein-carbon nanotube conjugates using guanidine hydrochloride. *Chemistry-A European Journal* 16 (40): 12221–12228.

Hirayama, K, M Kitamura, S Mayama, and K Umemura. 2023. Chirality-specific near-infrared photoluminescent responses of single-walled carbon nanotubes to detect papain adsorption assisted by the use of DNA and frustules. *Chemistry Letters* 52 (5): 408–411.

Hong, H, T Gao, and WB Cai. 2009. Molecular imaging with single-walled carbon nanotubes. *Nano Today* 4 (3): 252–261.

Horii, Y, S Ohura, and K Umemura. 2020. An efficient method to quantitatively detect competitive adsorption of DNA on single-walled carbon nanotube surfaces. *Analytical Biochemistry* 601: 113781.

Ishibashi, Y, M Ito, Y Homma, and K Umemura. 2018. Monitoring the antioxidant effects of catechin using single-walled carbon nanotubes: comparative analysis by near-infrared absorption and near-infrared photoluminescence. *Colloids and Surfaces B-Biointerfaces* 161: 139–146.

Ito, M, H Yajima, and Y Homma. 2016. Strain effect of cellulose-wrapped single-walled carbon nanotubes measured by photoluminescence and Raman scattering spectroscopy. *Japanese Journal of Applied Physics* 55 (7): 075101.

Jang, J, H Yeom, S Chae, and S Kee. 2025. Dispersion-stable carboxymethyl cellulose/single-walled carbon nanotube composite for water-processed organic thermoelectrics. *Materials* 18 (2): 337.

Johnson, ATC, C Staii, M Chen et al. 2006. DNA-decorated carbon nanotubes for chemical sensing. *Semiconductor Science and Technology* 21 (11): S17–S21.

Johnson, RR, ATC Johnson, and ML Klein. 2008. Probing the structure of DNA-carbon nanotube hybrids with molecular dynamics. *Nano Letters* 8 (1): 69–75.

Karachevtsev, VA, GO Gladchenko, MV Karachevtsev, VA Valeev, VS Leontiev, and OS Lytvyn. 2008. Adsorption of poly(rA) on the carbon nanotube surface and its hybridization with poly(rU). *Chemphyschem* 9 (14): 2010–2018.

Karachevtsev, VA, AY Glamazda, AM Plokhotnichenko, VS Leontiev, and AS Linnik. 2011. Comparative study on protection properties of anionic surfactants (SDS, SDBS) and DNA covering of single-walled carbon nanotubes against pH influence: luminescence and absorption spectroscopy study. *Materialwissenschaft Und Werkstofftechnik* 42 (1): 41–46.

Koh, B, JB Park, XM Hou, and W Cheng. 2011. Comparative dispersion studies of single-walled carbon nanotubes in aqueous solution. *Journal of Physical Chemistry B* 115 (11): 2627–2633.

Krasley, AT, EGE Li, JM Galeana, C Bulumulla, AG Beyene, and GS Demirer. 2024. Carbon nanomaterial fluorescent probes and their biological applications. *Chemical Reviews* 124 (6): 3085–3185.

Krupke, R, F Hennrich, O Hampe, and MM Kappes. 2003. Near-infrared absorbance of single-walled carbon nanotubes dispersed in dimethylformamide. *Journal of Physical Chemistry B* 107 (24): 5667–5669.

Kruss, S, AJ Hilmer, JQ Zhang, NF Reuel, B Mu, and MS Strano. 2013. Carbon nanotubes as optical biomedical sensors. *Advanced Drug Delivery Reviews* 65 (15): 1933–1950.

Lavis, LD 2017. Teaching old dyes new tricks: biological probes built from fluoresceins and rhodamines. *Annual Review of Biochemistry* 86: 825–843.

Lavis, LD, and RT Raines. 2008. Bright ideas for chemical biology. *ACS Chemical Biology* 3 (3): 142–155.

Li, H, G Gordeey, O Garrity, S Reich, and BS Flavel. 2019. Separation of small-diameter single-walled carbon nanotubes in one to three steps with aqueous two-phase extraction. *ACS Nano* 13 (2): 2567–2578.

Liang, XY, WT Shang, CW Chi et al. 2016. Dye-conjugated single-walled carbon nanotubes induce photothermal therapy under the guidance of near-infrared imaging. *Cancer Letters* 383 (2): 243–249.

Lin, WY, B Tian, PP Zhuang et al. 2016. Graphene-based fluorescence-quenching-related fermi level elevation and electron-concentration surge. *Nano Letters* 16 (9): 5737–5741.

Liu, M, HM Zhao, S Chen, HT Yu, YB Zhang, and X Quan. 2011. A "turn-on" fluorescent copper biosensor based on DNA cleavage-dependent graphene-quenched DNAzyme. *Biosensors & Bioelectronics* 26 (10): 4111–4116.

Liu, Z, S Tabakman, K Welsher, and HJ Dai. 2009. Carbon nanotubes in biology and medicine: in vitro and in vivo detection, imaging and drug delivery. *Nano Research* 2 (2): 85–120.

Lucente-Schultz, RM, VC Moore, AD Leonard et al. 2009. Antioxidant single-walled carbon nanotubes. *Journal of the American Chemical Society* 131 (11): 3934–3941.

Moon, SY, and WS Kim. 2019. Chirality control for predominant metallic or semiconducting single-walled carbon nanotubes prepared using a mild etchant. *Chemical Communications* 55 (92): 13888–13891.

Nakashima, N, S Okuzono, H Murakami, T Nakai, and K Yoshikawa. 2003. DNA dissolves single-walled carbon nanotubes in water. *Chemistry Letters* 32 (5): 456–457.

Nakayama, T, T Tanaka, K Shiraki, M Hase, and A Hirano. 2018. Suppression of single-wall carbon nanotube redox reaction by adsorbed proteins. *Applied Physics Express* 11 (7): 075101.

Nissler, R, L Kurth, H Li et al. 2021. Sensing with chirality-pure near-infrared fluorescent carbon nanotubes. *Analytical Chemistry* 93 (16): 6446–6455.

Nugraha, ART, GD Sanders, K Sato, CJ Stanton, MS Dresselhaus, and R Saito. 2011. Chirality dependence of coherent phonon amplitudes in single-wall carbon nanotubes. *Physical Review B* 84 (17): 174302.

Palwai, NR, DE Martyn, LFF Neves, Y Tan, DE Resasco, and RG Harrison. 2007. Retention of biological activity and near-infrared absorbance upon adsorption of horseradish peroxidase on single-walled carbon nanotubes. *Nanotechnology* 18 (23): 235601.

Pan, HL, LQ Liu, ZX Guo et al. 2003. Carbon nanotubols from mechanochemical reaction. *Nano Letters* 3 (1): 29–32.

Panwar, N, AM Soehartono, KK Chan et al. 2019. Nanocarbons for biology and medicine: sensing, imaging, and drug delivery. *Chemical Reviews* 119 (16): 9559–9656.

Petalcorin, MIR, VE Galkin, X Yu, EH Egelman, and SJ Boulton. 2007. Stabilization of RAD-51-DNA filaments via an interaction domain in Caenorhabditis elegans BRCA2. *Proceedings of the National Academy of Sciences of the United States of America* 104 (20): 8299–8304.

Petit, P, C Mathis, C Journet, and P Bernier. 1999. Tuning and monitoring the electronic structure of carbon nanotubes. *Chemical Physics Letters* 305 (5–6): 370–374.

Rao, CNR, R Voggu, and A Govindaraj. 2009. Selective generation of single-walled carbon nanotubes with metallic, semiconducting and other unique electronic properties. *Nanoscale* 1 (1): 96–105.

Sa, V, and KG Kornev. 2011. Analysis of stability of nanotube dispersions using surface tension isotherms. *Langmuir* 27 (22): 13451–13460.

Sanchez-Valencia, JR, T Dienel, O Gröning et al. 2014. Controlled synthesis of single-chirality carbon nanotubes. *Nature* 512 (7512): 61–64.

Satishkumar, BC, LO Brown, Y Gao, CC Wang, HL Wang, and SK Doorn. 2007. Reversible fluorescence quenching in carbon nanotubes for biomolecular sensing. *Nature Nanotechnology* 2 (9): 560–564.

Segawa, Y, A Yagi, K Matsui, and K Itami. 2016. Design and synthesis of carbon nanotube segments. *Angewandte Chemie-International Edition* 55 (17): 5136–5158.

Singh, I, PK Bhatnagar, PC Mathur, and LM Bharadwaj. 2008. Optical absorption spectrum of single-walled carbon nanotubes dispersed in sodium cholate and sodium dodecyl sulfate. *Journal of Materials Research* 23 (3): 632–636.

Sun, YW, I Hernández, J González et al. 2016. Significance of bundling effects on carbon nanotubes' response to hydrostatic compression. *Journal of Physical Chemistry C* 120 (3): 1863–1870.

Tachikawa, Y, M Ito, M Irita, T Harada, and K Umemura. 2024. Near-infrared photoluminescence responses of single-walled carbon nanotubes induced by biomolecules detected on a microbead surface. *ACS Omega* 9 (44): 44734–44740.

Tsyboulski, DA, SM Bachilo, and RB Weisman. 2005. Versatile visualization of individual single-walled carbon nanotubes with near-infrared fluorescence microscopy. *Nano Letters* 5 (5): 975–979.

Tu, XM, S Manohar, A Jagota, and M Zheng. 2009. DNA sequence motifs for structure-specific recognition and separation of carbon nanotubes. *Nature* 460 (7252): 250–253.

Umemura, K, Y Ishibashi, M Ito, and Y Homma. 2019. Quantitative detection of the disappearance of the antioxidant ability of catechin by near-infrared absorption and near-infrared photoluminescence spectra of single-walled carbon nanotubes. *ACS Omega* 4 (4): 7750–7758.

Umemura, K, S Sato, G Bustamante, and JY Ye. 2017. Using a fluorescence quenching method to detect DNA adsorption onto single-walled carbon nanotube surfaces. *Colloids and Surfaces B-Biointerfaces* 160: 201–206.

Umemura, K, Y Sato, Y Ishibashi, M Ito, and Y Homma. 2020. Various responses of single-walled carbon nanotubes with differing chirality: a suggestion for biosensing. *Journal of Near Infrared Spectroscopy* 28 (1): 51–56.

Umemura, K, K Ishizaka, D Nii, and K Izumi. 2016. Non-uniform binding of single-stranded DNA binding proteins to hybrids of single-stranded DNA and single-walled carbon nanotubes observed by atomic force microscopy in air and in liquid. *Applied Surface Science* 388: 381–384.

Umemura, K, K Izumi, and S Oura. 2016. Probe microscopic studies of DNA molecules on carbon nanotubes. *Nanomaterials* 6 (10): 14.

Venkataraman, A, EV Amadi, YD Chen, and C Papadopoulos. 2019. Carbon nanotube assembly and integration for applications. *Nanoscale Research Letters* 14: 220.

Wang, GF, G Xu, YH Zhu, and XJ Zhang. 2014. A "turn-on" carbon nanotube-Ag nanoclusters fluorescent sensor for sensitive and selective detection of Hg^{2+} with cyclic amplification of exonuclease III activity. *Chemical Communications* 50 (6): 747–750.

Wang, GJ, SZ Huang, Y Wang, L Liu, J Qiu, and Y Li. 2007. Synthesis of water-soluble single-walled carbon nanotubes by RAFT polymerization. *Polymer* 48 (3): 728–733.

Wang, X, YQ Feng, PP Dong, and JF Huang. 2019. A mini review on carbon quantum dots: preparation, properties, and electrocatalytic application. *Frontiers in Chemistry* 7: 671.

Welsher, K, Z Liu, SP Sherlock et al. 2009. A route to brightly fluorescent carbon nanotubes for near-infrared imaging in mice. *Nature Nanotechnology* 4 (11): 773–780.

Xu, Q, and TA Keiderling. 2004. Effect of sodium dodecyl sulfate on folding and thermal stability of acid-denatured cytochrome c: a spectroscopic approach. *Protein Science* 13 (11): 2949–2959.

Xu, Q, WJ Li, L Ding, WJ Yang, HH Xiao, and WJ Ong. 2019. Function-driven engineering of 1D carbon nanotubes and 0D carbon dots: mechanism, properties and applications. *Nanoscale* 11 (4): 1475–1504.

Xu, ZJ, XN Yang, and Z Yang. 2010. A molecular simulation probing of structure and interaction for supramolecular sodium dodecyl sulfate/single-wall carbon nanotube assemblies. *Nano Letters* 10 (3): 985–991.

Yamamoto, T, Y Miyauchi, J Motoyanagi et al. 2008. Improved bath sonication method for dispersion of individual single-walled carbon nanotubes using new triphenylene-based surfactant. *Japanese Journal of Applied Physics* 47 (4): 2000–2004.

Yamazaki, Y, and K Umemura. 2021. Sensing of epigallocatechin gallate and tannic acid based on near infrared optical spectroscopy of DNA-wrapped single-walled carbon nanotube hybrids. *Journal of Near Infrared Spectroscopy* 29 (2): 73–83.

Yang, F, M Wang, DQ Zhang, J Yang, M Zheng, and Y Li. 2020. Chirality pure carbon nanotubes: growth, sorting, and characterization. *Chemical Reviews* 120 (5): 2693–2758.

Yang, F, X Wang, DQ Zhang et al. 2014. Chirality-specific growth of single-walled carbon nanotubes on solid alloy catalysts. *Nature* 510 (7506): 522–524.

Yang, RH, ZW Tang, JL Yan et al. 2008. Noncovalent assembly of carbon nanotubes and single-stranded DNA: an effective sensing platform for probing biomolecular interactions. *Analytical Chemistry* 80 (19): 7408–7413.

Yi, HJ, D Ghosh, MH Ham et al. 2012. M13 phage-functionalized single-walled carbon nanotubes as nanoprobes for second near-infrared window fluorescence imaging of targeted tumors. *Nano Letters* 12 (3): 1176–1183.

Zhang, JQ, MP Landry, PW Barone et al. 2013. Molecular recognition using corona phase complexes made of synthetic polymers adsorbed on carbon nanotubes. *Nature Nanotechnology* 8 (12): 959–968.

Zhang, SC, LX Kang, X Wang et al. 2017. Arrays of horizontal carbon nanotubes of controlled chirality grown using designed catalysts. *Nature* 543 (7644): 234–238.

Zhao, EH, B Ergul, and W Zhao. 2015. Caffeine's antioxidant potency optically sensed with double-stranded DNA-encased single-walled carbon nanotubes. *Journal of Physical Chemistry B* 119 (10): 4068–4075.

Zhao, W, CH Song, and PE Pehrsson. 2002. Water-soluble and optically pH-sensitive single-walled carbon nanotubes from surface modification. *Journal of the American Chemical Society* 124 (42): 12418–12419.

Zhao, X, and JK Johnson. 2007. Simulation of adsorption of DNA on carbon nanotubes. *Journal of the American Chemical Society* 129 (34): 10438–10445.

Zheng, M, A Jagota, ED Semke et al. 2003. DNA-assisted dispersion and separation of carbon nanotubes. *Nature Materials* 2 (5): 338–342.

Zhou, J, Q Lu, Y Tong, W Wei, and SQ Liu. 2012. Detection of DNA damage by using hairpin molecular beacon probes and graphene oxide. *Talanta* 99: 625–630.

2 Evaluation of Modern Optical Diagnostics Tools

M. M. Balamurali
Vellore Institute of Technology

Kaushik Chanda
Rabindranath Tagore University

2.1 INTRODUCTION

Over the past three decades, optical techniques have played a pivotal role in the development of sensor devices that can serve as powerful analytical tools for detecting a wide range of biologically and chemically significant analytes.[1–5] Recently, efforts have been underway to simplify the fundamental principles of optical sensing. Either a bottom-up or top-down approach is being employed to design an optical sensor, with various optical parameters being used for sensing, followed by monitoring different response mechanisms for diverse classes of optical probes. Various approaches are available to translate molecular probes into optical sensor platforms, which form the current and future trends toward new sensor designs.[1,6–10]

The rapid evolution of new pathogenic microorganisms challenges the scientific society in their development of rapid detection and monitoring systems for disease-associated biomarkers and also toward elucidating the pathogenic mechanisms and hence toward the development of efficient diagnostics and therapeutics.[11–13]

Until now, traditional imaging methods, including computed X-ray tomography, magnetic resonance imaging, ultrasonic imaging, positron emission tomography, single photon emission computed tomography, fluorescence imaging technologies, *and so forth*, have been crucial in preclinical and basic research toward offering highly sensitive non-invasive real-time imaging with enhanced spatiotemporal resolution. Further, for improving the accuracy and sensitivity of optical imaging and broadening the application scope, new chemical tools are being developed to sense an array of disease-associated biomarkers and to screen key physiological processes. Recently, numerous luminescent probes were reported with applications in immunofluorescence,[14] live-cell imaging,[15–17] drug delivery,[18,19] and light-guided surgery.[20]

An efficient disease-targeting luminescent probe is expected to possess high target specificity, selectivity toward target biomarkers under physiological conditions, photo-stability, biocompatibility, and ambient wavelength excitation. The structural design of a fluorescent probe consists of three components—the target recognition moiety, the fluorophore, and the linker to connect them.[21–27] In addition, targeting groups are included in some probes to enable their interaction with specific tissues or organelles by localizing the fluorescence signals.

DOI: 10.1201/9781003570011-2

2.2 POINT-OF-CARE (POC) OPTICAL DEVICES

Recently, there has been an increasing interest in the research community on the development of PoC diagnostic tools for time-efficient detection of emerging pathogens for quick clinical decision making.[28–30] Moreover, such probes based on optical techniques are non-invasive, rapid, and low-cost, with sensitivity, selectivity, and specificity. Particularly in economically downtrodden countries, traditional diagnostic techniques like polymerase chain reaction (PCR), enzyme-linked immunosorbent assay, culture analyses, *and so forth*, are not available due to technological constraints. These are portable devices that use light to specifically target biomolecules and convert them into readable signals. Some of the characteristics of PoC devices are depicted in Figure 2.1.

In general, these PoC optical devices are appreciated for their miniaturized design and laboratory equivalent diagnostic efficiency. The diagnosis with these PoC devices is categorised under to three classifications—qualitative, quantitative, and semi-quantitative.

The development of PoC devices requires transfer of technology from scientists from various disciplines, including chemistry, biochemistry, optics, healthcare professionals, and so forth. Some of the PoC devices for specific applications available in the market are given in Figure 2.2, and different types of available methods, along with their advantages and disadvantages, are given in Table 2.1.

In recent years, the increase in illnesses caused by pathogens has resulted in increased morbidity and mortality. A variety of pathogens with increased virulence are the causative agents for the observed fatalities. The quick spread of such diseases has created a threat to the human population. This has necessitated efficient diagnostic tools that can supersede the conventional time-consuming and sophisticated laboratory-based diagnostic tools to detect microbial pathogens. The successful treatment of these infections employing simple, accurate, and time- and cost-effective detection tools has gained priority. Particularly, the optical biosensors have served the purpose with the evolution of a variety of efficacious PoC diagnostics. Currently, efforts to enhance the efficiencies

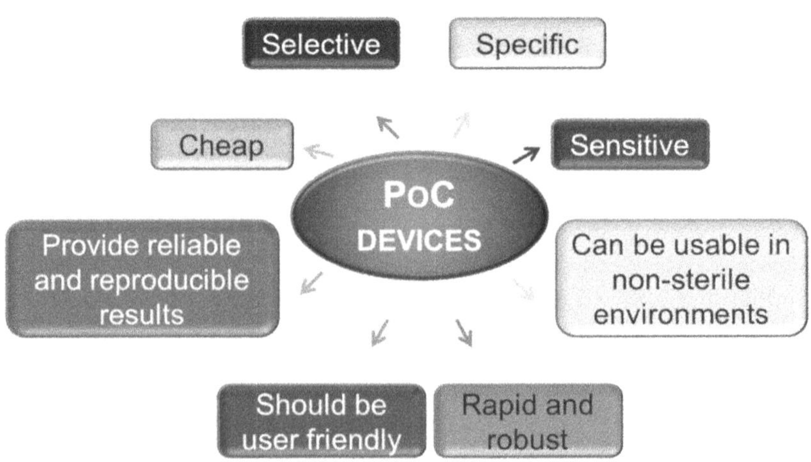

FIGURE 2.1 Expected characteristics of point-of-care (PoC) devices.

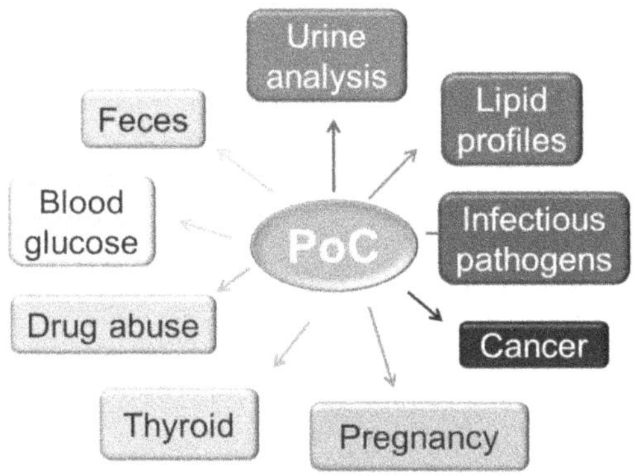

FIGURE 2.2 PoC devices available in the market for various purposes.

TABLE 2.1
Classifications of PoCs

	Method	Advantage/Disadvantage
Based on the Detection Techniques		
Label-free	The direct interaction between the biomolecules and the target measured	No modification or labeling required
Label-based	Interaction of luminescent molecules tagged to biomolecules with targets or bioreceptors	Involves multiple steps, labeling agents, and specific laboratory equipment
Based on the Optical Interrogation		
Vertical interrogation	The reflected or transmitted light is analysed as in a spectrophotometer, CCD camera, and so forth.	The optical setup is simple without sophistications
Horizontal interrogation	Waveguide-coupled light detects the signals before and after bio-recognition	Complex optical setup that requires an expensive grating coupler
Based on the Light Emission		
Fluorescence	Quantification and localization monitored by following the fluorescence emission upon specific wavelength excitation. In this, the dynamic energy of biomolecules is converted into fluorescence energy.	Results are reproducible, rapid, and low-cost. Suitable for PoC applications
Based on Scattering		
Raman scattering	Deviation of the incident light resulting in the change of the wavelength—elastic and inelastic scattering	Qualitative and quantitative recognition of biomolecules with a label-free and non-invasive method
Surface-enhanced Raman scattering	Enhancement of Raman signals upon interacting with surface-localized biomolecules.	High efficiency.

of these devices are underway with the integration of artificial intelligence, machine learning, Internet of Things, and so forth. The existing gap between bioinformatics and clinical validations can be filled with these smart technologies. Such advanced

PoC-based optical biosensors can help in monitoring the progression of diseases and the emergence of new microbial pathogens with increased virulence.[31]

Most of the recently developed PoCs are reported to follow either of the lateral flow assay, paper-based, and thread-based microfluidic devices, and so forth and adopt color-based transduction. The PoCs are promising in their applications, with cost-effectiveness, ease of use, and amenability. Conventional lateral flow methods employ target-specific antibodies attached to colloidal gold or silver nanoparticles.

2.3 POCS FOR DIAGNOSTIC APPLICATIONS

2.3.1 Virus Detection

There are numerous PoCs available for the effective detection of viral infections with very high specificity and sensitivity. The gold standard for screening utilized by these analytical devices is the reverse transcription PCR. Recently, a paper-based micro-analytical device was reported to screen the Ebola virus, which involves isothermal amplification of nucleic acids along with real-time recombinase polymerase reaction.[32] The device consists of positive, negative, and test sample sections wax-printed on paper (Figure 2.3). These sections were embossed with real-time recombinase polymerase amplification components for detection analysis. In this, the RNA matrix of the Ebola virus is used as the positive control, while the extracted RNA from the blood plasma of patients will be the test samples, and buffer alone will be used as a control sample.[32]

Similarly, for the detection of HIV, low-cost microfluidic miniaturized devices are available as PoC devices.[33] HIV remains a significant public health challenge, imposing a substantial economic burden on many countries and necessitating the availability of rapid self-testing for prompt diagnosis. Commercially available kits for the rapid diagnosis of HIV infection include lateral flow assays that involve the recognition of HIV-1/2 antibodies and/or HIV-1 p24 antigen. However, there are currently no assays,

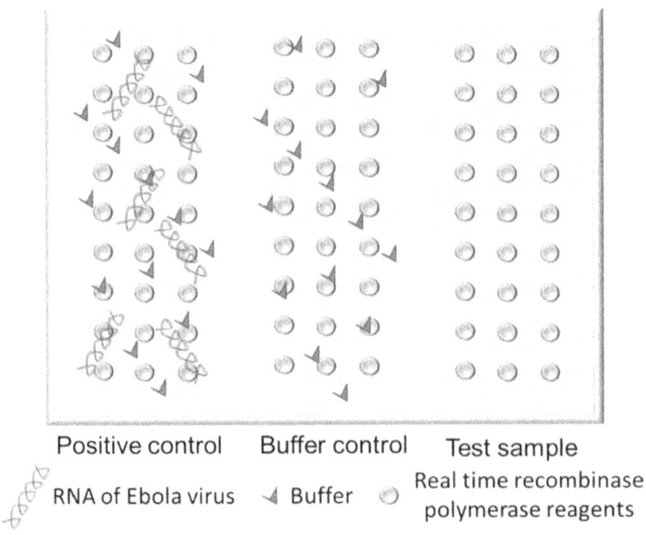

FIGURE 2.3 Schematic of a paper-based microanalytical device for the detection of viruses.

such as nucleic acid amplifications, that can track and quantify the development of anti-retroviral medication responses and the progression of infection. Recently, Phillips et al. have reported a rapid microfluidic analytical device to detect HIV-1 virus in whole blood samples of humans.[34] Briefly, the design of this device includes a temperature-controlled circuit along with a laminated microfluidic paper-based analytical device adopting a later flow assay strip for easy visualization. The assay time reported for the device was 90 min with a sensitivity of 2.3×10^7 copies of virus per ml of whole blood. This count is observed at the peak period of infection.

With the recent pandemic, SARS-CoV-2 caused COVID-19, and effective screening was the key priority to curb the rate of viral spread. RT-PCR was the primary tool for its diagnosis, along with CT scanning and hematological marker detection. But all these methods consume a lot of time for diagnosis in addition to the requirement of skilled personnel, intricate procedures, expensive equipment, and so forth. All these limitations paved the way for the development of a qualitative later flow assay by Hu et al.[35] This assay allowed for the detection of IgM and IgG in infected blood samples using colorimetric visualization with gold nanoparticles. During infections, IgM serves as the first line of defence, and the infection-specific IgG responses are the important determinants for anti-IgG and anti-IgM detection of the virus. The IgM antibody helps to detect recent exposure, while the IgG antibody shows a relatively longer time of virus exposure. With this later flow assay, it is possible to detect IgM and IgG in just 15 min, and hence, it served as a rapid diagnostic tool during the last stage of the pandemic period. This device contains five components—(1) the sample pad, (2) the conjugation pad, (3) the nitrocellulose membrane, (4) the absorption pad, and (5) the adhesive card. The anti-Human IgM and IgG are chemically immobilized in the device as detection sites. In the conjugation pad, the antigen of SARS-CoV-2 and rabbit IgG, both conjugated with gold nanoparticles, were sprayed. The assembly of the above components is depicted in Figure 2.4. The operation of this device relies on the capillary flow of the infected sample, which contains anti-SARS-CoV-2 IgG and IgM.

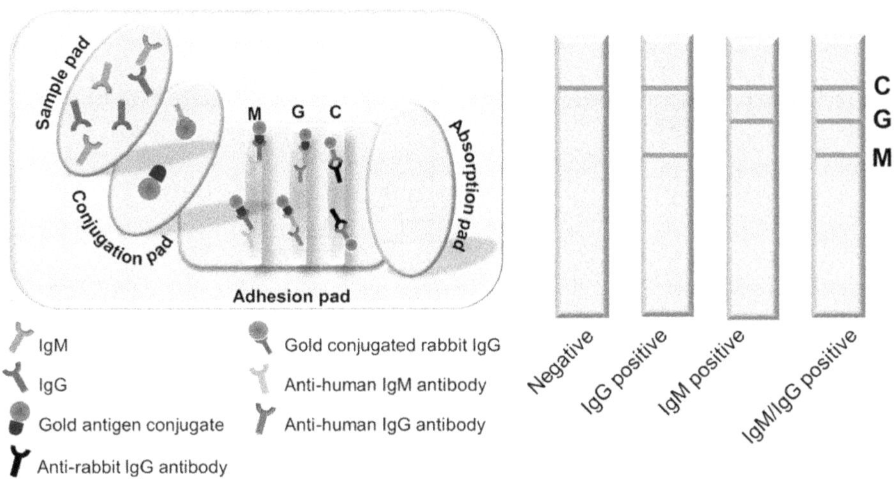

FIGURE 2.4 Schematic representation of a later flow-based microfluidic device to detect SARS-CoV-2.

As the sample moves through the device, the colloidal gold-tagged viral particles bind to the immobilized antibodies on the conjugation pad. Later, these conjugated samples are then bound to their respective antibodies at the respective designated detection lines. After flowing through the nitrocellulose membrane, the rabbit IgG conjugated with colloidal gold will bind with its respective antibody. The red lines will appear for positive samples with anti-SARS-CoV-2 IgM and IgG, while no lines will be detected for samples in the absence of any anti-SARS-CoV-2 IgG and IgM.

2.4 OPTICAL IMAGING INSTRUMENTATION TECHNIQUES

These probes have displayed promising potential by their accumulation at disease sites with a significant signal-to-noise ratio, along with concentration dependence and improved sensitivity. In general, these probes employ strategies including Foster's resonance energy transfer, photo-induced electron transfer, inter-/intramolecular charge/proton transfer, and so forth (Figure 2.5). On the other hand, there are certain probes known as single-lock activatable probes that are disease-based biomarker specific with optical signal turn-on/off characteristics.[36] Numerous such probes were reported for the detection of variations in the levels of reactive species, metabolites, pH, hypoxia, and so forth.[37]

More specifically, several optical probes are available for the detection of specific enzymes that play crucial roles in many essential biochemical pathways that

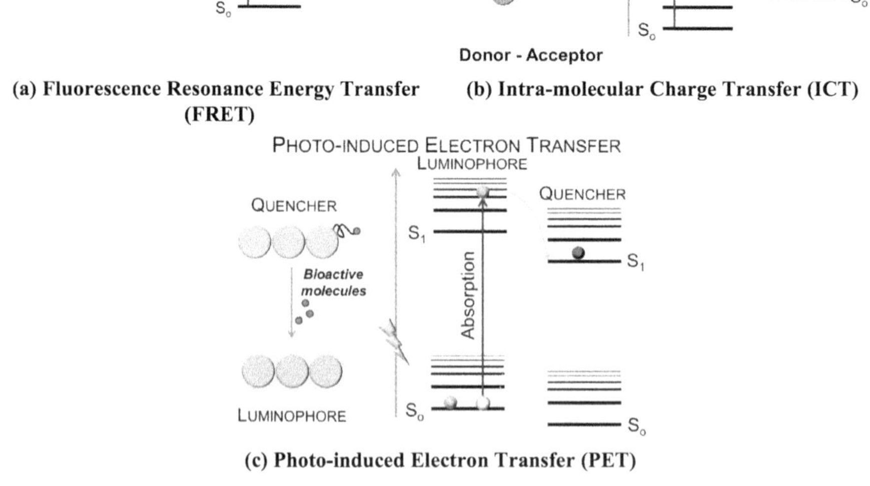

(a) Fluorescence Resonance Energy Transfer (FRET)

(b) Intra-molecular Charge Transfer (ICT)

(c) Photo-induced Electron Transfer (PET)

FIGURE 2.5 Various strategies employed in the development of optical detection probes.

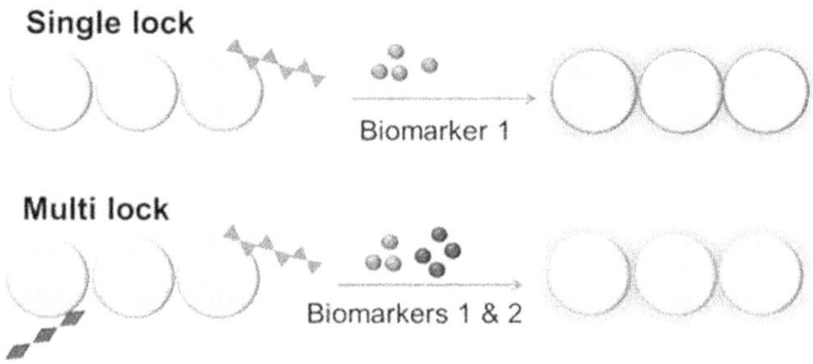

FIGURE 2.6 Schematic differentiating single and multi-lock luminescent probes.

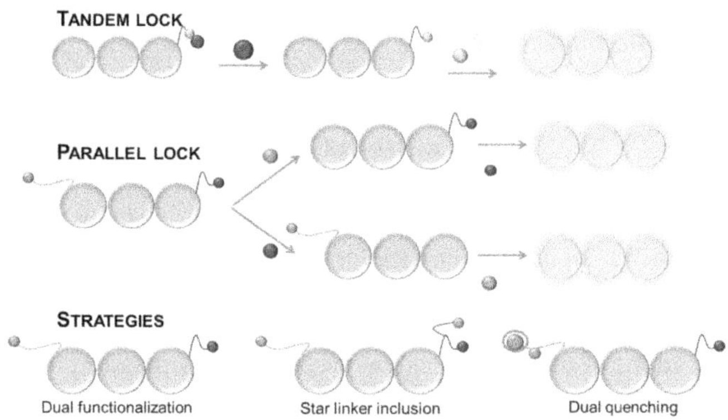

FIGURE 2.7 Sensing strategies with tandem-lock optical probes.

are known to influence disease progression. Under diseased conditions, these enzymes are expressed abnormally in physiological systems. Such probes are termed enzyme-responsive probes. These probes are limited in their applications under certain circumstances where significant signals are detected even from healthy tissues. To overcome this limitation, the concept of multi-lock optical probes has emerged, wherein the same probe can detect more than one enzyme or biomarker.[36] The multi-lock probes are reported with higher specificity toward target tissues and even possess the capability of differentiating disease subtypes (Figure 2.6).

2.5 MULTI-LOCK OPTICAL PROBES

The multi-lock optical probes are classified under two categories—(1) optical signals turned on when all the biomarkers are detected and (2) optical signals turned on when any of the biomarkers are detected. The former case is further classified as (1) tandem-lock optical probes and (2) parallel-lock optical probes. In tandem-lock probes, the sequential and successive biomarker-responsive elements undergo specific changes further to expose the other biomarker-responsive elements sequentially (Figure 2.7). The activation of such probes happens through the precise sensing of biomarker elements. However, the designs of these probes are limited by the projected difficulty in selecting highly specific substrates to enable sequential cleavage. Moreover, such probes can be enhanced by coupling them with external optical stimuli to enable their applications for molecular imaging in live cells with defined spatiotemporal control.

The molecular design strategies involved in the development of parallel-lock optical probes include multi-functionalization, star linker combination, and multi-quenching mechanisms.[38] In the first case, multiple biomarker-responsive elements are engineered on a tunable luminophore. In the second mechanism, multiple locking sites are combined through a star linker with multiple sites of functional groups for the incorporation of distinct biomarker-responsive elements. In the third approach, multiple fluorescence quenching mechanisms will be incorporated (Figure 2.8).

Some examples of tandem and parallel optical multi-lock probes for cancer diagnostics are listed in Table 2.2.

In the other case, where optical signals get turned on upon sensing of the biomarkers, based on the molecular design and sensing mechanisms, they are classified as parallel channel and alternate channel optical probes. In parallel channel optical probes, separate and simultaneous optical outputs are obtained upon sensing different biomarkers.[39,40] With alternate channel optical probes, the same luminescent probes give different optical outputs for different biomarkers under different stimuli. However, the real-time applications of these probes are limited by the inevitable crosstalk between the signals from different biomarkers, which leads to compromised sensitivity and specificity. Some examples of parallel channel and alternate channel optical probes are listed in Table 2.3.

The classifications of multi-lock optical probes are summarized in Figure 2.8.

TABLE 2.2
List of Multi-Lock Optical Probes for Specific Biomarker Sensing in Cancer Diagnostics[36]

Probe Design Strategy	Biomarkers	Triggering Signal	Applications
	β-galactosidase and nitroreductase	λ_{ex} 555 nm and λ_{em} 582 nm	To monitor the enzyme activity *via* intracellular imaging
	Esterase and peroxide	λ_{ex} 545 nm and λ_{em} 590 nm	
	Esterase and glutathione	λ_{ex} 615 nm and λ_{em} 754 nm	For the diagnosis of breast cancer
	Alkaline phosphatase and hydrogen sulfide	λ_{ex} 640 nm and λ_{em} 760 nm	For the diagnosis of breast cancer
	Hypochlorite and cathepsin B	bioluminescence: λ_{em} 600 nm	For the diagnosis of breast cancer
	Nitroreductase followed by light exposure	λ_{ex} 532 nm and λ_{em} 550 nm	To monitor the enzyme activity via intracellular imaging
	B-d-glucosidase	λ_{ex} 340 nm and λ_{em} 390 nm	To monitor the enzyme activity via intracellular imaging
Parallel	Phosphodiesterase I and β-D-glucosidase	λ_{ex} 340 nm and λ_{em} 460 nm	To monitor the enzyme activity via intracellular imaging

(Continued)

TABLE 2.2 (*Continued*)
List of Multi-Lock Optical Probes for Specific Biomarker Sensing in Cancer Diagnostics[36]

Probe Design Strategy	Biomarkers	Triggering Signal	Applications
	β-galactosidase and leucine aminopeptidase	λ_{ex} 360 nm and λ_{em} 530 nm	To monitor the enzyme activity via intracellular imaging
	Peroxide and tyrosine	λ_{ex} 550 nm and λ_{em} 588 nm	For the diagnosis of melanoma cancer
	Peroxide and tyrosine	λ_{ex} 665 nm and λ_{em} 684 nm	For the diagnosis of melanoma cancer
	Tyrosine and hydrogen sulfide	λ_{ex} 574 nm and λ_{em} 600 nm	For cell imaging of selective melanoma
	Nitroreductase	λ_{ex} 670 nm and λ_{em} 710 nm	For imaging of hypoxic mitophagy
	γ-glutamyl transferase followed by acidification	λ_{ex} 710 nm and λ_{em} 770 nm	For the diagnosis of metastatic ovarian cancer
	β-galactosidase followed by light exposure	Fluorescence: λ_{em} 600 nm	For the diagnosis of breast cancer
Tandem	β-galactosidase followed by light exposure	Chemiluminescence: λ_{em} 600 nm	For the diagnosis of breast cancer

TABLE 2.3
List of Parallel and Alternate Multi-Lock Optical Probes for Biomarker Sensing in Cancer Diagnostics

Design Strategy	Biomarkers	Triggering Signal $\lambda_{ex}/\lambda_{em}$ (nm)	Applications
Parallel	Intrinsic state	340/656	Intracellular imaging of enzyme activity
	Phosphodiesterase I	340/460	Intracellular imaging of enzyme activity
	β-D-glucosidase	340/390	Intracellular imaging of enzyme activity
	Phosphodiesterase I and β-D-glucosidase	340/460	Intracellular imaging of enzyme activity
Alternate	Intrinsic	340/656	Intracellular imaging of enzyme activity
	Esterase	340/656	Intracellular imaging of enzyme activity
	ALP	340/390	Intracellular imaging of enzyme activity
	Esterase & ALP	340/460	Intracellular imaging of enzyme activity
	Carboxylesterase	Ratiometric $\lambda_{ex} = 410$; $\lambda_{em} = \lambda_{550}/\lambda_{460}$	Mitochondrial acidification imaging
	Acidic pH	Ratiometric $\lambda_{ex} = 530$; $\lambda_{em} = \lambda_{620}/\lambda_{550}$	Mitochondrial acidification imaging

2.5.1 Probes to Detect Bio-Macromolecules

Numerous optical probes have been widely used in the diagnosis and prognosis of cancer and other diseases. The enzyme-responsive probes are commonly used for this purpose. Particularly, cancer theranostics employ such optical probe-based

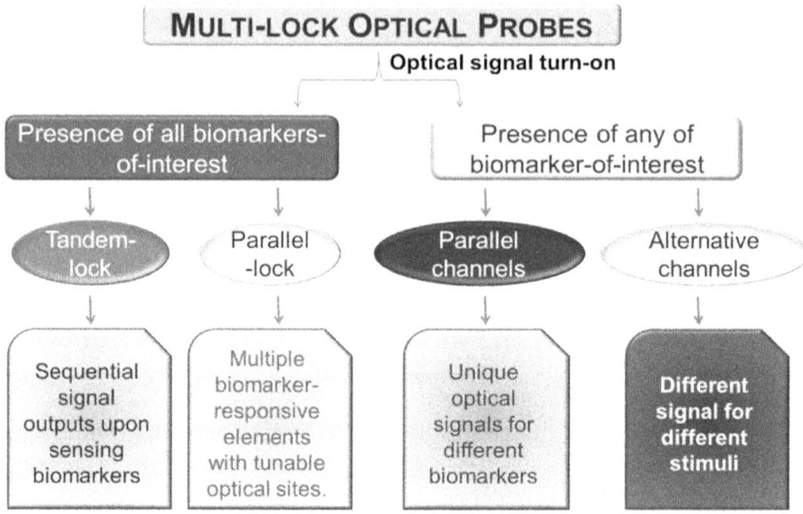

FIGURE 2.8 Classifications of multi-lock optical probes.

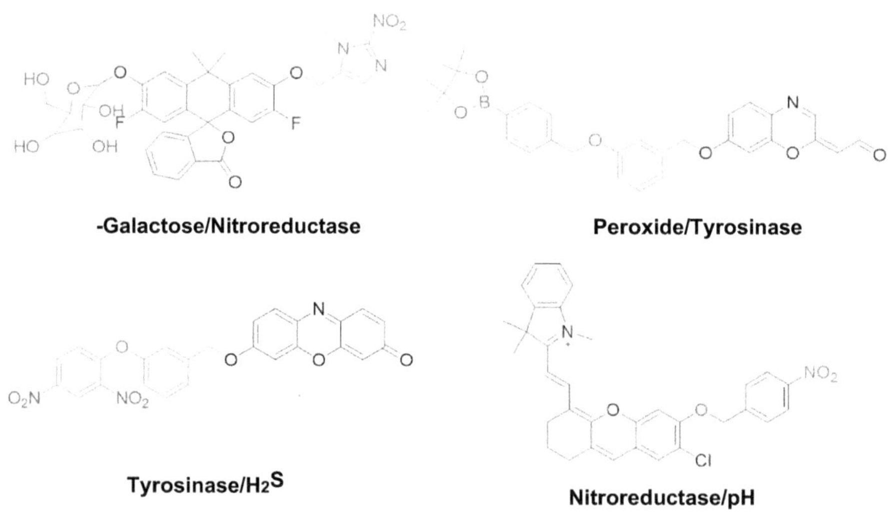

FIGURE 2.9 Chemical structures of multi-lock cancer diagnostic probes.

interventions for them to be included as precise medicine. These optical probes have proven their potential as enhanced cancer diagnostics with high sensitivity and specificity, along with high signal-to-noise ratios. Some of the reported multi-lock optical probes used in cancer diagnostics are shown in Figure 2.9.

There are optical probes available for the detection of enzymes that are overexpressed under cancer conditions—β-galactosidase, nitroreductase, leucine aminopeptidase, tyrosinase, and so forth. Apart from these, there are also probes that can sense small molecule biomarkers like H_2O_2, H_2S, glutathione, and so forth.

Taking advantage of the tyrosinase enzyme that catalyzes tyrosine to dopaquinone conversion further to produce melanin during melanogenesis, Zhan et al. have designed a probe for the detection of tyrosinase enzyme to facilitate melanoma imaging (Figure 2.10). This probe gets activated and releases fluorescent methylene blue upon sensing H_2O_2–Tyr cascade.[41,42]

The FRET quenching mechanism for cancer cell imaging was reported by Chen et al.[43] The design of this probe **A** included a covalent conjugation of the fluorophore to a quencher through a glutathione-responsive disulfide bond, along with an optically tunable hydroxyl group on the hemicyanine scaffold caged inside an esterase-responsive substrate. In normal cells, in the absence of esterase enzyme, the intra-molecular charge transfer is off ("ICT off"), while FRET is facilitated. In cancer cells, where esterase is overexpressed, fluorescence signals are tuned on in the presence of GSH to turn on intra-molecular charge transfer while turning off FRET. The above mechanism is schematically represented in Figure 2.11.

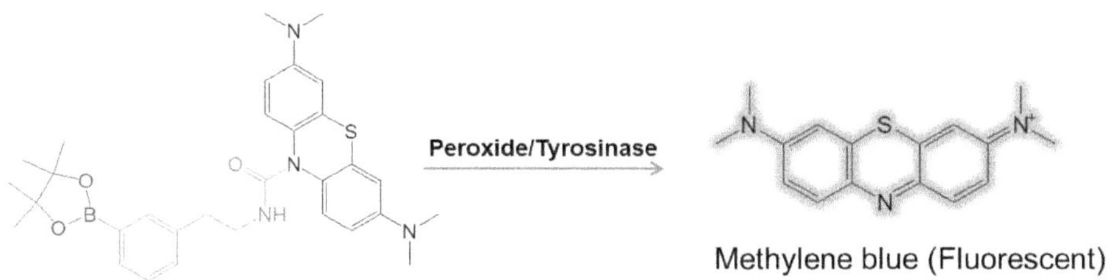

FIGURE 2.10 Optical imaging probe to detect melanogenesis by the release of fluorescent methylene blue.

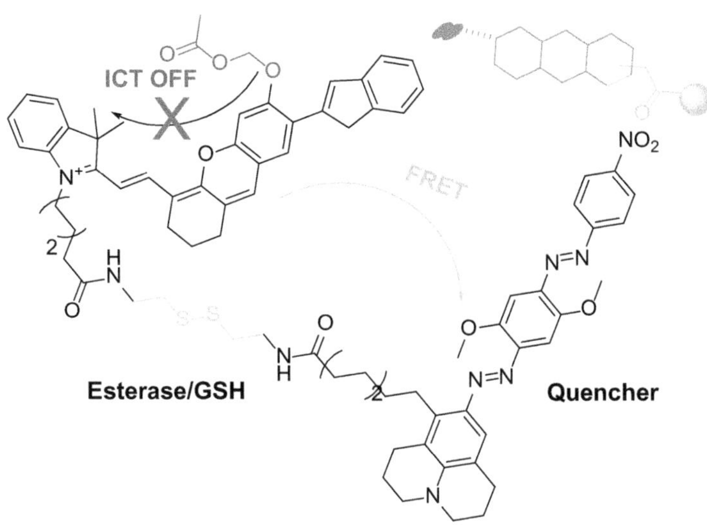

A

FIGURE 2.11 FRET quenching mechanism-based optical probes for cancer cell imaging.

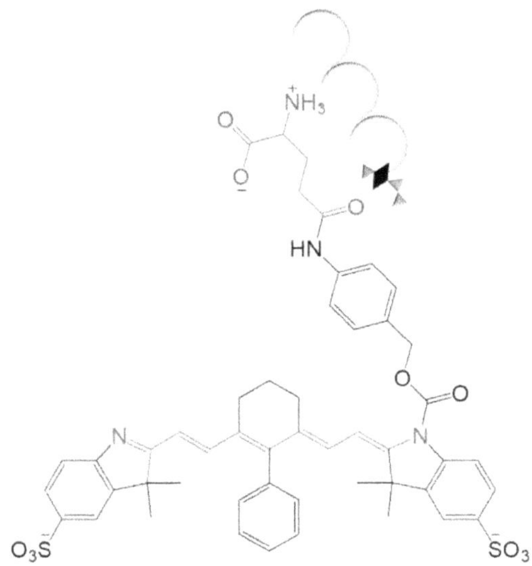

B

FIGURE 2.12 Structure of a chemiluminescent probe with luminophore–adamantylidene scaffold for cancer detection.

Another optical probe that emits fluorescence in the NIR region upon variation in pH was designed, along with the presence of overexpressed membrane protease γ-glutamyl transpeptidase in cancer cells.[44,45]

On the other hand, there are probes that are photo-activatable. Yao et al. have reported such a chemiluminescent probe containing a luminophore–adamantylidene scaffold **B** (Figure 2.12). The sequence of reactions that happen upon sensing β-galactosidase results in the accumulation of pre-chemiluminophores.[46] This, upon triggering with light, further generates 1,2-dioxetane, and the chemiluminescence is recovered.

2.6 OPTICAL PROBES FOR PARKINSON'S DISEASE DIAGNOSIS

Till date, the diagnostics and therapeutics for neurodegenerative Parkinson's disease are limited and demand significant improvement in their approaches. Development of fluorescent probes for imaging relevant biomarkers is being conducted. The pathogenesis of Parkinson's disease is associated with altered levels of reactive oxygen species (ROS) and inflammatory conditions, along with increased levels of HOCl in neurons. In this context, a probe to sense HOCl was recently developed and reported.[47,48] In this, a phenothiazine ring system and *o*-aminophenyl pyrilium group are used as the recognition elements for HOCl and non-HOCl reactive ROS-species. This probe is non-fluorescent under isolated conditions and becomes fluorescent when both pyrilium and phenothiazine groups are oxidized by ROS to lactone and sulfoxide, respectively. Probe **C** was designed to provide a rapid response with notable fluorescence enhancement and exceptional sensitivity to HOCl (Figure 2.13). Moreover, this probe was also demonstrated to be employed in the exploration of pathogenesis caused by HOCl-related diseases.

Peroxides are another ROS that plays a crucial role in Parkinson's disease. Liu et al. have developed a probe (**D**, Figure 2.13) having a boronate ester as the recognition moiety to sense the peroxides.[49] In this, the boronate ester influences the phenol functionality toward its electron-donating capability further to produce green fluorescence, by inhibiting the excited state intramolecular proton transfer (ESIPT)-induced NIR luminescence. Upon interaction of this probe with peroxides, the boronate ester is removed *via* oxidation to restore the phenol moiety and ESIPT-induced NIR fluorescence of the probe is observed. This probe possessed the ability to detect peroxides with the highest sensitivity and selectivity.

An NIR luminescent probe **E** to target lysosomes was reported by Quan et al.[50] This probe can detect another important biomarker, in Parkinson's disease, formaldehyde, using rhodamine derived as fluorophores, morpholines as targeting groups.

A dual-responsive and NIR-emissive luminescent probe **F** was developed to detect H_2S in mitochondria.[51] In this probe, a benzoindole salt is linked to a molecular rotor *N*,*N*-dimethyl-4(thiophen-2-yl)-aniline. Upon sensing H_2S, the C=C induces an enhancement in blue luminescence. Bei et al. developed another H_2S-sensing probe, Probe **G,** by incorporating 2-naphthyl benzothiazole fluorophore, a p-azide phenyl carbamate, as the reducible H_2S reaction site and a triphenylphosphonium moiety for mitochondrial targeting.[52] The carbamate moiety is cleaved upon reduction of the azide, further to release a fluorescence shifter amine. Zhang et al. developed a

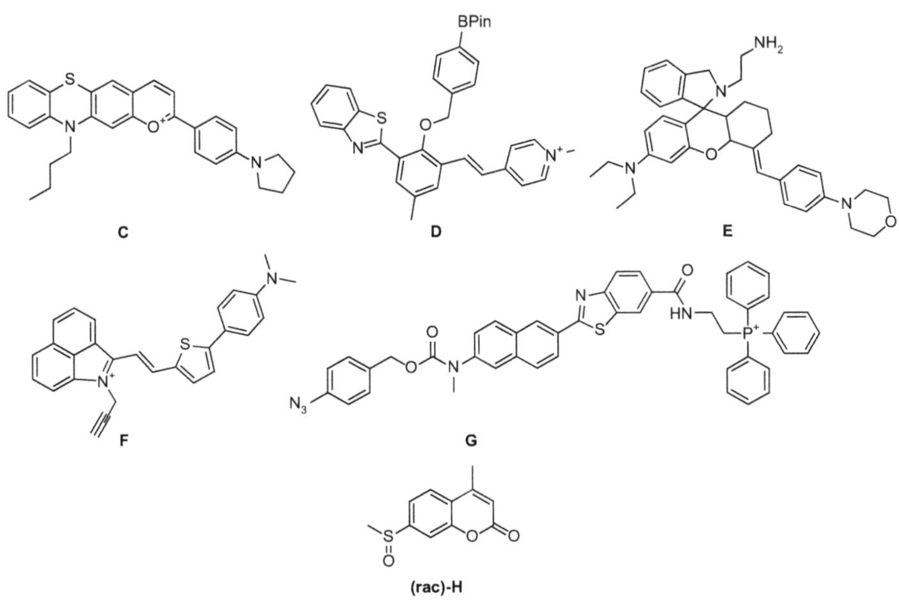

FIGURE 2.13 Structures of optical probes (**C–H**) for Parkinson's disease.

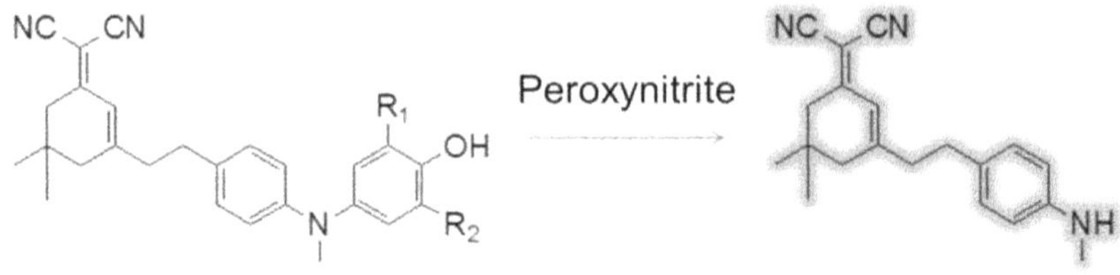

FIGURE 2.14 NIR-based probes for the detection of peroxynitrite for Parkinson's disease diagnostics.

luminescent optical probe, Probe **H**, designed to detect the methionine sulfoxide reductase enzyme, which facilitates the reduction of methionine sulfoxide to methionine.[53]

Peroxynitrite is another ROS linked to Parkinson's disease and displays its role as an oxidative stress inducer under diseased conditions, and hence its detection is also used as a biomarker in the diagnosis. A series of NIR emission-based probes (Figure 2.14) were developed by Xiong et al. to sense peroxynitrite.[54]

The design of these probes holding peroxynitrite reactive site adopts a typical sandwich structural design with (donor–π–acceptor), with dicyanoisophorone as fluorophore (donor) and a *p*-aminophenol as acceptor with different substitutions. This probe has proven to be highly selective and sensitive with a rapid response time for peroxynitrite detection.

2.7 OPTICAL PROBES FOR ALZHEIMER'S DISEASE DIAGNOSIS

The pathogenesis of progressive brain disorder, Alzheimer's disease, remains elusive to date. The abnormal accumulation of proteins forming amyloid plaques that induces tissue death and neural network breakdown is believed to be the cause of the disease. There are numerous biomarkers reported for Alzheimer's disease in the literature and are being detected with a variety of luminescent probes (Figure 2.15).[55]

In general, the polypeptide Aβ is recognized as the biomarker along with reactive oxygen/nitrogen species. The β- and γ- secretase enzymes hydrolyse the amyloid precursor protein to yield the polypeptide Aβ. The accumulation of these polypeptides in amyloid plaques is regarded as the key factor in the commencement and advancement of the disease.

Recently, a curcumin-based Aβ-specific fluorescent probe **I** (Figure 2.15) was reported by Yu et al.[56] with the capability of tomography and photo imaging of amyloidosis in the brain. This probe possessed the capability of quantitatively sensing Aβ fibrils with high specificity. It is also capable of differentiating the Aβ fibrils from their respective monomeric forms.

Gu et al. reported a "photo-triggered" luminescent probe **J** that can detect Aβ aggregates.[57] This probe, when the 6-nitroveratryl protecting group is irradiated, gets cleaved further to induce cyclization of the subsequent phenolic moiety onto the adjacent ester to produce a coumarin-based fluorescent core. This luminescence from this fluorophore is enhanced multi-fold upon binding to Aβ. In vivo experiments have revealed efficient blood–brain barrier permeability of this probe upon light

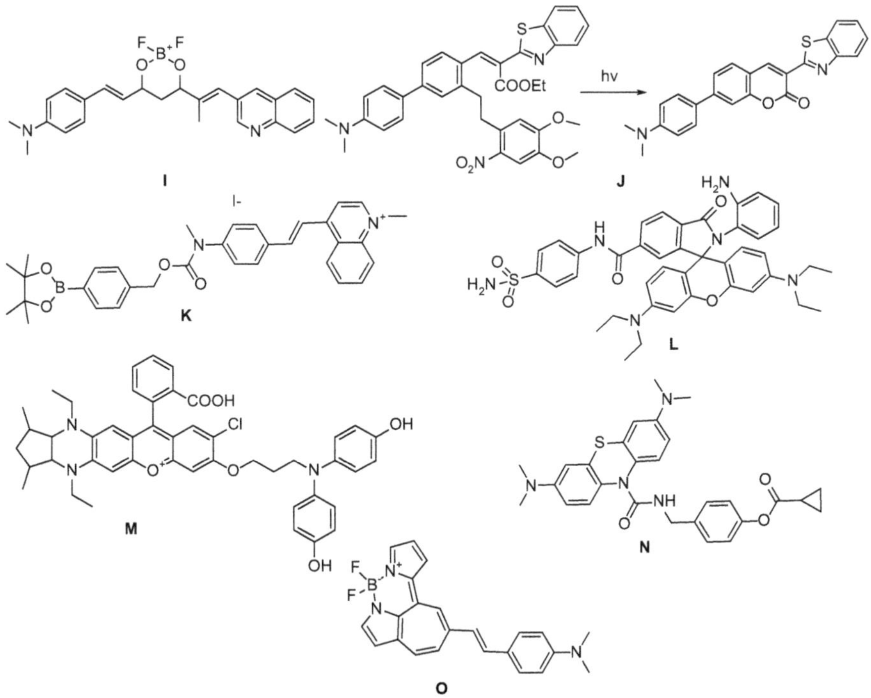

FIGURE 2.15 Structures of luminescent probes (**I–O**) for Alzheimer's disease diagnostics.

exposure. The photo-trigger strategy has demonstrated its potential to create various photo-triggered fluorophores, which can significantly aid in detecting specific biomolecules with a high signal-to-noise ratio.

Like in Parkinson's disease, oxidative stress is another biomarker, which is triggered by an abnormal production and accumulation of reactive oxygen/nitrogen species. Wang et al. developed a dual-sensing fluorescent probe **K** that targets both the Aβ peptide as well the alteration in hydrogen peroxide levels.[58] This probe is designed with a methylquinolinium moiety serving as the fluorescent moiety and a phenylboronate pinacol ester acting as the peroxide sensing unit. When sensing the plaque, this probe becomes susceptible to reacting with peroxide further to oxidize, which oxidizes the phenylboronate pinacol ester, resulting in the formation of a luminescent species that can quantitatively detect hydrogen peroxide at Aβ sites.

He et al. have reported a Golgi-targeting fluorescent optical probe **L** for the quantitative detection of nitric oxide.[59] The design of this probe includes 6-carboxyrhodamine B as the luminophore, 4-sulfamoylphenylamide as the Golgi-targeting moiety, and *o*-diaminobenzene as the nitric oxide sensing group.

Wang et al. have developed an NIR fluorescent probe **M** to sense peroxynitrite.[60] The design of this probe includes a diamino-substituted rhodol dye as the luminophore and a 4-aminophenol moiety as the peroxynitrite recognition moiety. This probe was used by Wang et al. to demonstrate the ability of peroxynitrite to serve as a biomarker for Alzheimer's disease.

The enzyme butyrylcholinesterase is considered a prominent target for Alzheimer's disease diagnosis. It is known that the levels of this enzyme progressively increase under diseased conditions. Zhang et al. designed and synthesized a "logic-gate"-based fluorescence probe **N** with a butyrylcholinesterase-specific cyclopropyl formate moiety and amide/ester bonds as ROS-reactive groups.[61] The fluorescence signals get turned on only when this probe senses both ROS and butyrylcholinesterase, projecting this as a highly sensitive and selective dual-analyte probe. Similarly, the neurofibrillary tangles, consisting of unusually hyper-phosphorylated tau proteins, are considered another hallmark of Alzheimer's disease. Utilizing the same, Xie et al. have developed a highly specific NIR-based probe **O** wherein a cycloheptatriene-BODIPY derivative is used.[62]

2.8 ENZYME-RESPONSIVE OPTICAL PROBES

In conventional clinical diagnostics, hepatic injuries are diagnosed by monitoring the levels of biomarkers in blood, like leucine aminopeptidase, monoamine oxidase, γ-glutamyl transpeptidase, caspase-3, glutathione S-transferase, nitroreductase, and so forth, which are detected at abnormal levels under diseased conditions.

During hepatic injuries, the cytoplasmic enzyme leucine aminopeptidase penetrates into the blood and hence gained its significance in clinical diagnosis for liver disease. For this purpose, a multi-lock NIR-based probe was developed that gets activated upon sensing enzyme-specific biomarkers of hepatopathy—leucine aminopeptidase and monoamine oxidase.[63] With this probe, it is possible to screen the extent of infiltration of these two enzymes in the blood stream. The principle of this probe involves the quenching of fluorescence upon the recognition and cleavage of the leucine-propylamino-p-hydroxybenzyl sequence by the overexpressed leucine aminopeptidase to expose the propylamino group, which is further oxidized by monoamine oxidase, followed by β-elimination to give NIR fluorescence emission. The enzymatic reactions, when followed through mass spectrometry and high-performance liquid chromatography under in vitro conditions, have revealed the second step as the rate-determining step in the reaction kinetics.

A high concentration of glutathione S-transferase in the liver catalyses the conjugation of glutathione and other toxic hepatic metabolites to reduce toxicity. Under conditions of hepatic damage, glutathione S-transferase and glutathione are released into the blood plasma from the liver, thus altering the levels of the enzyme in the liver. Taking advantage of this condition, Zhang et al. designed a naphthalimide-based two-photon fluorescent probe that gets activated by glutathione S-transferase to monitor the enzyme level.[64]

More recently, self-immolative probes were successfully employed to reveal the functions of numerous enzymes.[65] The rapid real-time diffusion of released fluorophores away from the active site limits the effectiveness of the existing optical chemosensors toward estimating the biological activity of enzymes. This challenge was addressed with the evolution of self-immolative probes, wherein the highly reactive quinone methide intermediates rapidly interact with the nearby nucleophiles and hence allow for the determination of enzyme activity locations at higher resolution.

A three-component chemosensor probe **P** to detect α-N-acetylgalactosaminidase enzyme was (Figure 2.16) reported by Kalidasan et al.[66] This includes α-N-acetylgalactosamine as the warhead, fluorescein as the luminescence reporter, and dabcyl moiety as the quencher. All the three components are linked *via* the mandelic acid core. When the above probe senses α-N-acetylgalactosaminidases, the fluorescein moiety linked with quinone methide is cleaved, which further reacts with other proximate nucleophiles. The mechanism of sensing is depicted in Figure 2.16. Qian et al. employed a similar mechanism to detect caspase-3/-7 activity in apoptosis-induced cancer cells.[67]

The potential of self-immolative probes was also demonstrated for the detection of β-galactosidase by Doura et al. (Figure 2.17).[68] They have designed a probe **Q** that can sense lacZ-positive cells by acting as the enzyme substrate. With these self-immolative probes, the enzyme activity in live cells can be precisely located through bio-imaging. Recently, numerous optical sensors for theranostic applications were reported, particularly for the diagnostics and therapeutics of cancer. These sensor probes act as prodrugs wherein a self-immolative spacer links the

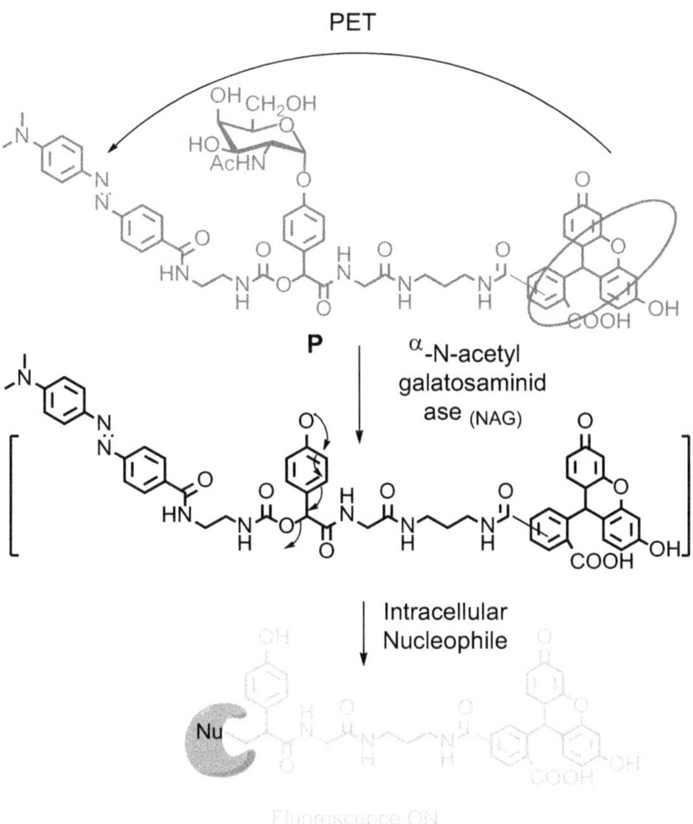

FIGURE 2.16 Sensing mechanism of the designed optical probe to sense a-N-acetylgalac -tosaminidases.

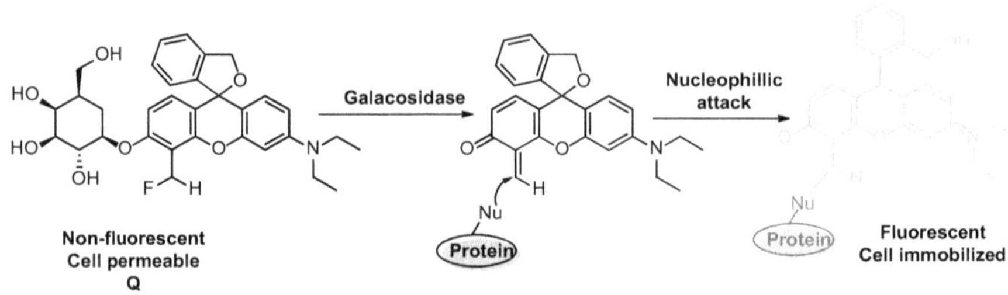

FIGURE 2.17 Sensing mechanism of probe **Q** (substrate for β-galactosidase) developed to detect lacZ-positive.

chemotherapeutic drug and a fluorophore. Generally, these sensors are triggered by overexpressed biomolecules like enzymes, metabolites, small molecules, or other reactive species in tumour cells. The drug and the fluorophore are released following a cascade of reactions where the drugs serve as therapeutics while the fluorophore serves as a diagnostic probe.

Roth-Konforti et al. developed another self-immolative probe **R** to detect cancer, through sensing of cathepsin B.[69] In this, the prodrug was triggered by overexpressed cathepsin B in cancer cells. Upon triggering the prodrug with cathepsin B, the therapeutic drug melphalan is released along with the fluorophore, as depicted in Figure 2.18a. Following the above, numerous enzymatic probes that can be triggered by their hydrolase activities were reported. In our recent report, we have revealed a self-immolative probe **I** that carries a NADH recognition moiety, a self-immolative spacer, and a fluorogenic agent. Upon recognizing NADH, this probe undergoes a sequence of reactions to finally release the fluorogenic drug moiety, as shown in Figure 2.18b.65

The development and progression of different cancers are associated with the abnormal activities of phosphatases, thus making them a significant biomarker for cancer diagnosis. Along with this requirement, a water-soluble fluorescent probe **T** based on the excited-state intramolecular proton transfer reaction (Figure 2.19) was reported to monitor the activity of alkaline phosphatases in live cells. Upon sensing alkaline phosphatase, this probe undergoes ESIPT reaction to display a deep red fluorescence. Numerous other probes were reported in the recent past for the detection of enzymes, including aminopeptidases like leucine aminopeptidase, which serves as an important biomarker to track the progression of cancer.

2.9 QUANTUM DOTS IN OPTICAL DIAGNOSTICS

Likewise, quantum dots doped with functionalized antibodies were employed for the detection of proteins that followed the FRET mechanism. To detect IgG in human samples, quantum dots doped with IgG antibodies from various human sources were immobilized in combination with gold nanoparticles. The luminescence of the quantum dots is quenched upon detecting IgG. In this approach, a complex is formed in which the donors and acceptors are sandwiched together, allowing for the monitoring of fluorescence quenching from the quantum dots while simultaneously detecting the FRET-induced emission, as shown in Figure 2.20.70

In the case of enzymes, the detection of trypsin was reported to employ doped quantum dots immobilized with arginine. In this method, quenching of fluorescence

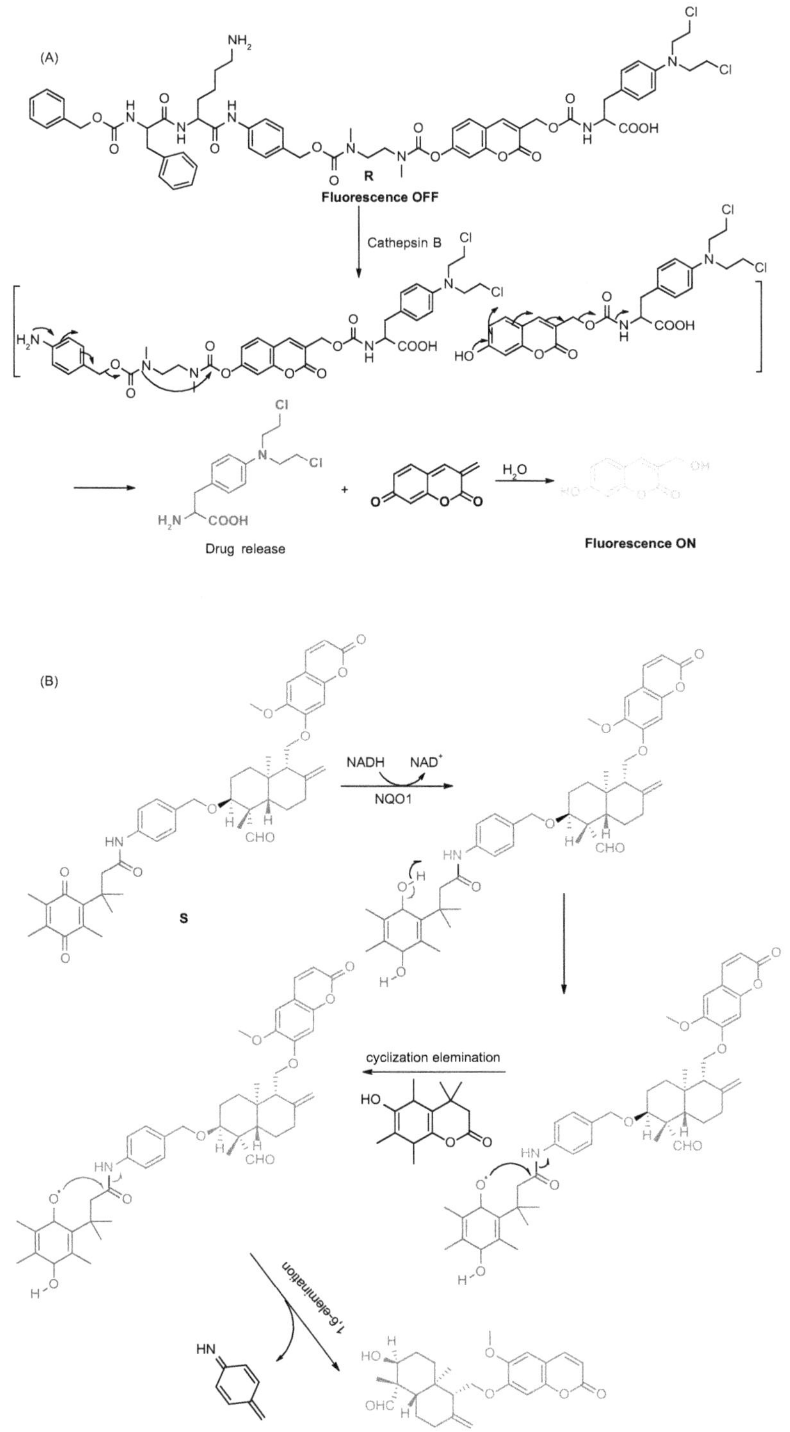

FIGURE 2.18 Theranostic mechanism of self-immolation (A) triggered with cathepsin B. (B) NADH recognition for cancer theranostic applications.

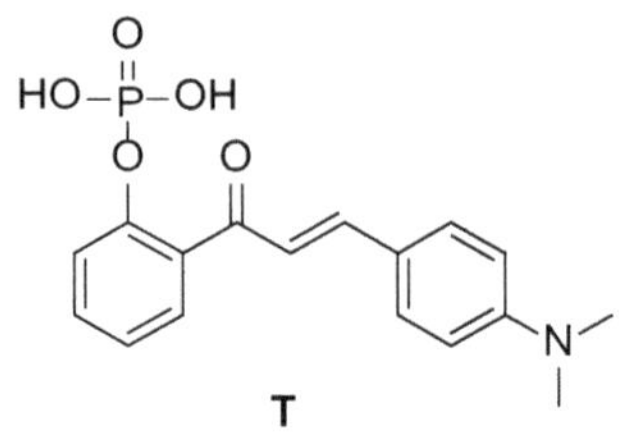

FIGURE 2.19 Structure of water-soluble fluorescent probe **T** for phosphatase enzyme detection.

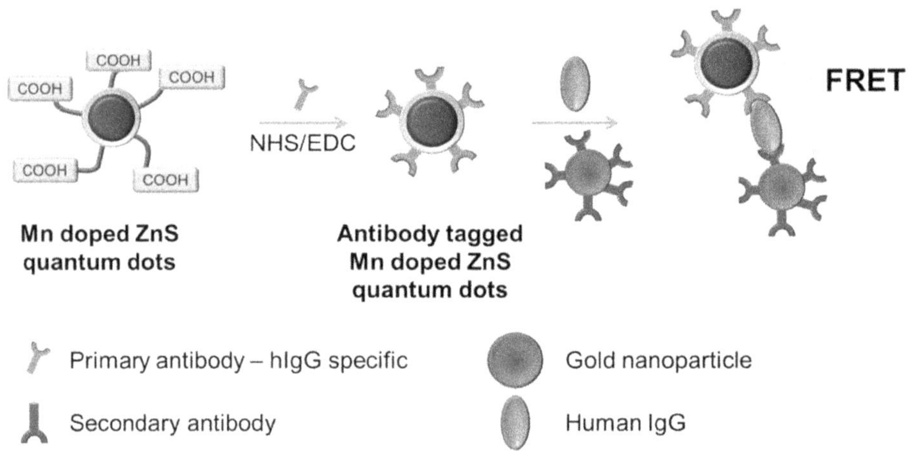

FIGURE 2.20 Schematic representing the use of antibody-tagged quantum dots to sense Immunoglobulin G.

is observed. In this mechanism, the restoration of luminescence from quantum dots is observed with the breakdown of arginine by trypsin. Similarly, in the case of methyl violet, doxorubicin, and other small molecules, the intercalation across the grooves of double-stranded DNA was successfully adopted in the detection of nucleic acids. The electrostatic interaction of these small molecules with doped quantum dots quenches the luminescence emission. On the other hand, the fluorescence is restored when the nucleic acids are detected.

2.10 PROBES TO DETECT PATHOGENS

The inhibition of microbial and human cell viability and their proliferation is known to be inhibited by numerous quantum dots that are functionalized and doped with different metals. The same was demonstrated in the case of ZnS doped with manganese and capped with histidine or L-cysteine for their proven ability towards inhibiting cell proliferation and viability. Similarly, chitosan-capped doped quantum dots were utilized to detect *E. coli*.[71] The positively charged quantum dots mentioned above can cause a blue-shifted emission upon interacting with the anionic cell wall of *E. coli.* These probes were employed in the development of biosensors to detect microbial pathogens in different sources, like water, food, and other samples that are more prone to contamination. In another experiment, it was shown that doped quantum dots that are chitosan-capped can easily infiltrate into the bacterial cell walls and exhibit their functions as internal fluorescent labels upon combining this strategy with fluorescence *in situ* hybridization techniques.

2.10.1 Potential of Optical Technology for UTI Diagnosis

The infections acquired in the urinary tract are a self-limiting disease, caused by uropathogenic microorganisms, including *E. coli, Klebsiella pneumonia, Proteus mirabilis,* and *Streptococcus agalactiae.* Traditional diagnosis of this infection is based on the assessment of urinary symptoms and the detection of pyuria and/or bacteriuria in the patients' urine culture. The gold standard will be the detection of bacteria itself, but the threshold for such diagnosis is limited by the bacterial count, which becomes insufficient in the preliminary stage of infection. Therefore, efforts to identify a more reliable and non-invasive diagnosis is being researched with the objective of early stage detection for systematic treatment of disinfection.

In this regard, particular focus is on the development of optical devices—colorimetric and fluorimetric optical systems that have shown promising results in the diagnosis of UTI. Recent investigations have revealed multiplex optical detection systems with enhanced specificity, selectivity, and sensitivity, as compared to traditional methods. Table 2.4 gives a list of available optical UTI diagnostic devices along with their detection limits.

2.10.2 PoCs for Co-Morbidity Diagnosis

In the modern lifestyle of human society, comorbidity has been a major issue in the development of novel therapies. For the prompt diagnosis of numerous co-existing diseases, efforts are taken to develop PoC devices for multi-disease detection. Recently, microfluidic-based POC devices employing optical techniques, including fluorescence, absorbance, and surface plasmon resonance for multi-disease detection, were developed.

2.10.3 Other Applications of Optical Probes

Various probes for applications in food processing, aimed at monitoring quality and safety, are available to meet the demand for simple, real-time, selective, and cost-effective techniques. For instance, cobalt phthalocyanine was utilized as a probe molecule in the development of an enzyme biosensor to monitor the aging of beer. Biosensors have been used to identify pathogens in food samples and to preserve cleanliness. Various chromophores are being created or manufactured for diagnostic applications in order to produce more dependable, efficient, and quick diagnostic instruments. RT-PCR was the sole and most reliable method for identifying the infection in the recent COVID-19 pandemic. Recently, a number of modified techniques

TABLE 2.4
Various Optical Diagnostic Devices for UTI-Related Pathogens

Device	Target	Sample Volume	Detection Limit	Time for Diagnosis
Paper-based analytical device	*E. coli*	20 μL	10^4 CFU/mL	6 h
Fluorescent sandwich immunoassay	*E. coli*	300 μL	240 CFU/mL	25 min
Dipstick	*E. coli*	500 μL	253 CFU/mL	10 h
Nucleic acid-based multicomponent assay with gold nanoparticles	*E. coli, Proteus mirabilis* and *Klebsiella pneumoniae*	6 μL	10^5 CFU/mL	3 h
Fluorescence sensor	*E. coli, Enterococcus faecium, K. pneumoniae, Enterococcus, Pseudomonas aeruginosa*	10 μL	10^7 CFU/mL	–
Fluorescence sensor	Sodium, potassium, calcium, chloride, and nitrite	2 μL	1.26 mM (sodium), 0.85 mM (potassium), 1.2 mM (calcium), 50 mM (chloride) 0.13 mM (nitrite)	2 h
Fidget spinner	*E. coli*	1,000 μL	10^5 CFU/mL	50 min
Double channel lateral flow immunoassay	*S. aureus and E. coli*	120 μL	9 cells/mL	20 min
Lateral flow immunoassay (colorimetric)	*E. coli*	20 μL	10^3 CFU/mL	45 min
Surface-enhanced Raman scattering chip based Bacteria-grasping assay	*E. coli, P. mirabilis and P. aeruginosa*	10 μL	10^5 cells/mL	60 min
Surface plasmon resonance-based biomimetic sensor	*E. coli*	5 mL	0.57 CFU/mL	20 min
Solution scattering imaging system	*E. coli*	70 μL	200 CFU/mL	10 min
Optical immunobiosensor	*S. aureus and E. coli*	3 μL	10^3 CFU/mL	40 min
Photoluminescence biosensor	*N-acyl homoserine lactone*	n/a	10 nM	–
Bioluminescence assays	*P. mirabilis, C. albicans, S. aureus and E. coli*	9 mL	10^5 CFU/mL	20 min

have also become popular, in which primers are fluorescently labeled for simple and rapid detection. Numerous therapeutic approaches may benefit from anti-biogram based on optical signals that determine the pathogen's susceptibility to different antibiotics. Drug discovery and cancer diagnosis both use fluorescence-based biosensors associated with imaging agents. It is also possible to get insights into the role and regulation of cellular enzymes. Further, genetically encoded FRET-based biosensors have also proven their vital roles in diagnostics.

In cancer therapeutics, photodynamic therapy is being extensively utilized as they are known to display minimal side effects. Numerous light-emitting probes are being adopted as nano-carriers for controlled drug delivery. The application of these optical probes extends toward monitoring gene expressions, localization of biomacromolecules or metabolites in sub-cellular fractions, and various signaling processes and cellular functions.

2.11 CONCLUSIONS

Despite the significant progress made toward the development of biosensors in the past few decades, rapid devices with precise specificity and selectivity have always been in demand with the emergence of complications in the pathogenicity levels of evading microorganisms. Moreover, there always exists a gap between the laboratory demonstrations and on-site applications as PoC devices. One of the most basic requirements of a sensor is its minimal invasive method adoption. Improvements are expected in the following areas of optical probes: (1) Most of the optical probes have a limited depth of penetrating the biological surfaces such as tissues, organs, or organelles. Improving the depth of penetration will provide knowledge on the specificity issues. (2) The main problem with optical probes, as compared to other existing probes, is their signal interference with endogenous luminophores. Therefore, optical probes with precise excitation wavelengths are now made possible with laser optics. This can also simplify the observations made in the diagnosis. (3) For any PoC device development, miniaturization is one of the fundamental requirements. With advancements in nanotechnology and the availability of a variety of semiconductors, this miniaturization has opened avenues in optical diagnostics. Moreover, the use of artificial intelligence has made a significant breakthrough in medical diagnostics. Herein, various optical probes for different diagnostic applications, their design strategies, and their sensing mechanisms were discussed. This chapter is expected to give a basic understanding of the developments made in optical sensors with the present time requirements, especially for beginners in this field.

ACKNOWLEDGMENTS

The authors thank the management of Vellore Institute of Technology for providing the necessary infrastructure to carry out this investigation. KC thanks Vice Chancellor of Rabindranath Tagore University, Hojai for his keen interest in this work.

REFERENCES

1. Dias, G. G.; Rodrigues, M. O.; Paz, E. R. S.; Nunes, M. P.; Araujo, M. H.; Rodembusch, F. S.; da Silva Junior, E. N., Aryl-phenanthro[9,10-d]imidazole: a versatile scaffold for the design of optical-based sensors. ACS Sens **2022**, *7* (10), 2865–2919.
2. Gupta, R.; Rahi Alhachami, F.; Khalid, I.; Majdi, H. S.; Nisar, N.; Mohamed Hasan, Y.; Sivaraman, R.; Romero Parra, R. M.; Al Mashhadani, Z. I.; Fakri Mustafa, Y., Recent progress in aptamer-functionalized metal-organic frameworks-based optical and electrochemical sensors for detection of mycotoxins. *Crit Rev Anal Chem* **2024**, *54* (6), 1707–1728.
3. Miele, E.; Dose, W. M.; Manyakin, I.; Frosz, M. H.; Ruff, Z.; De Volder, M. F. L.; Grey, C. P.; Baumberg, J. J.; Euser, T. G., Hollow-core optical fibre sensors for operando Raman spectroscopy investigation of Li-ion battery liquid electrolytes. *Nat Commun* **2022**, *13* (1), 1651.
4. Nejadmansouri, M.; Majdinasab, M.; Nunes, G. S.; Marty, J. L., An overview of optical and electrochemical sensors and biosensors for analysis of antioxidants in food during the last 5 years. *Sensors* **2021**, *21* (4), 1176.
5. Semenova, D.; Fernandes, A. C.; Bolivar, J. M.; Rosinha Grundtvig, I. P.; Vadot, B.; Galvanin, S.; Mayr, T.; Nidetzky, B.; Zubov, A.; Gernaey, K. V., Model-based analysis of biocatalytic processes and performance of microbioreactors with integrated optical sensors. *Nat Biotechnol* **2020**, 56, 27–37.
6. Ferrari, J. L.; Lima, K. O.; Goncalves, R. R., Refractive indexes and spectroscopic properties to design Er^{3+}-doped SiO_2-Ta_2O_5 films as multifunctional planar waveguide platforms for optical sensors and amplifiers. *ACS Omega* **2021**, 6 (13), 8784–8796.
7. Inglev, R.; Janting, J.; Bang, O., Annular cavity design for photoluminescent polymer optical fiber sensors. *Sensors* **2020**, *20* (18), 5199.
8. Iwanaga, M., A design strategy for surface nanostructures to realize sensitive refractive-index optical sensors. *Nanomaterials* **2023**, *13* (24), 3081.
9. Osorio, J. H.; Chesini, G.; Serrao, V. A.; Franco, M. A. R.; Cordeiro, C. M. B., Simplifying the design of microstructured optical fibre pressure sensors. *Sci Rep* **2017**, *7* (1), 2990.
10. Safarloo, S.; Nunez-Cascajero, A.; Sanchez-Gomez, R.; Vazquez, C., Polymer optical fiber plantar pressure sensors: design and validation. *Sensors* **2022**, *22* (10) 3883.
11. Arabi, M.; Chen, L., Technical challenges of molecular-imprinting-based optical sensors for environmental pollutants. *Langmuir* **2022**, *38* (19), 5963–5967.
12. Du, X.; Zhai, J.; Li, X.; Zhang, Y.; Li, N.; Xie, X., Hydrogel-based optical ion sensors: principles and challenges for point-of-care testing and environmental monitoring. ACS Sens **2021**, *6* (6), 1990–2001.
13. Tong, L., Micro/nanofibre optical sensors: challenges and prospects. *Sensors* **2018**, *18* (3), 903.
14. Chalmers, N. I.; Palmer, R. J., Jr.; Du-Thumm, L.; Sullivan, R.; Shi, W.; Kolenbrander, P. E., Use of quantum dot luminescent probes to achieve single-cell resolution of human oral bacteria in biofilms. *Appl Environ Microbiol* **2007**, *73* (2), 630–636.
15. Gupta, M.; Nagarajan, R.; Ramamurthy, C.; Vivekanandan, P.; Prakash, G. V., KLa((0.95-x))Gd(x)F(4):Eu(3+) hexagonal phase nanoparticles as luminescent probes for in vitro Huh-7 cancer cell imaging. *Dalton Trans* **2021**, 50 (15), 5197–5207.
16. Kougioumtzi, A.; Chatziathanasiadou, M. V.; Vrettos, E. I.; Sayyad, N.; Sakka, M.; Stathopoulos, P.; Mantzaris, M. D.; Ganai, A. M.; Karpoormath, R.; Vartholomatos, G.; Tsikaris, V.; Lazarides, T.; Murphy, C.; Tzakos, A. G., Development of novel GnRH and Tat(48–60) based luminescent probes with enhanced cellular uptake and bioimaging profile. *Dalton Trans* **2021**, 50 (26), 9215–9224.

17. Zhang, S.; Chen, T. H.; Lee, H. M.; Bi, J.; Ghosh, A.; Fang, M.; Qian, Z.; Xie, F.; Ainsley, J.; Christov, C.; Luo, F. T.; Zhao, F.; Liu, H., Luminescent probes for sensitive detection of pH changes in live cells through two near-infrared luminescence channels. ACS Sens **2017**, *2* (7), 924–931.
18. Comby, S.; Surender, E. M.; Kotova, O.; Truman, L. K.; Molloy, J. K.; Gunnlaugsson, T., Lanthanide-functionalized nanoparticles as MRI and luminescent probes for sensing and/or imaging applications. *Inorg Chem* **2014**, *53* (4), 1867–1879.
19. Gupta, B. K.; Narayanan, T. N.; Vithayathil, S. A.; Lee, Y.; Koshy, S.; Reddy, A. L.; Saha, A.; Shanker, V.; Singh, V. N.; Kaipparettu, B. A.; Marti, A. A.; Ajayan, P. M., Highly luminescent-paramagnetic nanophosphor probes for in vitro high-contrast imaging of human breast cancer cells. *Small* **2012**, *8* (19), 3028–3034.
20. Kaneko, Y.; Kobayashi, J.; Yamamoto, Y.; Tsuchiya, K., Light-guided surgery to repair coronary sinus orifice atresia with left superior vena cava. *Eur J Cardiothorac Surg* **2008**, *33* (5), 939–941.
21. Covaleda, G.; Gallego, P.; Vendrell, J.; Georgiadis, D.; Lorenzo, J.; Dive, V.; Aviles, F. X.; Reverter, D.; Devel, L., Synthesis and structural/functional characterization of selective m14 metallocarboxypeptidase inhibitors based on phosphinic pseudopeptide scaffold: implications on the design of specific optical probes. *J Med Chem* **2019**, *62* (4), 1917–1931.
22. Li, Y.; Cai, Z.; Liu, S.; Zhang, H.; Wong, S. T. H.; Lam, J. W. Y.; Kwok, R. T. K.; Qian, J.; Tang, B. Z., Design of AIEgens for near-infrared IIb imaging through structural modulation at molecular and morphological levels. *Nat Commun* **2020**, *11* (1), 1255.
23. Nabo, L. J.; Modzel, M.; Krishnan, K.; Covey, D. F.; Fujiwara, H.; Ory, D. S.; Szomek, M.; Khandelia, H.; Wustner, D.; Kongsted, J., Structural design of intrinsically fluorescent oxysterols. *Chem Phys Lipids* **2018**, 212, 26–34.
24. Price, S.; Mehta, R.; Tan, D.; Hinojosa, A.; Thomas, P. W.; Cummings, T.; Fast, W.; Que, E. L., Structural insights into the design of reversible fluorescent probes for metallo-beta-lactamases NDM-1, VIM-2, and IMP-1. *J Inorg Biochem* **2022**, *233*, 111869.
25. Rowland, R. J.; Chen, Y.; Breen, I.; Wu, L.; Offen, W. A.; Beenakker, T. J.; Su, Q.; van den Nieuwendijk, A.; Aerts, J.; Artola, M.; Overkleeft, H. S.; Davies, G. J., Design, synthesis and structural analysis of glucocerebrosidase imaging agents. *Chemistry* **2021**, *27* (66), 16377–16388.
26. Shi, Z.; Han, X.; Hu, W.; Bai, H.; Peng, B.; Ji, L.; Fan, Q.; Li, L.; Huang, W., Bioapplications of small molecule Aza-BODIPY: from rational structural design to in vivo investigations. *Chem Soc Rev* **2020**, 49 (21), 7533–7567.
27. Zhang, B.; Ge, C.; Yao, J.; Liu, Y.; Xie, H.; Fang, J., Selective selenol fluorescent probes: design, synthesis, structural determinants, and biological applications. *J Am Chem Soc* **2015**, 137 (2), 757–769.
28. Derichsweiler, C.; Herbertz, S.; Kruss, S., Optical bionanosensors for sepsis diagnostics. *Small* **2025**, *21*, e2409042.
29. Kruss, S.; Hilmer, A. J.; Zhang, J.; Reuel, N. F.; Mu, B.; Strano, M. S., Carbon nanotubes as optical biomedical sensors. *Adv Drug Deliv Rev* **2013**, 65 (15), 1933–1950.
30. Pashchenko, O.; Shelby, T.; Banerjee, T.; Santra, S., A comparison of optical, electrochemical, magnetic, and colorimetric point-of-care biosensors for infectious disease diagnosis. *ACS Infect Dis* **2018**, *4* (8), 1162–1178.
31. Shalu Yadav, A. P., Mohd Abubakar Sadique, Pushpesh Ranjan, Neeraj Kumar, Ayushi Singhal, Raju Khan, Emerging point-of-care optical biosensing technologies for diagnostics of microbial infections. *ACS Appl Optical Mater* **2024**, 1 (7), 1245–1262.
32. Magro, L.; Jacquelin, B.; Escadafal, C.; Garneret, P.; Kwasiborski, A.; Manuguerra, J. C.; Monti, F.; Sakuntabhai, A.; Vanhomwegen, J.; Lafaye, P.; Tabeling, P., Paper-based RNA detection and multiplexed analysis for Ebola virus diagnostics. *Sci Rep* **2017**, *7* (1), 1347.

33. Mauk, M.; Song, J.; Bau, H. H.; Gross, R.; Bushman, F. D.; Collman, R. G.; Liu, C., Miniaturized devices for point of care molecular detection of HIV. *Lab Chip* **2017**, *17* (3), 382–394.
34. Phillips, E. A.; Moehling, T. J.; Ejendal, K. F. K.; Hoilett, O. S.; Byers, K. M.; Basing, L. A.; Jankowski, L. A.; Bennett, J. B.; Lin, L. K.; Stanciu, L. A.; Linnes, J. C., Microfluidic rapid and autonomous analytical device (microRAAD) to detect HIV from whole blood samples. *Lab Chip* **2019**, *19* (20), 3375–3386.
35. Hu, Q.; Wei, Q.; Zhang, P.; Li, S.; Xue, L.; Yang, R.; Wang, C.; Zhou, L., An up-converting phosphor technology-based lateral flow assay for point-of-collection detection of morphine and methamphetamine in saliva. *Analyst* **2018**, *143* (19), 4646–4654.
36. Cheng, P.; Pu, K., Enzyme-responsive, multi-lock optical probes for molecular imaging and disease theranostics. *Chem Soc Rev* **2024**, 53 (20), 10171–10188.
37. Huang, J.; Li, J.; Lyu, Y.; Miao, Q.; Pu, K., Molecular optical imaging probes for early diagnosis of drug-induced acute kidney injury. *Nat Mater* **2019**, *18* (10), 1133–1143.
38. Gao, T.; Xiang, C.; Ding, X.; Xie, M., Dual-locked fluorescent probes for precise diagnosis and targeted treatment of tumors. *Heliyon* **2024**, *10* (18), e38174.
39. Ray, P.; Steckl, A. J., Label-free optical detection of multiple biomarkers in sweat, plasma, urine, and saliva. ACS Sens **2019**, *4* (5), 1346–1357.
40. Rockendorf, N.; Ramaker, K.; Gaede, K.; Tappertzhofen, K.; Lunding, L.; Wegmann, M.; Horbert, P.; Weber, K.; Frey, A., Parallel detection of multiple biomarkers in a point-of-care-competent device for the prediction of exacerbations in chronic inflammatory lung disease. *Sci Rep* **2024**, *14* (1), 12830.
41. Zhan, C.; Cheng, J.; Li, B.; Huang, S.; Zeng, F.; Wu, S., A fluorescent probe for early detection of melanoma and its metastasis by specifically imaging tyrosinase activity in a mouse model. *Anal Chem* **2018**, *90* (15), 8807–8815.
42. Peng, H.; Wang, T.; Li, G.; Huang, J.; Yuan, Q., Dual-locked near-infrared fluorescent probes for precise detection of melanoma via hydrogen peroxide-tyrosinase cascade activation. *Anal Chem* **2022**, *94* (2), 1070–1075.
43. Chen, N. T.; Cheng, S. H.; Liu, C. P.; Souris, J. S.; Chen, C. T.; Mou, C. Y.; Lo, L. W., Recent advances in nanoparticle-based Forster resonance energy transfer for biosensing, molecular imaging and drug release profiling. *Int J Mol Sci* **2012**, 13 (12), 16598–16623.
44. Scott, J. I.; Deng, Q.; Vendrell, M., Near-infrared fluorescent probes for the detection of cancer-associated proteases. *ACS Chem Biol* **2021**, 16 (8), 1304–1317.
45. Zhou, F.; Yang, S.; Zhao, C.; Liu, W.; Yao, X.; Yu, H.; Sun, X.; Liu, Y., Gamma-glutamyl transpeptidase-activatable near-infrared nanoassembly for tumor fluorescence imaging-guided photothermal therapy. *Theranostics* **2021**, *11* (14), 7045–7056.
46. Yao, Y.; Zhang, Y.; Yan, C.; Zhu, W. H.; Guo, Z., Enzyme-activatable fluorescent probes for beta-galactosidase: from design to biological applications. *Chem Sci* **2021**, 12 (29), 9885–9894.
47. Ashoka, A. H.; Ali, F.; Tiwari, R.; Kumari, R.; Pramanik, S. K.; Das, A., Recent advances in fluorescent probes for detection of HOCl and HNO. *ACS Omega* **2020**, 5 (4), 1730–1742.
48. Luo, L.; Yang, Y.; Chen, S.; Zhang, P.; Zeng, R., A photoelectrochemical sensor for the detection of hypochlorous acid with a phenothiazine-based photosensitizer. *Molecules* **2024**, *29* (3) 614.
49. Liu, Y.; Bai, L.; Li, Y.; Ni, Y.; Xin, C.; Zhang, C.; Liu, J.; Liu, Z.; Li, L.; Huang, W., Visualizing hydrogen peroxide in Parkinson's disease models via a ratiometric NIR fluorogenic probe. *Sens Actua B Chem* **2019**, 279, 38–43.
50. Quan, W.; Zhang, G.; Li, Y.; Song, W.; Zhan, J.; Lin, W., Upregulation of formaldehyde in Parkinson's disease found by a near-infrared lysosome-targeted fluorescent probe. *Anal Chem* **2023**, 95, 2925–2931.

51. Li, S.;. Huo, F.; Yin, C., NIR fluorescent probe for dual-response viscosity and hydrogen sulfide and its application in Parkinson's disease model. *Dyes Pigm* **2022**, *197*.
52. Bei, S. K.; Heo, C. H.; Choi, D. J.; Sen, D.; Joe, E. H.; Cho, B. R.; Kim, H. M., A ratiometric two-photon fluorescent probe reveals reduction in mitochondrial H_2S production in Parkinson's disease gene knockout astrocytes. *J Am Chem Soc* **2013**, 135, 9915–9923.
53. Zhang, L.; Peng, S.; Sun, J.; Yao, J.; Kang, J.; Hu, Y.; Fang, J., A specific fluorescent probe reveals compromised activity of methionine sulfoxide reductases in parkinson's disease. *Chem Sci* **2017**, 8, 2966–2972.
54. Xiong, J.; Wang, W.; Wang, C.; Zhong, C.; Ruan, R.; Mao, Z.; Liu, Z., Visualizing peroxynitrite in microvessels of the brain with stroke using an engineered highly specific fluorescent probe. ACS Sens **2020**, 5, 3237–3245.
55. Jun, Y. W.; Cho, S. W.; Jung, J.; Huh, Y.; Kim, Y.; Kim, D.; Ahn, K. H., Frontiers in probing Alzheimer's disease biomarkers with fluorescent small molecules. *ACS Cent Sci* **2019**, *5* (2), 209–217.
56. Yu, J.; Shao, C.; Ye, X.; Di, X.; Li, D.; Zhao, H.; Zhang, B.; Chen, G.; Liu, H. K.; Qian, Y., In Vivo brain imaging of amyloid-beta aggregates in Alzheimer's disease with a near-infrared fluorescent probe. ACS Sens **2021**, 6, 863–870.
57. Gu, Y.; Ding, Z.; Zheng, C.; Xu, Y.; Liu, T.; Mao, C.; Ran, C.; Yang, J.; Wang, P., Light-controlled fluorescent probes for precisely monitoring brain amyloid-β aggregates in Alzheimer's disease. *Chem Eng J* **2022**, *446*.
58. Wang, X.; Iyaswamy, A.; Xu, D.; Krishnamoorthi, S.; Sreenivasmurthy, S. G.; Yang, Y.; Li, Y.; Chen, C.; Li, M.; Li, H. W.; Wong M. S., Real-time detection and visualization of amyloid-beta aggregates induced by hydrogen peroxide in cell and mouse models of Alzheimer's disease. *ACS Appl Mater Interfaces* **2023**, 15, 39–47.
59. He, Z.; Liu, D.; Liu, Y.; Li, X.; Shi, W.; Ma, H., Golgi-targeted fluorescent probe for imaging NO in Alzheimer's disease. *Anal Chem* **2022**, 94, 10256–10262.
60. Wang, P.; Yu, L.; Gong, J.; Xiong, J.; Zi, S.; Xie, H.; Zhang, F.; Mao, Z.; Liu, Z.; Kim, J. S., An activity-based fluorescent probe for imaging fluctuations of peroxynitrite (ONOO–) in the Alzheimer's disease brain. *Angew Chem Int Ed* **2022**, *61*, e202206894.
61. Zhang, P.; Fu, C.; Liu, H.; Guo, X.; Zhang, Q.; Gao, J.; Chen, W.; Yuan, W.; Ding, C., AND-logic strategy for accurate analysis of Alzheimer's disease via fluorescent probe lighted up by two specific biomarkers. *Anal Chem* **2021**, 93, 11337–11345.
62. Xie, T.; Li, Y.; Tian, C.; Yuan, C.; Dai, B.; Wang, S.; Zhou, K.; Liu, J.; Tan, H.; Liang, Y.; Dai, J.; Chen, B.; Cui, M., Fused cycloheptatriene-BODIPY is a high-performance near-infrared probe to image tau tangles. *J Med Chem* **2022**, 65, 14527–14538.
63. Liu, Y.; Teng, L.; Xu, C.; Liu, H. W.; Xu, S.; Guo, H.; Yuan, L.; Zhang, X. B., A "Double-Locked" and enzyme-activated molecular probe for accurate bioimaging and hepatopathy differentiation. *Chem Sci* **2019**, *10* (47), 10931–10936.
64. Zhang, J.; Jin, Z.; Hu, X. X.; Meng, H. M.; Li, J.; Zhang, X. B.; Liu, H. W.; Deng, T.; Yao, S.; Feng, L., Efficient two-photon fluorescent probe for glutathione s-transferase detection and imaging in drug-induced liver injury sample. *Anal Chem* **2017**, *89* (15), 8097–8103.
65. Venkatesan, S.; Chanda, K.; Balamurali, M. M., An in silico approach to investigate the theranostic potential of coumarin-derived self-immolative luminescent probes. *Chem Biodivers* **2024**, *21* (2), e202301400.
66. Kalidasan, K.; Su, Y.; Wu, X.; Yao, S. Q.; Uttamchandani, M., Fluorescence-activated cell sorting and directed evolution of alpha-N-acetylgalactosaminidases using a quenched activity-based probe (qABP). *Chem Commun* **2013**, *49* (65), 7237–7239.
67. Qian, L.; Zhang, C. W.; Mao, Y.; Li, L.; Gao, N.; Lim, K. L.; Xu, Q. H.; Yao, S. Q., Two-photon enzymatic probes visualizing sub-cellular/deep-brain caspase activities in neurodegenerative models. *Sci Rep* **2016**, *6*, 26385.

68. Doura, T.; Kamiya, M.; Obata, F.; Yamaguchi, Y.; Hiyama, T. Y.; Matsuda, T.; Fukamizu, A.; Noda, M.; Miura, M.; Urano, Y., Detection of lacz-positive cells in living tissue with single-cell resolution. *Angew Chem Int Ed Engl* **2016**, *55* (33), 9620–9624.
69. Roth-Konforti, M. E.; Bauer, C. R.; Shabat, D., Unprecedented sensitivity in a probe for monitoring cathepsin B: chemiluminescence microscopy cell-imaging of a natively expressed enzyme. *Angew Chem Int Ed Engl* **2017**, *56* (49), 15633–15638.
70. Gao, X.; Tang, G.; Li, Y.; Su, X., A novel optical nanoprobe for trypsin detection and inhibitor screening based on Mn-doped ZnSe quantum dots. *Anal Chim Acta* **2012**, 743, 131–136.
71. Baruah, S.; Ortinero, C.; Shipin, O. V.; Dutta, J., Manganese doped zinc sulfide quantum dots for detection of *Escherichia coli*. *J Fluoresc* **2012**, 22 (1), 403–408.

3 Fluorescence Lifetime Imaging Data Analysis and Interpretation

Viktor V. Nikolaev, Tatiana B. Lepekhina, Hala Zuhayri, Alexey V. Borisov, Olga A. Zakharova, and Yury V. Kistenev
National Research Tomsk State University

3.1 MULTI-EXPONENT FLUORESCENCE LIFETIME IMAGING DATA PRESENTATION

Fluorescence is a kind of spontaneous emission of excited molecules (fluorophores) lasting on the timescale of several nanoseconds. Fluorescence lifetime τ is the average time that a fluorophore remains in its excited state and continues to emit. Taking into consideration that fluorescence intensity $I(t)$ is proportional to the concentration of excited molecules, which decreases after every act of fluorescence by a molecule, τ is defined as the interval of fluorescence decreasing in $1/e$ times relative to its initial value.

In general, the fluorescence lifetime imaging microscopy (FLIM) signal $I(t)$ is approximated by a multi-exponential decomposition as follows [1]:

$$I(t) = I_0 \sum\nolimits_{i=0}^{N} \alpha_i e^{-t/\tau_i} \tag{3.1}$$

where α_i and τ_i are the amplitudes and lifetimes of the fluorescence decay curve components, respectively, and I_0 is the intensity of the measured fluorescence signal at the initial moment $t = 0$. Usually, the fluorescence lifetime decay curve is approximated by two-exponential non-linear regression with fast (τ_1, a_1) and slow (τ_2, a_2) fluorescence lifetime and amplitude decay components:

$$I(t) = \alpha_1 e^{-t/\tau_1} + \alpha_2 e^{-t/\tau_2} \tag{3.2}$$

The mean fluorescence lifetime is defined as:

$$\tau_m = (a_1\tau_1 + a_2\tau_2) / (a_1 + a_2) \tag{3.3}$$

There are two ways of measuring fluorescence lifetime value. In the frequency domain method, pumping laser radiation is frequency-modulated. It causes a time

DOI: 10.1201/9781003570011-3

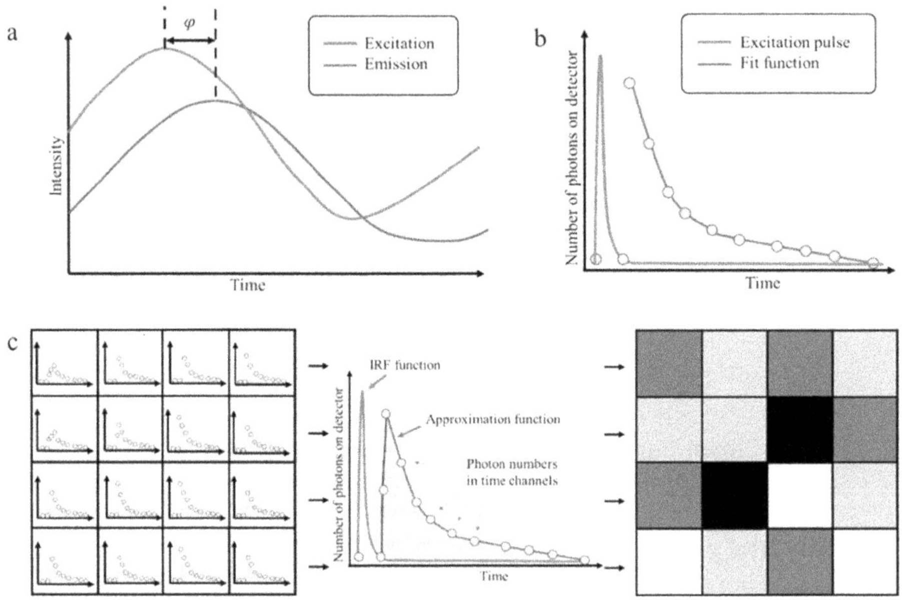

FIGURE 3.1 (a) Time dependence of laser radiation intensity and fluorescence signal in frequency domain method, (b) the signal on a detector from arrival of photons depending on their arrival time in time domain method, and (c) schematic view of graphical representation of FLIM data using fitting function of fluorescence lifetime decay.

delay between the time curve of laser radiation intensity and the fluorescence signal (Figure 3.1a). In the time-domain method, the sample is pumped by a short-duration laser pulse, and the signal on a detector from the arrival of photons is measured depending on their arrival time (Figure 3.1b). An application of multi-exponential fitting for graphical visualization is shown in Figure 3.1c.

Endogenous fluorophores, which can be detected by autofluorescence molecular imaging, can serve as biomarkers of various human diseases and pathological states [2]. This is due to the fact that the emission properties of endogenous fluorophores may depend on the metabolic state of the cell and its microenvironment. Fluorescence microscopy can be used to assess cellular metabolism by imaging endogenous fluorophores [3–5]. The metabolic status can be determined by the integral characteristic of fluorescence excited at different wavelengths and by estimating the fluorescence lifetime decay curve [6]. Typical endogenous fluorophores fluorescence lifetimes are shown in Table 3.1 (adapted from articles [1,2]).

3.2 EXAMPLES OF APPLICATION OF THE MULTI-EXPONENTIAL APPROACH

The papillary dermis is mainly composed of two protein molecules: collagen and elastin. This layer also contains capillaries, which are responsible for supplying the epidermis with nutrients and oxygen. Two-exponential fit of FLIM data obtained using two-photon microscopy allows for the assessment of the inner and outer areas

TABLE 3.1
Fluorescence Lifetime of Endogenous Fluorophores and their Spectral Characteristics

Endogenous Fluorophore	λ_{ex}, nm	λ_{em}, nm	Lifetime (ns)	References
Structural Proteins				
Collagen	280–350(1PE)	370–440	1.15–1.3 (Type III)	[1,2,7]
	760(2PE)		1.55–1.75 (Type I)	[1,2,7]
Elastin	300–370(1PE) 760(2PE)	420–460	0.2–2.5	[1,2]
Metabolic Coenzymes				
Free NAD(P)H	330–360(1PE)	460	0.4	[2,8]
Bound NAD(P)H	760(2PE)	440	1.3–9, 1–6.5	[2,8]
Free FAD	360–450(1PE)	520, 535 (max)	2.3–2.9	[9,10]
Bound FAD	800–900(2PE)		<0.1	[8]
Flavin mononucleotide (FMN)	444 (max)(1PE)	558 (max)	4.27–4.67	[1]
Amino Acids				
Tryptophan	280 (max)(1PE) 600(2PE) 760(3PE)	250–310, 310–360	0.11 or 3.03ns	[11–14]
Phenylalanine	258 (max)(1PE)	280 (max)	7.5 ns	[1]
Tyrosine	275 (max)(1PE)	300 (max)	2.5 ns	[1]
Pigments				
Keratin	227 (max)(1PE) 700–900(1PE)	382 (max)	1.4	[1]
Melanin	300(1PE) 760–800(2PE)	400, 520, 575	0.1–0.2, 0.5–1.8, 7.9	[1,2,15]
Eumelanin	355(1PE)	520	0.058, 0.51, 2.9, 7	[1]
Porphyrins	800 (2PE)	590–635	5.2, 18–20	[16–18]
Protoporhorphyrin IX	400–450(1PE)	630, 690, 710	9.7–16	[1]
Lipofuscin	340–395(1PE) 400–500(1PE) 755, 860(2PE)	540, 430–460	1.34, 0.4	[1,2]
Bilirubin	360–520(1PE)	480–650	0.02–0.09, 1–2	[1]
Vitamins				
Vitamin B6	330 (max)(1PE)	420 (max)	0.6–8.4	[1]
Retinol	327 (max)(1PE) 700–830 (free)(2PE)	~500 (free)	1.8, 5.0 (free), 0.7, 3.6, 12 (bound)	[19,20]
Riboflavin	420–500(1PE)	520–750	4.12	[1]

Source: Adapted from articles [1,2].
1PE, one-photon excitation; 2PE, two-photon excitation; 3PE, three-photon excitation.

of skin capillaries. The analysis of the fingernail bed and inner forearm areas of 6 healthy volunteers, which demonstrates the possibility of visualizing capillaries that are not visible without preliminary processing (capillaries have lower values of the τ_1 parameters compared to the surrounding area), was conducted (see Figure 3.2) [21]. Confirmation of the fact that capillaries have low fluorescence lifetime values was carried out using scanning of human erythrocytes and blood plasma. The blood plasma preparation protocol was chosen as standard for this type of procedure and is described in detail in the article (the blood was centrifuged for 10 min at 1,500 rpm). Two-photon measurements were performed with the Dermainspect (JenLab GmbH, Jena, Germany) device on a wavelength of 760 nm. The processing was based on the two-exponential approach described above (see formulas 3.2 and 3.3).

In Figure 3.2, the dermal papilla has been segmented into three regions—an inner (red), an outer (black), and an "intensity mask" region that corresponds to the pixels in the outer region with the highest autofluorescence intensity (blue). The results of the two-exponential fit of the autofluorescence signal decay curve for erythrocytes and blood plasma show that they correspond to fast fluorescence decay with a narrow lifetime distribution (100 ± 20 ps) and (170 ± 20 ps) for erythrocytes and blood plasma, respectively.

The metabolic activity of fibroblasts of the 3T3 line incubated on ceramics was analyzed using two-photon microscopy (MPTflex, Jena, Germany) [22]. In this study, fibroblasts were incubated on two types of ceramics: yttrium-stabilized zirconia ceramic hardened with aluminum oxide (ATZ) and stabilized zirconia ceramics (Y-TZP). Cover glass was a control group. Two-photon excitation wavelength for the

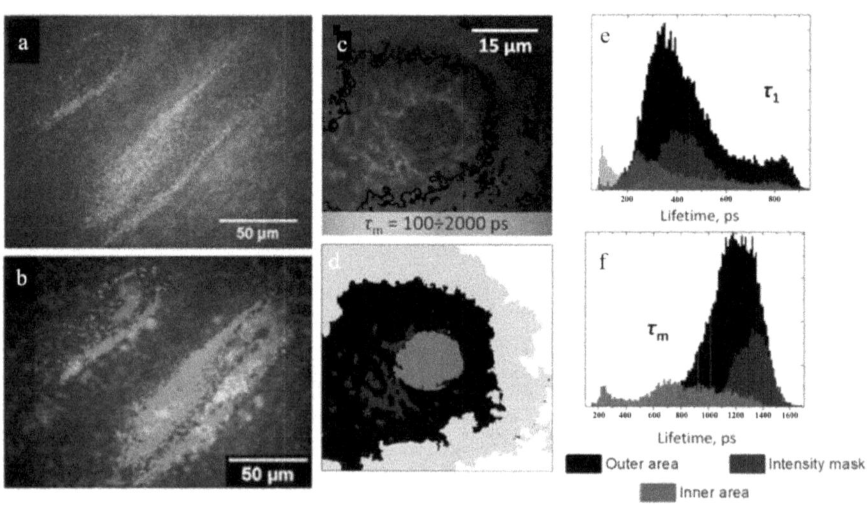

FIGURE 3.2 Two-photon imaging of the fingernail bed area: (a) merged second harmonic generation and autofluorescence signals, (b) merged second harmonic generation and FLIM images, (c) FLIM image of the papillary structure, colors correspond to the values of mean fluorescence lifetime in the 100–2,000 ps range, (d) segmentation of the FLIM image of papilla into the inner area (red), outer area (black) and intensity mask area (blue), the yellow area corresponds to the melanin-rich area around the papilla, (e) distributions of fast decay components (τ_1), and (f) mean fluorescence lifetime [21].

NAD(P)H fluorophore was 760 nm, and for the FAD, it was 830 nm. A two-exponential approximation was used to analyze the decay of the fluorescence lifetime of endogenous fluorophores NAD(P)H and FAD of fibroblasts 3T3 line. Optical characteristics of the 3T3 fibroblasts (for NAD(P)H and for FAD) are presented in Tables 3.2 and 3.3. The relationship between the metabolic state of cells and the fluorescence signal of fluorophores such as NAD(P)H and FAD has been previously shown using two-photon microscopy [23].

For Y-TZP, short (τ_1), long (τ_2), and mean (τ_m) fluorescence lifetimes were higher compared to ATZ and the control groups. The a_1 / a_2 ratio was also higher in the ATZ group.

For ATZ, short (τ_1), long (τ_2), and mean (τ_m) fluorescence lifetimes were higher compared to Y-TZP and the control groups. A higher a_1/a_2 ratio in the control group compared to ATZ and Y-TZP was measured. The amplitude of the short component a_1 has a greater contribution compared to the long component a_2 for both fluorophores.

Two-exponential approximation was used for the visualization the epidermis layers [24]. The authors evaluated the fluorescence decay components' lifetimes (τ_1, τ_2), their ratios (τ_1/τ_2), and the fluorescence decay components amplitude ratios (a_1/a_2) of the stratum corneum and viable epidermis.

Karrobi et al. [25] used an approach for FLIM data analysis to capture changes in the cellular metabolism of 3D breast spheroids. Spheroids were obtained from two cell lines (MCF10A, MD-MB-231) and embedded in a collagen-based gel in a 1–4 ratio. The studies were carried out over time (4 days) and at a depth of ~ 1–2 mm from the collagen gel surface. The measurements were conducted at two wavelengths: 760 and 880 nm for visualization of NAD(P)H or FAD, respectively. Collagen visualization was performed using a second harmonic generation signal with a laser at

TABLE 3.2
Optical Characteristics of the 3T3 Fibroblasts for NAD(P)H [22]

	Optical Characteristics (Mean ± STD)					
	a_1 (%)	a_2 (%)	a_1 / a_2	τ_1 (ps)	τ_2 (ps)	τ_m (ps)
Control	74 ± 11	26 ± 11	3.48 ± 2.1	459 ± 93	2,595 ± 762	774 ± 161
ATZ	75 ± 4	25 ± 4	3.82 ± 1.6	494 ± 110	2,498 ± 409	766 ± 178
Y-TZP	73 ± 9	27 ± 9	3.55 ± 1.8	552 ± 134	2,728 ± 517	881 ± 260

TABLE 3.3
Optical Characteristics of the 3T3 Fibroblasts for FAD [22]

	Optical Characteristics (Mean ± STD)					
	a_1 (%)	a_2 (%)	a_1 / a_2	τ_1 (ps)	τ_2 (ps)	τ_m (ps)
Control	83 ± 6	17 ± 5	4.3 ± 2.1	245 ± 117	1,512 ± 491	438 ± 161
ATZ	82 ± 13	18 ± 13	5.1 ± 2.3	299 ± 120	1,804 ± 408	594 ± 243
Y-TZP	87 ± 10	13 ± 9	4.6 ± 2.2	214 ± 145	1,364 ± 538	434 ± 245

a wavelength of 1,050 nm. Filters were installed in front of the detectors to block residual effects of wavelengths of 720 nm and above, and a 440 ± 40 bandpass filter was used to register NAD(P)H. A 550 ± 50 nm filter was used to register FAD. The first step of data processing was to create a binary mask, followed by 20 × 20 binning and time grouping (the number of time points was reduced from 256 to 64) immediately after applying the binary mask [26]. The next step was to use a two-exponential approximation of the measured signals at wavelengths of 760 and 880 nm. It is noted that τ_1 and a_1 correspond to free NAD(P)H, τ_2 and a_2 correspond to enzyme-bound NAD(P)H, and for FAD, there is an inverse correspondence [27]. To analyze metabolic activity, the fluorescence lifetime imaging redox ratio (FLIRR) computed for each masked pixel was used as follows:

$$\text{FLIRR} = \frac{\text{fraction of enzyme bound NAD}(\text{P})\text{H}}{\text{fraction of enzyme bound FAD}} = \frac{a_2\text{NAD}(\text{P})\text{H}}{a_1\text{FAD}}$$

The lower FLIRR values are known to correspond to glycolytic metabolism, and higher FLIRR values correspond to a more oxidative phosphorylation (OXPHOS) phenotype [27].

Cells located closer to the edge of the MCF10A spheroid showed fluorescence lifetime changes associated with shift to OXPHOS, while the spheroid core showed changes indicative of shift to glycolysis. MDA-MB-231 spheroids showed a significant shift indicative of an increase in OXPHOS, with this change being more pronounced at high collagen concentrations. Over time, MDA-MB-231 spheroids invaded the collagen gel with cells that invaded furthest, showing the greatest changes consistent with a shift toward OXPHOS.

3.3 PHASOR-PLOT FLUORESCENCE LIFETIME IMAGING DATA PRESENTATION

The phasor approach is an alternative to the multi-exponential approximation, which provides a graphical representation of multivariate data. In essence, the idea of this approach was proposed a long time ago, and the original source is quite difficult to find. For example, such an approach was used to subtract background fluorescence in phase modulation spectroscopy [28]. For the analysis of FLIM data in the frequency domain, this approach was first proposed [29,30]. Later, this approach was applied in the time domain [31].

3.3.1 Frequency Domain Fluorescence Lifetime Imaging

Interpretation of fluorescence signal in the frequency domain method can be based on the relationship between a segment rotating at a constant angular velocity ω and a harmonic oscillation:

$$g(\varphi) = A\cos(\omega t)$$

$$s(\varphi) = A\sin(\omega t)$$

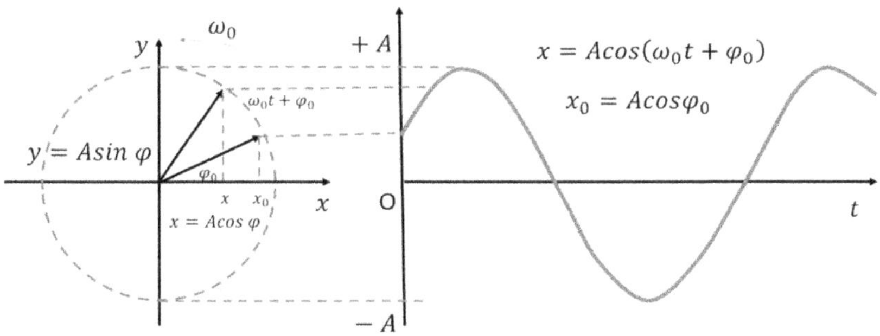

FIGURE 3.3 Method of vector diagrams.

where m is the length of the segment; g, s is the projection of the segment on x and y axes, respectively. If we have two segments rotating with the same angular velocity ω but with different initial phases (Figure 3.3), then the angle between segments does not vary in time, and projections of one segment on the axis coinciding with another segment and on the axis perpendicular to the latter will be:

$$g(\omega) = A\cos(\varphi)$$

$$s(\omega) = A\sin(\varphi)$$

We can interpret values $g(\omega)$, $s(\omega)$ as phasor-plot coordinates of a fluorescence signal measured at each pixel location in frequency domain fluorescence lifetime imaging. The parameter A should be considered as the amplitude modulation ratio and φ is the phase delay between the amplitude-modulated pumping laser radiation and a molecule's fluorescence radiation [32].

3.3.2 Time Domain Fluorescence Lifetime Imaging

The phasor-plot approach can be implemented in time domain fluorescence lifetime imaging through the Fourier transform of a fluorescence signal measured at each pixel location as follows [31,33]:

$$G = \frac{\int_0^T I(t)\cos(\omega t)\,dt}{\int_0^T I(t)\,dt} \tag{3.4}$$

$$S = \frac{\int_0^T I(t)\sin(\omega t)\,dt}{\int_0^T I(t)\,dt} \tag{3.5}$$

where $T = 2\pi / \omega$, ω—the laser radiation angular frequency. The idea of the phasor transformation is presented in Figure 3.4.

The fluorescence decay curve of a single fluorophore is presented by a point on the semi-circle of the phasor plot [31]. In this case, the fluorescence lifetime is directly related to coordinate values of the point on the phasor plot [32]:

$$\tau = \frac{1}{\omega}\frac{s}{g} \tag{3.6}$$

It follows from this formula that the position of a FLIM data point on the phasor depends on the modulation frequency. When a fluorescence decay curve is presented by a point inside the semi-circle, it is better approximated by the two-exponential model (see Figure 3.5).

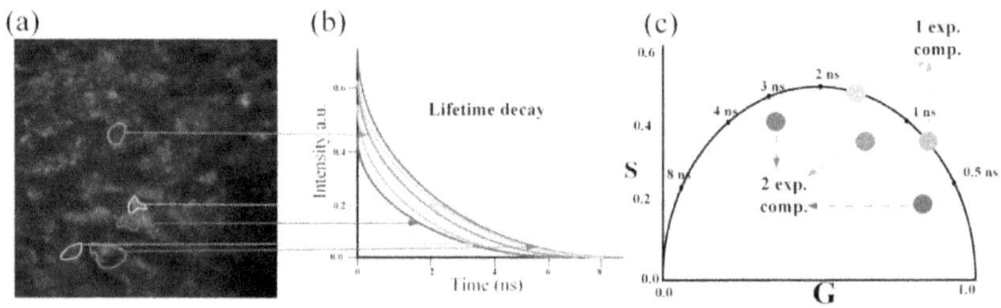

FIGURE 3.4 The phasor-plot presentation of FLIM data. (a) A fluorescence image, (b) fluorescence lifetime decay curves associated with the fluorescence image, and (c) the phasor-plot FLIM data presentation [33].

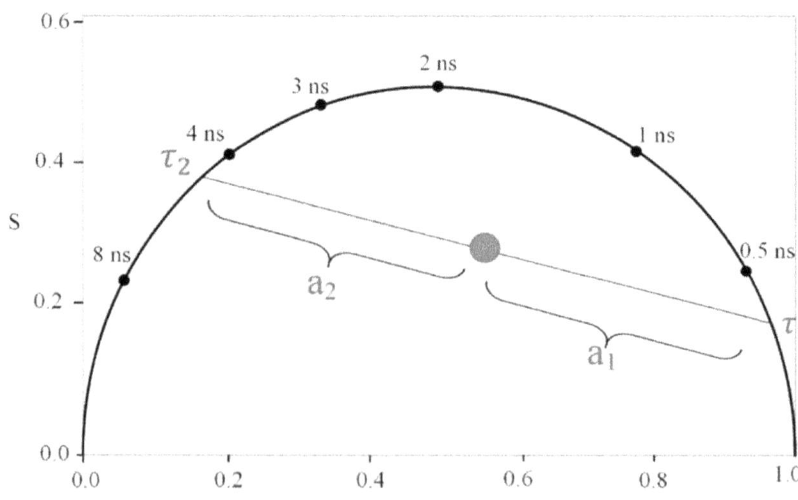

FIGURE 3.5 Decomposition of the fluorescence lifetime curve on two exponents through its phasor-plot presentation for a 80 MHz repetition rate. (Adapted from [33].)

As mentioned above, the phasor approach is described by two parameters, phase and modulation degree [34]. The phase is used as the pointer angle and the modulation degree as the amplitude. Some features of the phasor graph are as follows:

1. The points lying on the universal curve are described by single-exponential decay profiles.
2. The phasor angle, that is, the location on the semicircle, depends on the fluorescence lifetime.
3. The fluorescence lifetime located on the universal curve depends on the pulse repetition rate, so the images taken with different parameters will differ.
4. The phasors of multi-exponential fluorescence decay profiles lie inside the universal curve and are described by the sums of several decay components.

3.4 EXAMPLES OF APPLICATION OF THE PHASOR APPROACH

Glucose stimulation and insulin secretion in specialized mammalian pancreatic cells, the β-cells, are linked and have long been studied to understand biochemical processes that arepivotal to the maintenance of glucose systemic homeostasis. Non-invasive and rapid strategies capable of monitoring the glucose response in living cells are important in studying such processes. The study [35] presents results on the determination of a combination of quantitative indices between protein-bound and free nicotinamide adenine dinucleotide (phosphate) residues in their reduced form (NAD(P)H), as well as insulinoma cell lines INS-1E as a β-like cell form using a phasor-plot approach. INS-1E and A549 cell cultures were grown according to a standard protocol described in detail by the authors; stimulation with glucose was done by incubating the cells for 45 min in Krebs solution with different glucose concentrations of 11.1 and 30 mmol/L (hyperglycemic conditions) immediately before FLIM measurements (Figure 3.6).

The authors established a shift in the phasor plane associated with glucose stimulation, and they conclude that these changes are related to oxidative stress. This conclusion was based on an experiment with treating cells cultured under standard conditions with 100 μmol/L hydrogen peroxide solution.

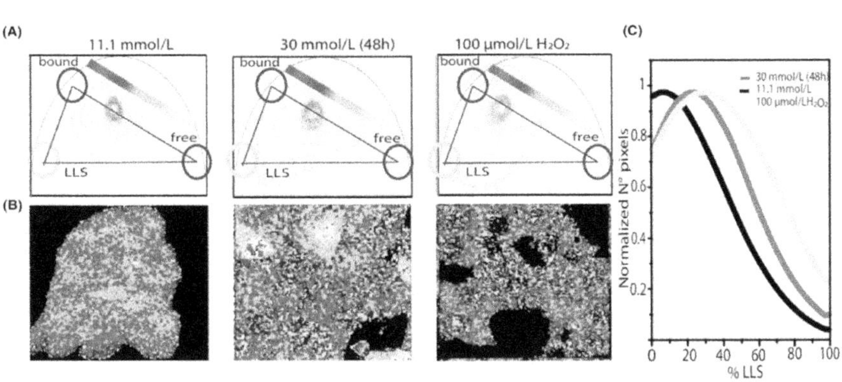

FIGURE 3.6 Results of the cell examination after stimulation with glucose [35].

Wound healing is a complex physiological process that occurs at the cellular level [36]. However, this process slows down and becomes more complicated in the presence of diabetes mellitus. The study [33] presents results of the application of the FLIM-phasor approach in the evaluation of the wound state of laboratory animals with and without model diabetes (mice, CD1). A two-week protocol of low-dose Streptozotocin injections was implemented to induce diabetes mellitus in the experimental group. The skin surgery wound model was used. Also, a low-intensity photodynamic therapy (LDPDT) of skin wounds was implemented using topically administered 5-aminolevulinic acid (5-ALA) and methylene blue (MB) directly to the wound for 30 min of incubation. After that, the wounds were illuminated by an AlGaInP laser ($\lambda = 630$ nm, $P = 5$ mW) with two optical energy densities: 1 and 4 J/cm^2 for 3 min 45 s and 15 min, respectively. The 25 healthy mice were randomly divided into five groups (five mice per group): LDPDT-5-ALA/1 J/cm^2, LDPDT-MB/1 J/cm^2, LDPDT-5-ALA/4 J/cm^2, and LDPDT-MB/4 J/cm^2 groups and the healthy control group. The 25 diabetes mellitus mice were randomly divided on five groups (five mice per group): dB LDPDT-5-ALA/1 J/cm^2, dB LDPDT-MB/1 J/cm^2, dB LDPDT-5-ALA/4 J/cm^2, and dB LDPDT-MB/4 J/cm^2 groups and the diabetes mellitus control group.

The skin state was analyzed using a MPTflex two-photon microscope (Jenlab GmbH, Germany, Jena). The phasor-plot FLIM data presentation was used to emphasize the changes that occur during the wound healing process. To extract the details about FLIM data spatial distribution on the phasor plane, the authors analyzed S-axis and G-axis projections of data clusterizations (see Figure 3.7).

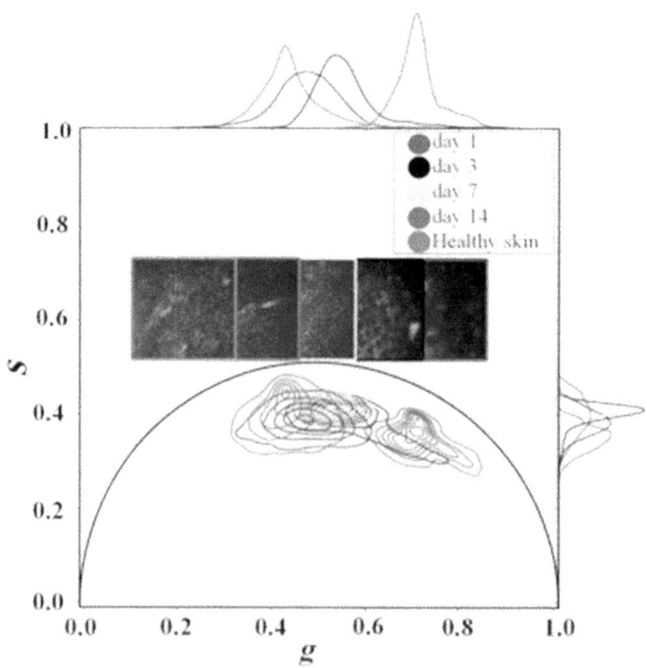

FIGURE 3.7 The phasor-plot clusters for the control group [33].

In order to evaluate quantitatively the wound state relative to the healthy skin, the density distribution of the phasor-plot cluster G-coordinate was analyzed. Its box diagram presentation is shown in Figure 3.8.

In particular, these results demonstrate the superiority of the 4 J/cm^2 regime over 1 J/cm^2. One of the reasons can be connected with NADH metabolism. NADH is involved in redox metabolic reactions and exists in two distinctive states: free and bound, which differ in their fluorescence lifetimes (see Table 3.1). The results presented in Figure 3.8 show a shift of fluorescence lifetime to smaller values. The protein-bound form of NADH is responsible for the OXPHOS [37]. Free form of NADH is responsible for glycolysis [23]. Therefore, abovementioned shift can be corresponded to increasing the free form of NADH, which can be interpreted as a change in the balance between OXPHOS and glycolysis.

Colorectal cancer (CRC) is the second leading cause of cancer-associated deaths worldwide [38,39]. Standard CRC tests like colonoscopy are hard to tolerate by patients. It is a reason of keen interest to diagnostics based on the analysis of molecular biomarkers in biofluids. For example, about 30 specific blood-containing CRC molecular biomarkers were discovered [40]. Exosomes released by cells in various biofluids have biochemical cargo corresponding to a specific cell type. It is the reason exosomes are potential CRC biomarkers [41,42].

The exosomes extracted from blood samples of 11 colon polyps (CPs) patients and 13 CRC patients with stages T2-4 N0-3 M0-1 were analyzed using a two-photon MPTflex microscope (Jenlab GmbH, Jena, Germany) and FLIM-phasor approach [43]. Comparative analysis in the phasor plane of FLIM data from a person with CRC and a person with CPs established that there are two classes of data (Figure 3.9). Both classes are presented in FLIM images of the person with CRC, and the only class was in FLIM data of a person with CPs. The authors interpreted the class occurring in FLIM data of a person with CPs as associated with exosomes released by healthy cells. Another class was interpreted to be associated with exosomes released by cancer cells.

To estimate the radii of classes R_j, the averaged distance from each point of the class to the center plus its tripled standard deviation (the 3σ rule) was used:

$$R_j = \frac{1}{N_j}\sum_{i=1}^{N_j} r_{i,j} + 3\sqrt{\frac{1}{N_j}\sum_{i=1}^{N_j}\left(r_{i,j} - \frac{1}{N_j}\sum_{i=1}^{N_j} r_{i,j}\right)^2} \tag{3.7}$$

where $r_{i,j} = \sqrt{\left(g_{c,j} - g_{i,j}\right)^2 + \left(s_{c,j} - s_{i,j}\right)^2}$, $g_{i,j}$ and $s_{i,j}$ are the coordinates of a data point in the jth class; $g_{c,j}, s_{c,j}, N_j$ are the coordinates of the center and the number of data points in jth class. Using these circle masks, fluorescence lifetime distributions for cancer-associated and healthy exosomes were extracted (see Figure 3.10a).

Taking it into account, a quantitative criterion for CRC diagnosis was suggested by the authors as follows:

$$N_{ch} = \frac{\sum N_c}{\sum N_h + \sum N_c} \tag{3.8}$$

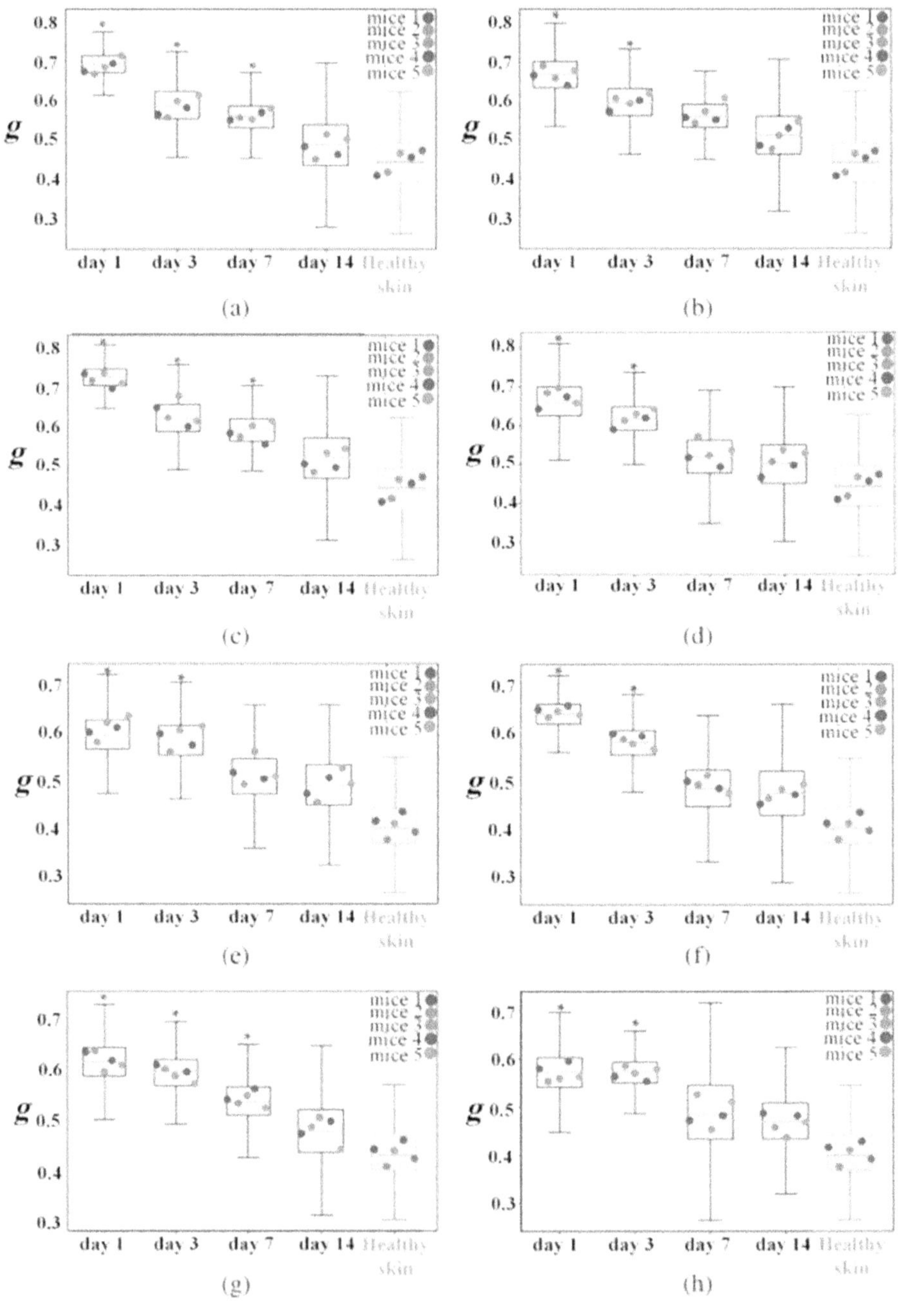

FIGURE 3.8 The box diagram of the g coordinate of phasor-plot density distributions for: (a) dB LDPDT 5ALA-1 J/cm², (b) dB LDPDT 5ALA-4 J/cm², (c) dB LDPDT MB-1 J/cm², (d) dB LDPDT MB-4 J/cm², (e) LDPDT 5ALA-1 J/cm², (f) LDPDT 5ALA-4 J/cm², (g) LDPDT MB-1 J/cm², and (h) LDPDT MB-4 J/cm² [33]

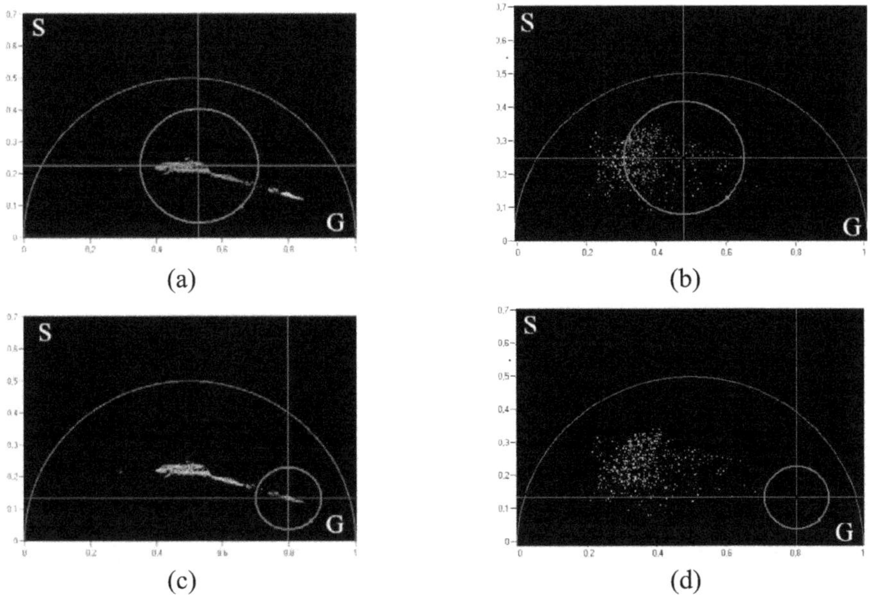

FIGURE 3.9 The class areas on phasor plots: the first class position on the phasor plot for a CRC patient (a) and a CP patient (b), the second class position on the phasor plot for a CRC patient (c), and a CP patient (d) [43].

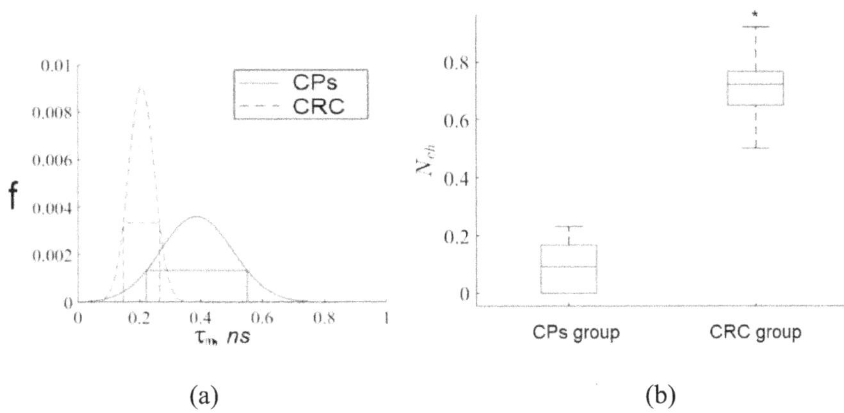

FIGURE 3.10 (a) Interval approximations of the distribution functions for the cancer-associated and healthy exosomes FLIM data fluorescence lifetimes presented in Figure 3.9 and (b) evaluation of N_{ch} values for CPs and CRC groups, *p-value was 3×10^{-5} [43].

where N_c is the number of CRC-associated exosomes, N_h is the number of "non-cancerous" (CPs-associated) exosomes. The results of applying this criterion to CPs and CRC patients' groups are presented in Figure 3.10b. Thus, the FLIM-phasor approach allows detecting "cancerous" exosomes in the blood.

Since the phasor approach is a variant of graphical data visualization, most of the work on using this approach is aimed at finding visual differences in data from different groups. A phasor plot can be interpreted through changes in the point cloud position. A graphical representation of FLIM microscopy results of different cell types allowed for their differentiation without additional staining. In addition, it was possible to observe different metabolic states of macrophages and mast cells (see Figure 3.11). In vivo studies have been conducted for some cell types [44–46]. The authors demonstrated the separability of the data in additional materials using machine learning methods. The accuracy of group separation was shown to be more than 90%.

The importance of studying the metabolic activity of bacteria is defined by their ability to modulate metabolic processes in response to adverse conditions. Such possibilities of bacteria give them unique advantages. In practice, this affects the drug stability in the treatment of infectious diseases. In the example of several bacterial cultures (Escherichia Coli, Salmonella Enterica Serovar Typhimurium, Pseudomonas Aeruginosa, Bacillus Subtilis, and Staphylococcus Epidermidis), the possibility of their differentiation was shown using methods of two-photon microscopy and phasor approach. The phasor plot varied depending on environmental conditions such as antibiotic exposure and growth phase, suggesting that the observed shifts in phasor are representative of metabolic changes within the bacteria [47].

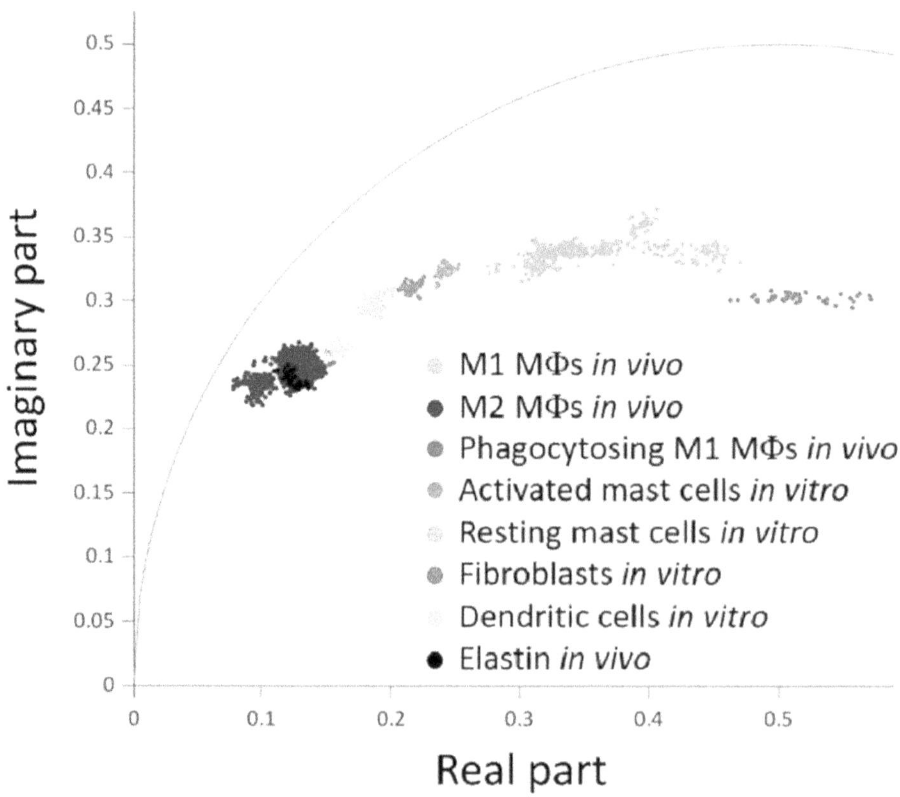

FIGURE 3.11 Various skin cell types fluorescence curves data phasor presentation.

3.5 CONCLUSION

The use of multi-exponential fluorescence lifetime imaging data presentation makes it possible to determine the lifetime of various endogenous fluorophores. The two-exponential approach allows for the analysis of the metabolic activity of cells, skin capillaries, and the distribution of erythrocytes in blood plasma.

The phasor approach is a kind of graphical presentation of FLIM data, which makes data interpretation more convenient. The phasor approach can be used for the determination of a combination of quantitative indices between protein-bound and free NAD(P)H, as well as in insulinoma cell lines INS-1E as a β-like cell. The FLIM data phasor approach also allows for the estimation of the skin wound state and can be used for CRC diagnosis.

ACKNOWLEDGMENTS

The authors thank M. E. Darvin (independent researcher from Berlin, Germany), M. S. Snegerev (National Research Tomsk State University), A. P. Kokhanenko and K. A. Lozovoy (Department of Quantum Electronics and Photonics, Faculty of Radiophysics, National Research Tomsk State University), and A. S. Lozhkomoev and E. I. Senkina (Institute of Strength Physics and Materials Science of the Siberian Branch of the Russian Academy of Sciences) for their great contribution and assistance in writing an article on the study of metabolic activity of fibroblasts for the diagnostics of osteoimplant survival.

The authors thank H. Baalbaki, A. I. Knyazkova, and N. A. Krivova (National Research Tomsk State University) for their help in preparing the article on the study of diabetic and non-diabetic skin wound healing.

The authors wish to thank A. A. Samarinova (National Research Tomsk State University), Natalia V. Yunusova, and Olga V. Cheremisina (Laboratory of Tumor Biochemistry, Cancer Research Institute, Tomsk National Research Medical Center, Russian Academy of Sciences) for their invaluable contribution to the preparation of the article on colorectal cancer diagnosis using exosomes.

REFERENCES

1. R. Datta, T. M. Heaster, J. T. Sharick, A. A. Gillette, M. C. Skala, Fluorescence lifetime imaging microscopy: fundamentals and advances in instrumentation, analysis, and applications. *J. Biomed. Opt.* 25(7), 071203, 2020. doi:10.1117/1.JBO.25.7.071203.
2. E. A. Shirshin, B. P. Yakimov, M. E. Darvin et al., Label-free multiphoton microscopy: the origin of fluorophores and capabilities for analyzing biochemical processes. *Biochem. Mosc.* 84, 69–88, 2019.
3. S. Huang, A. A. Heikal, W. W. Webb, Two-photon fluorescence spectroscopy and microscopy of NAD(P)H and flavoprotein. *Biophys. J.* 82, 2811–2825, 2002.
4. M. C. Skala, K. M. Riching, A. Gendron-Fitzpatrick et al., In vivo multiphoton microscopy of NADH and FAD redox states, Fluorescence lifetimes, and cellular morphology in precancerous epithelia. *Proc. Natl. Acad. Sci. USA* 104, 19494–19499, 2007.
5. A. Zoumi, A. Yeh, B. J. Tromberg, Imaging cells and extracellular matrix in vivo by using second-harmonic generation and two-photon excited fluorescence. *Proc. Natl. Acad. Sci. USA* 99, 11014–11019, 2002.

6. F. Yu, S. Zhuo, Y. Qu et al., On chip two-photon metabolic imaging for drug toxicity testing. *Biomicrofluidics* 11, 034108, 2017.
7. S. Ranjit, A. Dvornikov, M. Stakic et al., Imaging fibrosis and separating collagens using second harmonic generation and phasor approach to fluorescence lifetime imaging. *Sci. Rep.* 5, 13378, 2015.
8. B. Torrado, S. Malacrida, S. Ranjit, Linear combination properties of the phasor space in fluorescence imaging. *Sensors* 22, 999, 2022.
9. O. I. Kolencm, K. P. Quinn, Evaluating cell metabolism through autofluorescence imaging of NAD(P)H and FAD. *Antioxi. Redox Sig.* 20(30), 875–889, 2019.
10. M. Monici, Cell and tissue autofluorescence research and diagnostic applications. *Biotechnol. Ann. Rev.* 11, 227–256, 2005.
11. C. Li, R. K. Pastila, C. Pitsillides et al., Imaging leukocyte trafficking in vivo with two-photon-excited endogenous tryptophan fluorescence. *Opt. Express* 18(2), 988, 2010.
12. S. R. Alam, H. Wallrabe, Z. Svindrych et al., Investigation of mitochondrial metabolic response to doxorubicin in prostate cancer cells: an NADH, FAD and tryptophan FLIM assay. *Sci. Rep.* 7(1), 10451, 2017.
13. V. Jyothikumar, Y. Sun, A. Periasamy, Investigation of tryptophan–NADH interactions in live human cells using three-photon fluorescence lifetime imaging and Förster resonance energy transfer microscopy. *J. Biomed. Opt.* 18(6), 060501, 2013.
14. J. T. Vivian, P. R. Callis, Mechanisms of tryptophan fluorescence shifts in proteins. *Biophys. J.* 80(5), 2093–2109, 2001.
15. E. Dimitrow, I. Riemann, A. Ehlers et al., Spectral fluorescence lifetime detection and selective melanin imaging by multiphoton laser tomography for melanoma diagnosis. *Exp. Dermatol.* 18(6):509–515, 2009. doi:10.1111/j.1600-0625.2008.00815.x.
16. H. Schneckenburger, K. König, K. Kunzi-Rapp, C. Westphal-Frösch, A. Rück, Time-resolved in-vivo fluorescence of photosensitizing porphyrins. *J. Photochem. Photobiol. B Biol.* 21(2–3), 143–147, 1993.
17. R. Cubeddu, A. Pifferi, P. Taroni, A. Torricelli, G. Valentini, E. Sorbellini, Fluorescence lifetime imaging: an application to the detection of skin tumors. *IEEE J. Sel. Top. Quantum Electron.* 5(4), 923–929, 1999.
18. K. Konig, Multiphoton microscopy in life sciences. *J. Microsc.* 200(2), 83–104, 2000.
19. M. Collini, B. Leo, G. Bal, Probing protein aggregation by time-resolved fluorescence during β-lactoglobulin crystal growth. *Eur. Biophys. J.* 31(2), 111–117, 2002.
20. W. R. Zipfel, R. M. Williams, R. Christie, A. Y. Nikitin, B. T. Hyman, W. W. Webb, Live tissue intrinsic emission microscopy using multiphoton-excited native fluorescence and second harmonic generation. *Proc. Natl. Acad. Sci.* 100(12), 7075–7080, 2003.
21. E. A. Shirshin, Y. I. Gurfinkel, A. V. Priezzhev et al. Two-photon autofluorescence lifetime imaging of human skin papillary dermis in vivo: assessment of blood capillaries and structural proteins localization. *Sci. Rep.* 7, 1171, 2017. doi:10.1038/s41598-017-01238-w
22. T. B. Lepekhina, V. V. Nikolaev, M. E. Darvin et al., Two-photon-excited FLIM of NAD(P)H and FAD—metabolic activity of fibroblasts for the diagnostics of osteoimplant survival. Int. *J. Mol. Sci.* 25, 2257, 2024. doi:10.3390/ijms25042257.
23. M. M. Lukina, M. V. Shirmanova, T. F. Sergeeva, E. V. Zagaynova, Metabolic imaging in the study of oncological processes (review). *Sovrem. Teh. Med.* 8, 113–126, 2016. https://cyberleninka.ru/article/n/metabolic-imaging-in-the-study-of-oncological-processes-review
24. M. S. Roberts , Y. Dancik, T. W. Prow et al. Non-invasive imaging of skin physiology and percutaneous penetration using fluorescence spectral and lifetime imaging with multiphoton and confocal microscopy. *Eur. J. Pharmaceut. Biopharmaceut.* 77(3), 469–488, 2011. doi:10.1016/j.ejpb.2010.12.023.
25. K. Karrobi, A. Tank, M. A. Fuzail et al. Fluorescence lifetime imaging microscopy (FLIM) reveals spatial-metabolic changes in 3D breast cancer spheroids. *Sci. Rep.* 13, 3624, 2023. doi:10.1038/s41598-023-30403-7.

26. S. J. DeCamp, V. M. K. Tsuda, J. Ferruzzi et al. Epithelial layer unjamming shifts energy metabolism toward glycolysis. *Sci. Rep.* 10, 18302, 2020.
27. H. Wallrabe, Z. Svindrych, S. R. Alam et al. Segmented cell analyses to measure redox states of autofluorescent NAD(P)H, FAD & Trp in cancer cells by FLIM. *Sci. Rep.* 8, 79, 2018.
28. J. R. Lakowicz, R. Jayaweera, N. Joshi, I. Gryczynski, Correction for contaminant fluorescence in frequency-domain fluorometry. *Anal. Biochem.* 160, 471–479, 1987.
29. A. H. A. Clayton, Q. S. Hanley, P. J. Verveer, Graphical representation and multicomponent analysis of single-frequency fluorescence life-time imaging microscopy data. *J. Microsc. Oxf.* 213, 1–5, 2004.
30. G. I. Redford, R. M. Clegg, Polar plot representation for frequency-domain analysis of fluorescence lifetimes. *J. Fluoresc.* 15(5), 805–815, 2005.
31. M. A. Digman, V. R. Caiolfa, M. Zamai, E. Gratton, The phasor approach to fluorescence lifetime imaging analysis. *Biophys.* J. 94(2), L14–L16, 2008.
32. S.-C. Liao, Y. Sun, U. Coskun, *FLIM Analysis Using the Phasor Plots.* ISS Technical Notes No. 61822. ISS, Champaign, IL, USA, 2014.
33. H. Zuhayri, T. B. Lepekhina, V. V. Nikolaev et al., The quantitative in vivo assessment of diabetic and non-diabetic skin wound healing using phasor-FLIM approach. *J. Biophoton.* 2024, e202400227. doi:10.1002/jbio.202400227.
34. W. Becker, Fluorescence lifetime imaging–techniques and applications. *J. Microsc.* 247(2), 119–136, 2012.
35. G. Ferri, M. Tesi, F. Massarelli, L. Marselli, P. Marchetti, F. Cardarelli, Metabolic response of insulinoma 1E cells to glucose stimulation studied by fluorescence lifetime imaging. *FASEB BioAdv.* 2, 409–418, 2020. doi:10.1096/fba.2020-00014.
36. H. N. Wilkinson, M. J. Hardman, Wound healing: cellular mechanisms and pathological outcomes. *Open Biol.* 10(9), 200223, 2020. doi:10.1098/rsob.200223.
37. A. Rück, C. Hauser, S. Mosch, S. Kalinina, Spectrally resolved fluorescence lifetime imaging to investigate cell metabolism in malignant and nonmalignant oral mucosa cells. *J. Biomed. Opt.* 19, 096005, 2014. doi:10.1117/1.JBO.19.9.096005.
38. H. Sung, J. Ferlay, R. L. Siegel, et al., Global cancer statistics 2020: GLOBOCAN estimates of incidence and mortality worldwide for 36 cancers in 185 countries. *CA Cancer J. Clin.* 71, 209–249, 2021. doi:10.3322/caac.21660.
39. M. S. Hossain, H. Karuniawati, A. A. Jairoun et al., Colorectal cancer: a review of carcinogenesis, global epidemiology, current challenges, risk factors, preventive and treatment strategies. *Cancers* 14, 1–25, 2022. doi:10.3390/cancers14071732.
40. J. Rao, X. Wan, F. Tou et al., Molecular characterization of advanced colorectal cancer. *Front. Mol. Biosci.* 8, 687229, 2021.
41. Y. Xiao, J. Zhong, B. Zhong et al., Exosomes as potential sources of biomarkers in colorectal cancer. *Cancer Lett.* 476, 13–22, 2020. doi:10.1016/j.canlet.2020.01.033.
42. D. Lucchetti, F. Litta, C. Ratto, A. Sgambato, The role of extracellular vesicles as biomarkers in colorectal cancer. *Tech. Coloproctol.* 22, 989–990, 2018. doi:10.1007/s10151-018-1880-4.
43. A. V. Borisov, O.A. Zakharova, A.A. Samarinova, N.V. Yunusova, O.V. Cheremisina, Y. Kistenev, A criterion of colorectal cancer diagnosis using fluorescence-lifetime imaging exosomes analysis. *Diagnostics* 12, 1272, 2022.
44. M. Kröger, J. Scheffel, V.V. Nikolaev et al., In vivo non-invasive staining-free visualization of dermal mast cells in healthy, allergy and mastocytosis humans using two-photon fluorescence lifetime imaging. *Sci. Rep.* 10, 14930, 2020. doi:10.1038/s41598-020-71901-2.
45. M. Kröger, J. Scheffel, E. A. Shirshin et al., Label-free imaging of M1 and M2 macrophage phenotypes in the human dermis in vivo using two-photon excited FLIM. *Elife.* 11, e72819, 2022. doi:10.7554/eLife.72819.

46. V. V. Nikolaev, Y. V. Kistenev, M. Kröger, H. Zuhayri, M. E. Darvin, Review of optical methods for noninvasive imaging of skin fibroblasts-From in vitro to ex vivo and in vivo visualization. *J. Biophoton.* 17(1), e202300223, 2024. doi:10.1002/jbio.202300223.
47. A. Bhattacharjee, R. Datta, E. Gratton et al., Metabolic fingerprinting of bacteria by fluorescence lifetime imaging microscopy. *Sci. Rep.* **7**, 3743, 2017. doi:10.1038/s41598-017-04032-w.

4 Fluorescence Lifetime Imaging Microscopy

Pietro Ricci
University of Florence
Lab. Europeo di Spettroscopie Non-Lineari (LENS)

Luca Pesce
University of Pisa

Valeria Vetri and Giuseppe Sancataldo
University of Palermo

4.1 INTRODUCTION

Over the past few decades, fluorescence has become a cornerstone of biological and medical sciences, enabling advances in various fields, including clinical chemistry, DNA sequencing, genetic analysis, flow cytometry, and cellular imaging (Lakowicz 2006; Stockert and Blázquez-Castro 2017). Within this framework, fluorescence microscopy has become useful for investigating a wide range of cellular and tissue properties, from spatial morphology to fine molecular interactions (Kubitscheck 2013). However, despite its widespread use, steady-state fluorescence intensity imaging suffers from several inherent limitations. These include susceptibility to artifacts caused by variations in fluorophore density, excitation intensity, detector settings, optical path losses, and photobleaching. Moreover, it cannot distinguish fluorophores with overlapping excitation and emission spectra, severely limiting its specificity in complex systems. To address these challenges, fluorescence lifetime imaging microscopy (FLIM) has emerged as a powerful method. Unlike intensity-based imaging, FLIM relies on the measurement of fluorescence decay times, an intrinsic property of fluorophores that reflects their interactions with the surrounding microenvironment. This approach provides an additional dimension of contrast and yields quantitative, highly reliable data, even in challenging imaging conditions.

FLIM is particularly valuable for examining the physicochemical properties of biological samples, including molecular interactions, pH, polarity, viscosity, and ion concentrations (Marcu et al. 2015). An illustrative comparison between fluorescence intensity imaging and FLIM is presented in Figure 4.1. While intensity imaging quantifies photon counts per pixel (Figure 4.1a), FLIM measures fluorescence decay times (Figure 4.1b), which are independent of fluorophore concentration and excitation intensity. This independence enhances the robustness of FLIM measurements, making them less prone to artifacts commonly associated with traditional

DOI: 10.1201/9781003570011-4

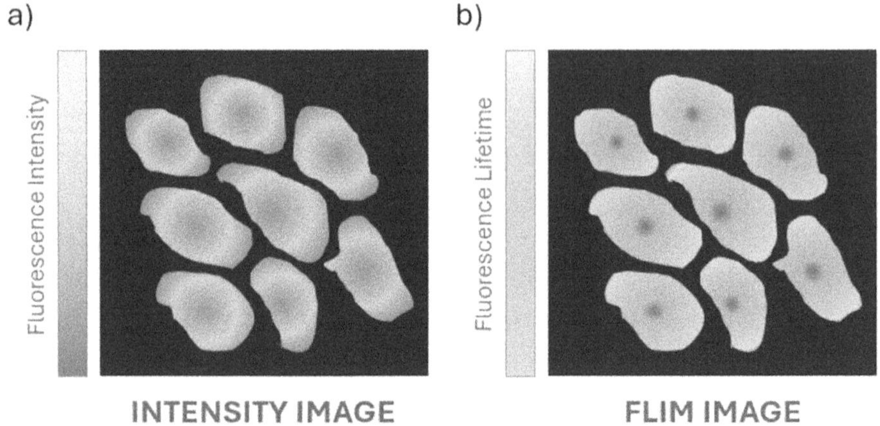

FIGURE 4.1 Schematic comparison of intensity-based and FLIM-based imaging. Intensity imaging (a) quantifies photon counts per pixel, whereas FLIM. (b) Measures fluorescence decay times, offering a contrast parameter independent of fluorophore concentration and excitation intensity.

intensity-based methods. For instance, variations in excitation light penetration, which are especially problematic in thick tissues or complex biological samples, have reduced impact on FLIM data. The unique capabilities of FLIM extend beyond artifact reduction. Its sensitivity to microenvironmental changes enables it to detect subtle variations in fluorophore behavior, offering insights into dynamic processes that are otherwise difficult to observe (Suhling et al. 2015; Datta et al. 2020). This advantage is particularly evident when distinguishing regions with similar fluorescence intensities but differing decay times, providing high contrast and revealing previously inaccessible details. Notably, another significant strength of FLIM lies in its ability to differentiate intrinsic autofluorescence from specific fluorescence signals. Autofluorescence, a common source of noise in biological samples, often complicates traditional fluorescence imaging, particularly in tissues with high background signals. FLIM can effectively resolve this issue by detecting differences in decay times between autofluorescent components and target fluorophores. This capability enhances image quality and contrast, making FLIM particularly useful for studying highly autofluorescent samples, such as thick animal specimens, plant tissues, or engineered constructs with complex fluorescent profiles.

This chapter provides a comprehensive overview of FLIM, beginning with an exploration of the fundamental principles and mechanisms underlying fluorescence. The focus will be on fluorescence decay and fluorescence lifetime, which form the foundation of FLIM measurements. We will then examine the properties of fluorophores and their spectroscopic characteristics, crucial for accurate FLIM imaging. Following this, the chapter delves into the most commonly used FLIM techniques, with a particular emphasis on time-domain and frequency-domain approaches. Next, we explore how FLIM integrates with other established imaging methodologies, enhancing its capabilities and expanding its range of applications. A detailed discussion on phasor analysis will follow, highlighting its advantages in FLIM data interpretation and providing insights into its widespread use for quantitative analysis.

Finally, the chapter concludes with an in-depth look at the application of FLIM in cellular and tissue imaging, emphasizing its significant contributions to biological research and clinical diagnostics.

4.2 FLUORESCENCE

4.2.1 Fluorescence Principles

Fluorescence is a photophysical process that occurs when a fluorophore absorbs photons, causing the molecule to shift to a higher energy state, and successively emits light as it returns to its ground state (Figure 4.2a). The absorption and emission processes are properly represented in an energy level scheme (Jablonski diagram), as illustrated in Figure 4.2b, where electronic states are arranged vertically by energy, and vertical arrows indicate transitions between levels. When a molecule in its ground state (S_0) absorbs light with sufficient energy (occurring in 10^{-15} s time scale), an electron is excited to higher energy states (S_1, S_2,...). The electron then can return to the ground state through radiative emission, where a fluorescence photon is released, or non-radiatively, by dissipating the excited state energy as heat. Ground state and excited states contain multiple vibrational and rotational sub-levels, and, after excitation, the electron will undergo vibrational relaxation to the lowest vibrational level of the excited state by a nonradiative process (internal conversion).

Fluorescence specifically refers to the radiative return of the electron from the lowest vibrational level of the first excited singlet state (S_1) to the ground state. Electronic transitions typically occur in the ultraviolet-visible (UV-Vis) region of the electromagnetic spectrum. An example of the absorption and emission spectra of a model fluorophore, fluorescein, is presented in Figure 4.2c. The absorption spectrum reflects the excitation of molecules to higher electronic states, while the emission spectrum corresponds to the radiative relaxation back to the ground state. Rotational and vibrational modes contribute to the broadening of the fluorescence emission band, leading to a characteristically smooth and wide spectral profile. Importantly, fluorescence emission generally originates from the lowest vibrational level of the

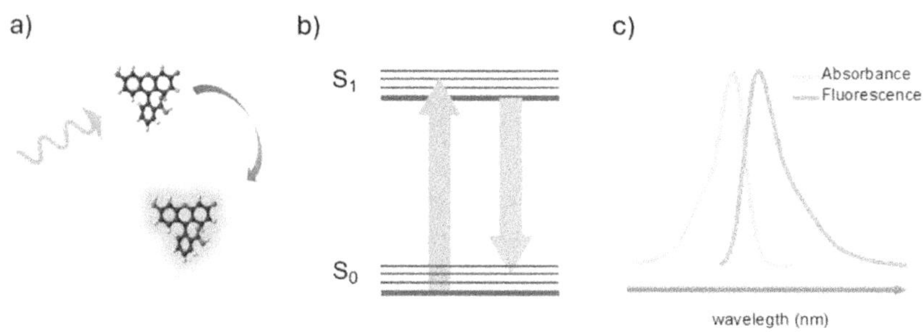

FIGURE 4.2 Fluorescence photophysics. (a) Schematic representation of the excitation process, where a photon promotes a molecule from the ground state to an excited electronic state. (b) Energy-level diagram (Jablonski diagram) illustrating the absorption of a photon, internal relaxation processes, and subsequent emission of fluorescence. (c) Absorption and emission spectra showing the characteristic wavelengths associated with the excitation and fluorescence emission of the molecule.

first electronically excited singlet state (S_1), irrespective of the specific vibrational level initially populated during excitation. This behavior, known as Kasha's rule, ensures that the fluorescence emission spectrum remains largely stable and independent of the excitation wavelength. Moreover, due to the partial loss of energy through non-radiative processes such as vibrational relaxation and internal conversion before photon emission, the energy of the emitted light is lower than that of the absorbed light. Consequently, the emission occurs at a longer wavelength, a phenomenon known as the Stokes shift. The magnitude of the Stokes shift provides valuable information about the molecular environment and dynamics surrounding the fluorophore.

4.2.2 Fluorescence Lifetime

In thermal equilibrium, the population of molecules across various energy levels follows a Boltzmann distribution, which governs how molecules are distributed between different states. Specifically, the ratio of molecules in the excited energy state (S_1) to those in the lower energy state (S_0) is determined by the energy difference between these states, $\Delta E = E_1 - E_0$, according to the equation:

$$\frac{N_{S1}}{N_{S0}} = \frac{g_{s1}}{g_{s0}} e^{-\frac{\Delta E}{kT}} \tag{4.1}$$

Here, g_{S1} and g_{S0} represent the degeneracies (number of accessible quantum states) of the upper and lower energy levels. In the case of singlet states, these degeneracies are typically equal to 1. At room temperature ($T = 293$ K), where $kT = 0.0252$ eV, and for an absorption wavelength of 500 nm (corresponding to an energy difference $\Delta E = 2.46$ eV), the exponential factor becomes extremely small. This value indicates that the number of molecules in the excited state (S_1) is only a negligible fraction of those in the ground state (S_0). As such, at room temperature, nearly all molecules reside in the ground electronic state, with very few achieving excitation. Following light excitation, fluorescent molecules are promoted to an electronically excited state, where they remain for a finite period before dissipating the absorbed energy through the emission of light or through non-radiative processes, such as internal conversion and vibrational relaxation. The time spent in the excited state is a critical parameter that reflects both the intrinsic properties of the fluorophore and its interactions with the surrounding environment. The average time a fluorophore remains in the electronic excited state S_1, before returning to the ground state, is called fluorescence lifetime (τ). This is defined as the inverse of the sum of all possible depopulation rates, which include both radiative and non-radiative pathways:

$$\tau = \frac{1}{k_r + k_{nr}} \tag{4.2}$$

where k_r is the radiative rate constant, governing the emission of fluorescence photons, and k_{nr} is the non-radiative rate constant, which accounts for internal conversion and intersystem crossing. After excitation, the number of fluorophores N_{S1} in the excited state S_1 decreases over time according to the following rate equation:

$$\frac{dN_{S1}(t)}{dt} = -(k_r + k_{nr})N_{S1}(t) \tag{4.3}$$

By integrating this equation and considering that the fluorescence intensity is directly proportional to the population of molecules in the excited state ($F(t) \propto N_{s1}(t)$), the solution gives:

$$F(t) = F_0 e^{-(k_r + k_{nr})t} = F_0 e^{-t/\tau} \tag{4.4}$$

Here, F_0 is the initial fluorescence intensity at $t=0$. This equation shows that fluorescence intensity decays exponentially over time, following a simple exponential decay law. This behavior is schematically illustrated for different regions of Figure 4.3a–c. When plotted on a logarithmic scale (Figure 4.3c), the monoexponential decay of fluorescence appears as a straight line, making it easy to visually assess the decay characteristics. This plotting method simplifies the analysis and provides a clear, intuitive way to evaluate the time-dependent behavior of fluorescence intensity.

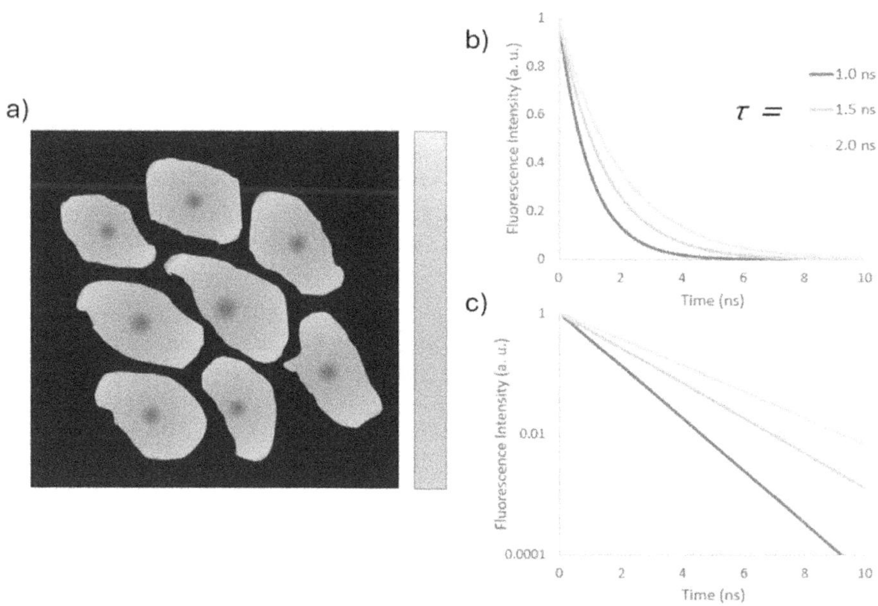

FIGURE 4.3 Schematic representation of FLIM Data. (a) Schematic of a typical FLIM image, where each pixel encodes information on the fluorescence lifetime rather than intensity. (b) Representative fluorescence decays of a fluorophore exhibiting different lifetimes, shown on a linear time scale. (c) The same fluorescence decays are plotted on a semi-logarithmic scale, facilitating the visualization of lifetime differences and the identification of multi-exponential decay behaviors.

4.3 FLUOROPHORES AND THEIR SPECTROSCOPIC CHARACTERISTICS

Fluorescence spectroscopy grants exceptional sensitivity to changes in the structural and dynamics properties of molecules and molecular complexes. Multiple methods can be employed, ranging from steady-state fluorescence emission measurements to complex time-resolved techniques. These methods can be applied in cuvettes or under a microscope, providing access to micrometric and nanometric scale resolution images.

The content of information increases with the combined analysis of various fluorescence observables. Key fluorescence observables include fluorescence emission peak intensity, shape, and position, fluorescence lifetime, fluorescence polarization, quantum yield, emission and excitation wavelengths, anisotropy, Förster resonance energy transfer efficiency, fluorescence correlation spectroscopy, and so on (Lakowicz 2006). The significance and the amount of evidence in each experiment vary depending on the specific fluorophore being analyzed. Fluorophores are molecular groups responsible for fluorescence, characterized by their ability to absorb photons at specific wavelengths and subsequently emit photons at longer ones. Among them, fluorescent dyes (or probes) represent a class of invaluable compounds whose properties can be effectively harnessed in the design of experiments that illuminate the nature of various phenomena or systems under investigation. These molecules can change their fluorescence in response to binding with a specific target, chemical reaction, or change in their immediate environment.

It is challenging to pinpoint the exact origins of studies on natural chromophores, but they likely began in the late 14th century with the observation of color variations in certain substances under illumination. Early investigations focused on the properties of pigments in plants and minerals, as well as their interactions with light. The term "fluorescence" was introduced by George Gabriel Stokes in 1852 during his studies on natural fluorophores and their applications in various fields (Stokes 1852).

Once the scientific question is defined, the investigation of multiple aspects of a certain phenomenon is made possible by using fluorescent dyes as molecular reporters. Fluorescent dyes can be used to sense pH, viscosity, and charge, as well as to quantify the presence of ions. Moreover, these molecules may sense temperature, solvent polarity, molecular interactions, and the presence of specific analytes (Langhals 2020). Within the cellular environment, dyes can be used to identify specific cellular components, monitor metabolic processes, track particle dynamics, assess changes in membrane potential, and monitor cell death among other applications (Specht et al. 2017).

Notably, fluorescent dyes can be either naturally occurring or artificially synthesized for specific applications. Natural fluorophores include small molecules such as aromatic amino acids, chlorophyll, NADH, NADPH, anthocyanins, flavins, and fluorescent proteins derived from marine organisms. Among these, green fluorescent protein (GFP) from the jellyfish Aequorea Victoria is one of the most well-known and widely used. Wild-type GFP is a 238-amino acid protein (~27 kDa) with a cylindrical β-barrel structure formed by 11 β-strands, capped by short α-helices (Zimmer 2002). Its central chromophore (Ser65-Tyr66-Gly67) forms through autocatalytic cyclization and is shielded from solvent, protecting it from quenching (Craggs 2009). When excited by UV light, GFP emits strong green fluorescence. GFPs are highly biocompatible, soluble, and resistant to aggregation, making them ideal for diverse biotechnological applications (Wang et al. 2023). Through protein engineering and organic synthesis, variants with enhanced properties have been developed, enabling signal-switching capabilities and fluorescence across the entire visible spectrum (Tian et al. 2023). As a result, GFPs have become essential tools in the design of advanced fluorescent biosensors.

Alongside fluorescent proteins, organic dyes are essential components in modern fluorescence microscopy. Examples of synthetic dyes include fluorescein, rhodamine, sulfonamides, Alexa and Atto dyes, naphthalimides, cyanine, and BODIPY dyes. Dyes can be covalently bound to specific targets or interact with them via specific non-covalent interactions, playing a crucial role in the analysis of local interactions. Typically, organic dyes are covalently attached through chemical reactions to biological vectors, such as antibodies (Abs). Abs are the most commonly used affinity probes due to their high specificity and the wide variety currently available to researchers. They are typically functionalized with organic dyes to label target biological structures. However, their relatively large size (~150 kDa in molecular weight and ~15 nm in length) and bivalent binding nature can lead to significant experimental artifacts (Fornasiero and Opazo 2015). In recent years, several alternatives have been developed. Among these, Fabs, nanobodies, and aptamers are the most widely used. These affinity probes are characterized by a smaller size (ranging from 10 to 50 kDa) and high specificity.

Nowadays, to enhance the ability to study multiple physicochemical processes with precision and versatility, several advancements have been made in tailoring the properties of new probes. These innovations have expanded the range of applications for synthetic fluorophores, enabling more effective exploration of complex phenomena. Quantum dots, which are semiconducting nanostructures, that exhibit size-dependent fluorescence, are another significant development in this field (Tandale et al. 2021), and this list has grown longer with the introduction of carbon nano-dots, which are biocompatible, soluble nanoparticles whose fluorescence properties can be modulated by acting on the surface (Lim et al. 2014).

A wide world of molecules or molecular structures can be used based on the selected application in the following few examples will be described for natural compounds, synthetic fluorescent probes. Fluorescent dyes can be categorized according to their chemical structure and excitation and emission properties, and often the information on the application is given. Although these latter are commonly specifically defined for a probe, by understanding the fundamental mechanisms governing the process and the chemical properties of the molecules, new and diverse uses for these probes can be explored.

4.4 TECHNIQUES FOR FLUORESCENCE LIFETIME MEASUREMENT

FLIM enables the measurement of the fluorescence decay of fluorophores, usually over a time range that goes from sub-nanoseconds to tens of nanoseconds. Various techniques allow for capturing these very short lifetimes (Liu et al. 2019). This section primarily discusses the main methods of fluorescence lifetime measurements, focusing on time-domain and frequency-domain techniques (Gratton 2003; Becker 2012).

4.4.1 Time-Domain Techniques

In time-domain approaches, a short light pulse, usually in the picosecond range and shorter than the sample's lifetime, is employed to excite the sample. The fluorescence response of the sample is essentially the convolution of the excitation pulse with the decay dynamics of the molecule's excited state, typically defining an exponentially decaying signal. The fluorescence decay can then be observed by using time-resolved electronics, which detect photons according to their arrival time, or by using a gated detector.

One of the most widely used time-domain methods is the time-correlated single photon counting (TCSPC) (Becker et al. 2004). In this approach, a highly precise "stop and watch" system measures the interval between the arrival of the excitation photon and the consequent fluorescence photon emission. Thus, the time of arrival for each emitted photon is measured with great accuracy. A time–amplitude converter circuit subsequently converts this timing into an analog voltage, which is then recorded. However, at high photon count rates, many photons remain undetected due to the system's dead time, leading to a phenomenon called pile-up, where only photons with shorter arrival times are measured per excitation pulse. This leads to an incorrect photon distribution, artificially shortening the observed fluorescence lifetime. To minimize such an effect, the photon count rate at the detector should be kept low, ideally below 10% of the excitation rate. For this reason, time-domain approaches usually detect just one photon per excitation pulse, meaning that multiple pulses, often in the thousands, are needed to compile a photon histogram and enhance the signal-to-noise ratio. It is evident that the slow data acquisition speed is a key limitation of these methods. However, the recent integration of high-speed digitizers (with subnanosecond sampling) has improved data collection rates by reducing dead times to two nanoseconds or less, thus speeding up the generation of fluorescence decay profiles.

Another time-domain approach is time-gating (Soloviev et al. 2007). Unlike TCSPC, time-gating does not detect the arrival time of individual photons but instead selects specific intervals in which photons are recorded. This method allows the detector to activate only during particular windows following the excitation pulse. By adjusting the gate's timing across multiple experiment repetitions, a fluorescence decay curve can be generated. Time-gated photon counting has been combined with laser scanning approaches, enabling efficient fluorescence lifetime imaging with high photon detection rates, while when paired with position-sensitive detectors, it has allowed for wide-field imaging with single-photon sensitivity at high temporal resolution.

4.4.2 Frequency-Domain Techniques

In frequency-domain methods, instead of using pulsed light, a continuous light source with intensity modulation at a specific frequency excites the sample. The equation describing the modulated excitation signal is as follows:

$$E(t) = E(0)\left[1 + M_E \sin(\omega t)\right] \tag{4.5}$$

where $E(t)$ and $E(0)$ represent the intensity at times t and 0, respectively, M_E is the excitation modulation factor, and ω is the angular frequency of modulation. In the hypothetical case of instantaneous emission (i.e., zero fluorescence lifetime), the emission signal would match the excitation in both frequency and waveform. However, due to finite fluorescence lifetimes, the emitted signal exhibits a phase shift and reduced modulation amplitude compared to the excitation waveform, which can be expressed as:

$$F(t) = F(0)\left[1 + M_F \sin(\omega t + \phi)\right] \tag{4.6}$$

where $F(t)$ and $F(0)$ are the fluorescence intensity at times t and 0, respectively, M_F is the emission modulation factor, and ϕ denotes the phase delay between excitation and emission.

For single exponential decays, the phase lifetime τ_ϕ is related to the phase shift as:

$$\tau_\phi = \frac{\tan(\phi)}{\omega} \tag{4.7}$$

This indicates that the phase shift is proportional to the excited state's lifetime: the smaller the shift, the shorter the lifetime. The modulation factor can be determined by the ratio of the AC to DC components of the excitation and emission signals. Practically, it's used to define the relative modulation $M = M_F/M_E$ as:

$$M = \frac{1}{\sqrt{1 + (\omega \tau_M)^2}} \tag{4.8}$$

where τ_M is the modulation lifetime. For single exponential decay, phase and modulation lifetimes are equivalent, and a single modulation frequency can be sufficient to calculate the lifetime.

When the fluorescence decay is not single-exponential but consists of multiple lifetime components, the phase lifetime and the modulation lifetime represent different types of averages of those components, providing complementary information about the underlying decay dynamics. In this case, a more complex analysis is needed, typically requiring measurements at several modulation frequencies.

To measure lifetimes in the nanosecond range, modulation frequencies from 10 to 100 MHz are required. Fast data acquisition can be achieved using gain-modulated cameras for wide-field systems or gain-modulated detectors for point-scanning systems. Specifically, the high-frequency signals emitted are transformed into low-frequency or static phase-dependent signals through heterodyne and homodyne detection, respectively. In heterodyne detection, the detector gain is modulated at a frequency slightly different from the excitation source, resulting in a low-frequency signal that retains the original modulation and phase characteristics. Fourier analysis of this signal then retrieves the relevant data. In homodyne detection, the detector

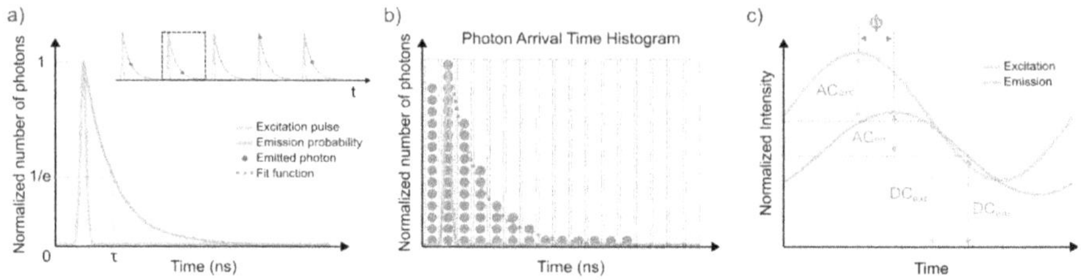

FIGURE 4.4 Schematic representation of time- and frequency-domain FLIM. (a) In time-domain FLIM using TCSPC, short excitation pulses generate fluorescence decays (the inset shows detected single photons across multiple excitation pulses). (b) Histogram of photon arrival times constructed from single-photon detections, where the green curve is the instrument response function (IRF) and the dotted red line is the fitted decay. (c) In frequency-domain FLIM, sinusoidally modulated excitation produces a phase-shifted and demodulated emission signal, with AC and DC components indicated. (d) Modulation and phase plots as a function of frequency for different lifetimes, highlighting modulation lifetime (TM) and phase lifetime (TP).

gain matches the excitation frequency, filtering the signal based solely on the phase difference between the emission and reference waveform.

Figure 4.4 reports schematically the physics and working principle of time-domain and frequency-domain FLIM. Figure 4.4a presents a schematic of the excitation and measurement process in the time domain, where each excitation pulse is followed by a corresponding fluorescence emission. Figure 4.4b illustrates the method for constructing the temporal decay curve through the acquisition of individual photons emitted after excitation in TCSPC. Figure 4.4c shows the modulation of the excitation and emission signals in the frequency domain, highlighting how phase and amplitude variations can provide detailed insights into the dynamic properties of the studied system.

As a matter of comparison, frequency-domain methods offer faster acquisition speeds compared to time-domain techniques like TCSPC, making them ideal for capturing rapid cellular processes. However, TCSPC's ability to accurately record times of individual photons allows it to distinguish multiple decay components more effectively. On the other hand, to resolve multiexponential decays, frequency-domain methods require multi-frequency measurements or digital heterodyning combined with phasor analysis. However, even if TCSPC provides superior temporal resolution and signal-to-noise ratios for dimmer samples, it struggles with bright samples and high photon rates.

As a final observation, frequency-domain FLIM is often more cost-effective since it can be implemented using modulated continuous-wave (CW) sources, which are generally less expensive and complex than pulsed lasers required for TCSPC.

4.5 LASER SCANNING AND WIDE-FIELD FLIM MICROSCOPY

FLIM can be implemented using two main imaging systems, scanning systems and widefield systems, each offering distinct advantages depending on the experimental needs.

Scanning systems, including confocal and multiphoton microscopes, operate by sequentially scanning a focused laser beam across the sample, measuring fluorescence lifetime on a pixel-by-pixel basis, usually in combination with TCSPC. This method yields high spatial resolution, particularly important in applications requiring fine structural detail, such as tissue imaging or subcellular analysis. Moreover, scanning systems offer better optical sectioning and depth penetration, which makes them ideal for 3D imaging in thick or optically dense samples. They also support advanced techniques such as spectral FLIM and polarization-resolved FLIM, which can provide additional layers of information on the molecular state of the sample. However, one of the main limitations of scanning systems is their slower acquisition speed, which can be problematic for studying fast dynamic processes like protein conformational changes or calcium signaling.

Widefield FLIM systems, on the other hand, employ widefield illumination, typically using pulsed LEDs or lasers in combination with fast-gating camera devices. This approach captures fluorescence lifetimes across the entire field of view (quasi)simultaneously, allowing for rapid acquisition of FLIM data. Widefield systems are thus highly suited for real-time imaging of dynamic processes, particularly in live cell imaging, where fast temporal resolution can be critical. However, widefield systems often face challenges related to spatial resolution, as they can suffer from out-of-focus light and background noise. Despite these limitations, modern developments in widefield optics, along with computational deconvolution methods, are steadily enhancing the quality of widefield FLIM measurements, bringing them closer to the resolution typically associated with scanning systems.

Ultimately, the choice between scanning and widefield FLIM depends on the specific demands of the experiment. For studies requiring high spatial resolution, deep tissue penetration, and detailed molecular analysis, scanning systems are often preferred. Conversely, for high-speed imaging of dynamic events in live cells or other fast processes, widefield FLIM is advantageous due to its rapid acquisition capabilities. Both systems remain critical tools in advancing the understanding of biological and material systems through fluorescence lifetime analysis.

4.6 PHASOR ANALYSIS

FLIM captures two- or three-dimensional data, integrating both spatial and temporal information. Analyzing FLIM data is inherently more complex than processing intensity-based measurements, as it requires specialized methods to extract fluorescence lifetime information from time-resolved signals. One such method is phasor analysis, a graphical and fit-free approach that has become widely adopted for FLIM data interpretation (Ranjit et al. 2018). In phasor analysis, the fluorescence decay curve of each pixel is transformed into a point on a phasor plot, with its position determined by the phase and modulation characteristics of the fluorescence signal. This visualization simplifies the interpretation of complex lifetime distributions, as distinct fluorophores or environmental conditions appear as clusters of points on the plot. This method is particularly effective for studying heterogeneous environments, distinguishing free and bound states of fluorophores, and identifying molecular interactions. The phasor approach offers a visually intuitive and computationally efficient alternative to traditional multi-exponential fitting. This transformation, based on Fourier analysis of the fluorescence intensity decay curve, provides a straightforward and robust way to analyze complex fluorescence data, making it a powerful tool in FLIM applications.

For a fluorescence decay curve *I(t)*, acquired in the time domain for each pixel of a FLIM image (Figure 4.5a), the phasor coordinates are calculated using the Fourier transform at a specific modulation frequency $\omega = 2\pi f$, where f represents the laser repetition frequency. The coordinates G (the real part) and S (the imaginary part) of the phasor are given by the following equations:

$$G = \frac{\int_0^\infty I(t)\cos(\omega t)\,dt}{\int_0^\infty I(t)\,dt} \tag{4.9}$$

$$S = \frac{\int_0^\infty I(t)\sin(\omega t)\,dt}{\int_0^\infty I(t)\,dt} \tag{4.10}$$

The integrals in the numerator represent the weighted sum of the cosine and sine components of the fluorescence decay, while the denominator normalizes the phasor, ensuring that all points fall within a semicircular arc, known as the universal semicircle, in the phasor plot (Figure 4.5b). Each pixel of the image gives a point on the phasor plot. For a single-exponential fluorescence decay characterized by a lifetime τ, the phasor coordinates G and S are mathematically related to the lifetime as:

$$G = \frac{1}{1+(\omega\tau)^2} \tag{4.11}$$

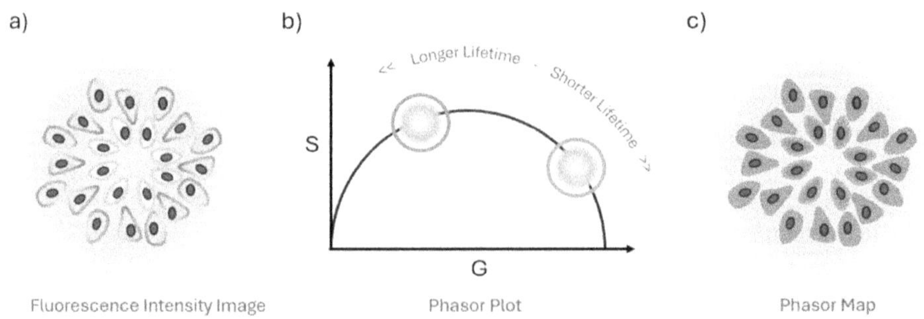

FIGURE 4.5 (a) Schematic representation of a FLIM image of a model sample. (b) Corresponding phasor plot displaying the distribution of lifetime components. (c) Reconstructed phasor maps based on the regions identified in the phasor plot, highlighting the spatial distribution of different fluorescence lifetimes within the sample.

$$S = \frac{\omega\tau}{1+(\omega\tau)^2} \tag{4.12}$$

On the phasor plot, all single-exponential decays fall along the universal semicircle. The position of the point along this semicircle depends on the fluorescence lifetime. Longer lifetimes are located near the origin (left), while shorter lifetimes move toward the lower right, closer to the edge of the unit circle. For example, a lifetime approaching zero would result in a point near $(G, S) = (0, 1)$, while an infinitely long lifetime would approach $(G, S) = (0, 0)$.

For pixels that contain multi-exponential decays or a mixture of fluorophores with different lifetimes, the resulting phasor lies inside the semicircle. In this case, the phasor coordinates represent a weighted average of the phasors corresponding to the individual components. For a bi-exponential decay with lifetimes τ_1 and τ_2 and their respective fractional contributions f_1 and f_2, the resulting phasor (G, S) is given by:

$$G = f_1G_1 + f_2G_2 \tag{4.13}$$

$$S = f_1S_1 + f_2S_2 \tag{4.14}$$

where G_1, S_1, G_2, and S_2 are the phasor coordinates for the individual lifetimes τ_1 and τ_2. This linear combination enables the phasor approach to capture complex lifetime behaviors in biological samples.

In the phasor plot, lifetime distributions can be selected using colored cursors, which assign corresponding colors to the image pixels, resulting in the so-called "lifetime maps" (Figure 4.5c). One of the key advantages of the phasor method is its ability to bypass complex curve-fitting algorithms. Instead of requiring manual fitting of multi-exponential models to fluorescence decay data, the phasor plot offers an intuitive visual representation of decay dynamics. Mixtures of fluorophores, varying biochemical states, or environmental differences appear as distinct clusters or regions on the phasor plot, enabling rapid identification of molecular interactions or environmental changes. Another major benefit is the phasor method's capability to separate fluorescent species with overlapping emission spectra but distinct lifetimes. By visually mapping different regions of the phasor plot, it becomes straightforward to isolate areas corresponding to specific fluorophore lifetimes or distinct cellular microenvironments. Furthermore, because each point on the phasor plot represents a linear combination of lifetime components, it is possible to easily differentiate between free and bound states of a fluorophore or identify regions undergoing specific biochemical reactions.

4.7 CELL AND TISSUE IMAGING

FLIM has become a useful tool in modern biological and medical research, offering insights into the functional and molecular landscape of cells and tissues. Building upon the foundational principles discussed earlier in this chapter, including the mechanistic details of fluorescence lifetimes, the practical implementation of FLIM, and its visualization via phasor plots, this section delves into its applications and broader significance. FLIM extends far beyond the mere imaging of cellular structures, enabling the investigation of dynamic biochemical environments, protein interactions, and metabolic states with unparalleled precision and specificity (Lakowicz, 2006). At its core, FLIM leverages the principle that the fluorescence lifetime of a fluorophore is influenced by its local environment, such as pH, polarity, and the presence of quenching agents or Förster resonance energy transfer (FRET) partners.

One of the most transformative applications of FLIM is in metabolic imaging. By exploiting the intrinsic fluorescence lifetimes of NADH and FAD, two coenzymes central to cellular metabolism, researchers can map the metabolic states of cells with remarkable resolution (Skala et al. 2007; Stringari et al. 2011). FLIM-based metabolic imaging has been applied to study cancer metabolism, stem cell differentiation, oxidative stress, and immune cell activation, providing insights into how metabolic shifts underlie pathological and physiological processes (Figure 4.6a) (Pugliese et al. 2023). For example, cancer cells often exhibit a preference for glycolysis over oxidative

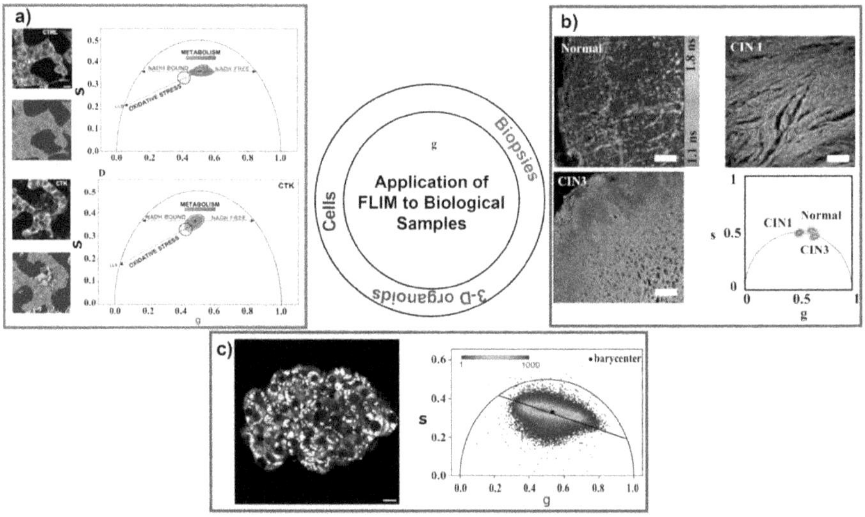

FIGURE 4.6 FLIM is primarily used to investigate cellular metabolism and oxidative stress in culture models. In the example shown in (a), FLIM is applied to a beta-cell model (INS-1 cells) by analyzing the NAD(P)H metabolism and long-lifetime fluorescence species after treatment with pro-inflammatory cytokines. (adapted from https://doi.org/10.1038/s41598-023-40272-9). (b) FLIM can also be applied to fixed tissue biopsies to distinguish between healthy and tumor regions. For instance, in cervical tissue, FLIM can differentiate stages of cervical intraepithelial neoplasia (CIN). Normal: is the healthy region; CIN1, which is low-risk and often resolves without treatment, and CIN3, which indicates severe dysplasia (adapted from https://doi.org/10.1364/BOE.386999). (c) FLIM can be combined with two-photon microscopy to study complex 3D models such as multicellular spheroids or organoids. (b) Displays an intensity image of human-derived Langerhans islets alongside the corresponding phasor plot, illustrating the metabolic trajectory. (Adapted from https://doi.org/10.1038/s42003-022-04215-w.)

phosphorylation, even in the presence of oxygen (Xu et al. 2022). FLIM can detect this metabolic reprogramming by distinguishing between free and protein-bound NADH based on their distinct lifetimes, offering a non-invasive approach to characterize tumor heterogeneity and monitor therapeutic responses. FLIM has also been used to distinguish tumor from healthy regions in unstained tissue biopsies, with results later correlated to Hematoxylin and Eosin (H&E) staining (Wang et al. 2020). In this context, FLIM revealed the relationship between metabolic changes and cancer progression, enabling the differentiation of various stages of cervical lesions, from low-risk to high-risk (Figure 4.6b), as well as identifying metabolic differences between normal and tumoral mammary carcinoma (Conklin et al. 2009). FLIM has also been applied to various 3D organoid models to investigate key questions in developmental and cancer biology, advanced disease modeling, therapeutic evaluation, and other bioengineering applications (Figure 4.6c) (Barroso et al. 2023).

Beyond metabolism, FLIM has improved the study of protein-protein interactions through FRET-FLIM. In FRET, the energy transfer between a donor and an acceptor fluorophore is highly distance-dependent, typically occurring within 1–10 nm, a scale relevant for molecular interactions. By measuring the reduction in the donor's fluorescence lifetime, FLIM provides a quantitative and spatially resolved readout of FRET efficiency, facilitating the study of dynamic biomolecular interactions in live cells (De Los Santos et al. 2015). Applications range from elucidating signaling pathways to monitoring conformational changes in proteins, with implications for understanding disease mechanisms and drug development. In tissue-level imaging, FLIM offers unique advantages in visualizing and characterizing the extracellular matrix (ECM), a critical component in tissue homeostasis and pathology. Collagen, a major ECM protein, exhibits intrinsic second-harmonic generation (SHG) signals that can be analyzed via Phasor FLIM (Ranjit et al. 2015). In neuroscience, FLIM is used to study metabolic variations within neurons and glial cells, providing information on brain function, plasticity, and disease states like Alzheimer's or Parkinson's disease, where altered metabolic profiles can serve as early biomarkers (Yasuda 2019). FLIM is used to map local biochemical (Bastiaens and Squire 1999) and biophysical information in real-time, such as intracellular pH (Orte et al. 2013) or examine oxygen distribution (Alspaugh et al. 2021), distinguishing between oxygenated and hypoxic regions within cells. This ability enables researchers to differentiate between ischemic and healthy regions in cardiac or brain tissue, identifying areas suffering from oxygen deprivation (Marcu 2012).

In clinical applications, FLIM has shown promise as a diagnostic and prognostic tool (Picart et al. 2024). In cancer diagnostics, FLIM can distinguish malignant from benign tissues based on differences in their metabolic signatures or the distribution of endogenous fluorophores such as porphyrins and flavins. Additionally, exploiting the fluorescence lifetime of fluorophores has a significant impact on fluorescence-guided imaging. For example, the near-infrared dye indocyanine green (ICG) has proven to be an effective tool for distinguishing healthy from cancerous tissues. Tumor tissue typically exhibits a longer fluorescence lifetime than non-cancerous tissue (Pal et al. 2023). This tumor-specific shift in lifetime enables discrimination between tumor and normal tissue with an accuracy exceeding 97% across multiple tumor types. As FLIM technology continues to advance, its potential in precision medicine is becoming increasingly apparent, enabling patient-specific diagnostics and monitoring. Despite its numerous advantages, FLIM is not without challenges. Recent developments in machine learning have further enhanced the utility of phasor analysis, enabling automated classification and analysis of FLIM data in high-throughput settings (Héliot and Leray 2021).

In conclusion, FLIM imaging has emerged as a powerful and versatile technique for exploring the molecular underpinnings of cellular and tissue biology. Its ability to provide quantitative, environment-sensitive, and label-free measurements has positioned FLIM as a cornerstone of modern fluorescence microscopy. By bridging the gap between molecular specificity and spatial resolution, FLIM enables a deeper understanding of complex biological systems and paves the way for innovative applications in research, diagnostics, and therapy. As this field continues to evolve, the convergence of technological advancements and interdisciplinary collaboration will undoubtedly unlock new possibilities, further cementing FLIM's role as a transformative tool in life sciences.

REFERENCES

Alspaugh G, Roarke B, Chand A et al. (2021) Developing analysis protocols for monitoring intracellular oxygenation using fluorescence lifetime imaging of myoglobin-mcherry. *Methods Mole Biol* 2304:315–337. https://doi.org/10.1007/978-1-0716-1402-0_17.

Barroso M, Monaghan MG, Niesner R, Dmitriev RI (2023) Probing organoid metabolism using fluorescence lifetime imaging microscopy (FLIM): the next frontier of drug discovery and disease understanding. *Adv Drug Deliv Rev* 201:115081. https://doi.org/10.1016/J.ADDR.2023.115081.

Bastiaens PIH, Squire A (1999) Fluorescence lifetime imaging microscopy: spatial resolution of biochemical processes in the cell. *Trends Cell Biol* 9:48–52. https://doi.org/10.1016/S0962-8924(98)01410-X.

Becker W (2012) Fluorescence lifetime imaging: techniques and applications. *J Microsc* 247:119–136. https://doi.org/10.1111/J.1365-2818.2012.03618.X.

Becker W, Bergmann A, Hink MA et al. (2004) Fluorescence lifetime imaging by time-correlated single-photon counting. *Microsc Res Tech* 63:58–66. https://doi.org/10.1002/JEMT.10421.

Conklin MW, Provenzano PP, Eliceiri KW et al. (2009) Fluorescence lifetime imaging of endogenous fluorophores in histopathology sections reveals differences between normal and tumor epithelium in carcinoma in situ of the breast. *Cell Biochem Biophys* 53:145–157. https://doi.org/10.1007/S12013-009-9046-7.

Craggs TD (2009) Green fluorescent protein: structure, folding and chromophore maturation. *Chem Soc Rev* 38:2865–2875. https://doi.org/10.1039/B903641P.

Datta R, Heaster TM, Sharick JT et al. (2020) Fluorescence lifetime imaging microscopy: fundamentals and advances in instrumentation, analysis, and applications. *J Biomed Opt* 25:1. https://doi.org/10.1117/1.jbo.25.7.071203.

De Los Santos C, Chang C, Mycek M, Cardullo RA (2015) FRAP, FLIM, and FRET: detection and analysis of cellular dynamics on a molecular scale using fluorescence microscopy. *Mol Reprod Dev* 82:587–604. https://doi.org/10.1002/MRD.22501.

Fornasiero EF, Opazo F (2015) Super-resolution imaging for cell biologists: concepts, applications, current challenges and developments prospects & overviews. *BioEssays* 37:436–451. https://doi.org/10.1002/BIES.201400170.

Gratton E (2003) Fluorescence lifetime imaging for the two-photon microscope: time-domain and frequency-domain methods. *J Biomed Opt* 8:381. https://doi.org/10.1117/1.1586704.

Héliot L, Leray A (2021) Simple phasor-based deep neural network for fluorescence lifetime imaging microscopy. *Sci Rep* 11:1–13. https://doi.org/10.1038/S41598-021-03060-X.

Kubitscheck U (2013) *Fluorescence Microscopy: From Principles to Biological Applications*. Wiley-VCH, Weinheim, p. 410.

Lakowicz JR (2006) *Principles of Fluorescence Spectroscopy*. Springer, New York.

Langhals H (2020) Fluorescence and fluorescent dyes. *Phys Sci Rev* 5. https://doi.org/10.1515/PSR-2019-0100.

Lim SY, Shen W, Gao Z (2014) Carbon quantum dots and their applications. *Chem Soc Rev* 44:362–381. https://doi.org/10.1039/C4CS00269E.

Liu X, Lin D, Becker W et al. (2019) Fast fluorescence lifetime imaging techniques: a review on challenge and development. *J Innov Opt Health Sci* 12:12. https://doi.org/10.1142/S1793545819300039.

Marcu L (2012) Fluorescence lifetime techniques in medical applications. *Ann Biomed Eng* 40:304. https://doi.org/10.1007/S10439-011-0495-Y.

Marcu L, French PMW, Elson DS (2015) *Fluorescence Lifetime Spectroscopy and Imaging: Principles and Applications in Biomedical Diagnostics*. CRC Press, Boca Raton, p. 554.

Orte A, Alvarez-Pez JM, Ruedas-Rama MJ (2013) Fluorescence lifetime imaging microscopy for the detection of intracellular pH with quantum dot nanosensors. *ACS Nano* 7:6387–6395. https://doi.org/10.1021/NN402581Q.

Pal R, Lwin TM, Krishnamoorthy M et al. (2023) Fluorescence lifetime of injected indocyanine green as a universal marker of solid tumours in patients. *Nat Biomed Eng* 7: 12 7:1649–1666. https://doi.org/10.1038/s41551-023-01105-2.

Picart T, Gautheron A, Caredda C et al. (2024) Fluorescence-guided surgical techniques in adult diffuse low-grade gliomas: state-of-the-art and emerging techniques: a systematic review. *Cancers* 16:2698. https://doi.org/10.3390/CANCERS16152698.

Pugliese LA, De Lorenzi V, Bernardi M et al. (2023) Unveiling nanoscale optical signatures of cytokine-induced β-cell dysfunction. *Sci Rep* 2023 13:1 13:1–14. https://doi.org/10.1038/s41598-023-40272-9.

Ranjit S, Dvornikov A, Stakic M et al. (2015) Imaging fibrosis and separating collagens using second harmonic generation and phasor approach to fluorescence lifetime imaging. *Sci Rep* 5:1–10. https://doi.org/10.1038/SREP13378.

Ranjit S, Malacrida L, Jameson DM, Gratton E (2018) Fit-free analysis of fluorescence lifetime imaging data using the phasor approach. *Nat Protoc* 13:1979–2004. https://doi.org/10.1038/S41596-018-0026-5.

Skala MC, Riching KM, Gendron-Fitzpatrick A et al. (2007) In vivo multiphoton microscopy of NADH and FAD redox states, fluorescence lifetimes, and cellular morphology in precancerous epithelia. *Proc Natl Acad Sci USA* 104:19494–19499. https://doi.org/10.1073/PNAS.0708425104.

Soloviev VY, Tahir KB, McGinty J et al. (2007) Fluorescence lifetime imaging by using time-gated data acquisition. *Appl Opt* 46(30):7384–7391. https://doi.org/10.1364/AO.46.007384.

Specht EA, Braselmann E, Palmer AE (2017) A critical and comparative review of fluorescent tools for live-cell imaging. *Annu Rev Physiol* 79:93–117. https://doi.org/10.1146/ANNUREV-PHYSIOL-022516-034055.

Stockert JC, Blázquez-Castro A (2017) *Fluorescence Microscopy in Life Sciences*. Elsevier, Amsterdam, p. 751.

Stokes GG, Stokes GG (1852) On the change of refrangibility of light. *RSPT* 142:463–562.

Stringari C, Cinquin A, Cinquin O et al. (2011) Phasor approach to fluorescence lifetime microscopy distinguishes different metabolic states of germ cells in a live tissue. *Proc Natl Acad Sci USA* 108:13582–13587. https://doi.org/10.1073/PNAS.1108161108.

Suhling K, Hirvonen LM, Levitt JA et al. (2015) Fluorescence lifetime imaging. *Scanning* 18:1–50. https://doi.org/10.1007/978-94-007-6174-2_13-2.

Tandale P, Choudhary N, Singh J et al. (2021) Fluorescent quantum dots: an insight on synthesis and potential biological application as drug carrier in cancer. *Biochem Biophys Rep* 26:100962. https://doi.org/10.1016/J.BBREP.2021.100962.

Tian F, Xu G, Zhou S et al. (2023) Principles and applications of green fluorescent protein-based biosensors: a mini-review. *Analyst* 148:2882–2891. https://doi.org/10.1039/D3AN00320E.

Wang M, Da Y, Tian Y (2023) Fluorescent proteins and genetically encoded biosensors. *Chem Soc Rev* 52:1189–1214. https://doi.org/10.1039/D2CS00419D.

Wang X, Wang X, Wang Y et al. (2020) Discriminating different grades of cervical intraepithelial neoplasia based on label-free phasor fluorescence lifetime imaging microscopy. *Biomed Opt Express* 11(4):1977–1990. https://doi.org/10.1364/BOE.386999.

Xu T, Junge JA, Delfarah A et al. (2022) Bladder cancer cells shift rapidly and spontaneously to cisplatin-resistant oxidative phosphorylation that is trackable in real time. *Sci Rep* 12:5518. https://doi.org/10.1038/S41598-022-09438-9.

Yasuda R (2019) Principle and application of fluorescence lifetime imaging for neuroscience: monitoring biochemical signaling in single synapses using fluorescence lifetime imaging. *Neurophoton Biomed Spectrosc* 53–64. https://doi.org/10.1016/B978-0-323-48067-3.00003-2.

Zimmer M (2002) Green fluorescent protein (GFP): applications, structure, and related photophysical behavior. *Chem Rev* 102:759–781. https://doi.org/10.1021/CR010142R.

5 Quantitative Phase Imaging in Biomedical Research

Varad Mahesh Phadke
Manipal Academy of Higher Education (MAHE)

Hemanth Noothalapati
Shimane University

Ajeetkumar Patil
Manipal Academy of Higher Education (MAHE)

5.1 INTRODUCTION

5.1.1 Overview of Quantitative Phase Imaging

Optical imaging techniques have been essential tools for the investigation of biological systems for a long time. The capability to record the images of biological systems with better resolution reveals the morphological and dynamic parameters [1]. Among several biomedical imaging techniques, QPI techniques have developed as powerful tools [2]. Also, QPI is a vital optical imaging technique that is also non-invasive, which provides label-free images of live samples at the subcellular level [3]. Additionally, it shows the competence to study living cells and tissue samples without the use of exogenous fluorophores, that is, fluorescent dyes, quantum dots (QDs), fluorescent proteins and so forth, which may affect cellular processes and lead to toxicity. QPI follows the mechanism of interferometry, which calculates the phase variations of a light beam passing through a specimen, which relates the optical path difference (OPD) of the sample, with high sensitivity [4]. Also, QPI becomes an important imaging method in biomedical research and diagnosis [5].

QPI acts as a powerful optical method that improves the contrast in biomedical imaging, specifically for transparent or inadequately absorbing biological specimens like live cells, thin tissue samples and subcellular assemblies. QPI also permits precise and detailed measurement of exterior structure in non-biological samples such as lithographic patterns, fibres, latex dispersions and glass fragments [6,7].

Compared to traditional bright-field microscopy, which works by determining the variations in amplitude of the light beam, QPI measures the phase shifts of the optical wave while passing through a sample to generate high-contrast images.

DOI: 10.1201/9781003570011-5

In QPI, in an interferometric setup, the incident light beam acts as the reference field, while the beam scattered through the sample acts as the object field. The interaction between these two fields allows the whole optical field in such a way that it measures the phase interruption produced by the sample. To boost image or interferogram contrast, typically a phase shift of $\pi/2$ is introduced between the reference and scattered fields. Such adjustments give rise to the destructive interference by converting the phase variations into amplitude variations, which can be easily detected as intensity variations in the resultant images. Later, the measured phase information, which is not directly manageable by standard optical detectors, gets converted into quantifiable amplitude variations. Finally, the phase image is a reconstruction of numerous intensity quantities at different phase shifts, which allows precise determination of the phase delay.

In QPI, the high contrast produced in the recorded images can be obtained from the phase delays induced by transparent or turbid biological specimens [8]. This phenomenon permits the quantification of very thin specimens with improved resolution and sensitivity [9]. The contrast in QPI images is directly dependent on the variations in the refractive index within the sample. The refractive index is related to the dry mass of biological material. Thus, QPI can directly measure dry mass without the need for external staining. This ability is mainly valuable for studying living cells and active biological developments by providing the high-resolution, label-free image recording of cellular morphology and activity.

This chapter provides detailed knowledge about QPI, along with its mechanism and application in various fields to enhance the precision and capability of biomedical research and diagnosis.

5.1.2 Types of QPI

Depending on the dimensionality of the resultant image obtained from QPI techniques, QPI can be divided into two types: two-dimensional (2D) QPI techniques and three-dimensional (3D) QPI techniques. The working mechanism of the QPI techniques can also be classified as 2D and 3D QPI techniques:

5.1.2.1 Two-Dimensional QPI

The 2D QPI technique works by calculating the phase shift in the x-y plane, which is produced in the light beam while passing through a transparent or turbid sample, such as biological cells [10]. This phase shift arises due to variations in the thickness and refractive index of the samples, which vary the speed of light compared to its path through the adjacent medium [10,11].

The basic imaging system contains an incident light beam, which is split into a sample beam that travels toward the specimen and a reference beam. Later, these beams are then recombined, generally via interferometry [12] or holography [13,14] to produce an interference pattern, also called an interferogram, recorded by a digital camera [15,16]. This recorded pattern encrypts the phase and amplitude data of the sample. Then, computational algorithms are used to reconstruct a quantitative phase image, which directly represents the variations in optical path length throughout the sample. Finally, these path length variations can be further converted to measure thickness [17], refractive index [18], or dry mass [19]. Depending on the technique,

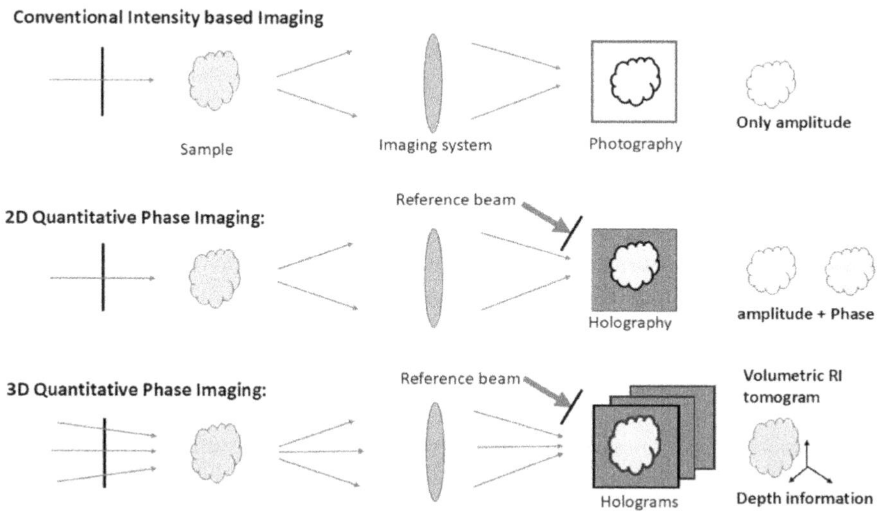

FIGURE 5.1 Mechanism schematics of conventional imaging, 2D QPI and 3D QPI.

phase variation can be extracted using methods such as temporal phase-shifting, off-axis holography, or wavefront sensing with diffraction gratings (Figure 5.1).

5.1.2.2 Three-Dimensional QPI

The 3D-QPI provides amplitude and phase information along with volumetric distribution of the sample's refractive index (RI), effectively creating a 3D tomogram of the internal structure of the sample. Additionally, it delivers detailed depth information of the specific organelles, compartments, proteins, or functional groups within a cell, which is not possible with 2D QPI.

In 3D-QPI, we can capture and reconstruct the phase variation while a light beam passes through different depths of the sample, by visualising the internal structure of the sample in three dimensions [20]. Generally, conventional 3D QPI techniques include multiple 2D phase images recorded at several focal planes, which can be used to obtain volumetric phase maps by using computational reconstruction. This method needs complex algorithms to extract the 3D phase information from the acquired data. Recent inventions have employed optical approaches, such as the wavelength-multiplexed diffractive optical processor [21]. In this method, a specially designed diffractive element transforms the phase information from various axial planes into discrete intensity patterns, where every pattern is fixed with a single wavelength. These intensity patterns are recorded instantaneously on an image sensor, allowing quick, label-free 3D images without digital algorithms. One more method uses the transport of intensity equation (TIE) [20], where multiple defocused intensity images are mathematically processed to gain the 3D phase distribution of the sample.

In summary, 2D QPI delivers rapid, label-free images of cell morphology and quantitative features [22]. But it is restricted to surface or projected information. However, 3D QPI reconstructs the volumetric refractive index distribution by visualisation and quantification of subcellular structures. Additionally, 2D QPI is suitable for rapid analysis of a large number of samples, while 3D QPI is essential to study the depth information of cellular morphology and processes in complex biological samples.

5.2 MECHANISM OF QPI

QPI measures the variations in optical phase shift of a light beam while travelling through transparent or semi-transparent specimens for the quantification of physical parameters. QPI depends on interferometry for the detection of phase variations produced by the light beam while slowing down in the region having a high refractive index. The phase shift ϕ can be measured using the following formula [20].

$$\phi = \frac{2\pi}{\lambda}\int_{z=0}^{h} n(z)\,dz$$

where ϕ denotes the phase shift caused by the sample with refractive index n and height h, in the z direction. This calculated phase shift is relative to the dry mass in the sample [21].

As we consider advancements in QPI techniques, all of the QPI techniques follow two different types of mechanisms. One is the interferometric QPI techniques, which work by producing interference of the reference and sample beams. This interference design encodes the phase information, permitting highly precise phase measurements. However, these methods need accurate alignment, high stability and complex optical setups [23,24]. On the other hand, non-interferometric techniques retrieve phase data without using interference between the reference and sample beams. Instead, these techniques use computational algorithms for the reconstruction of phase from intensity measurements by analysing images recorded at various focal lengths or incident angles. Generally, these techniques are simple and robust against the noise and environmental factors and can be executed in wide-field, real-time imaging techniques.

5.2.1 Interferometric Techniques

Interferometric techniques are at the core of QPI, which enables the accurate, label-free quantification of phase shifts produced by transparent or semi-transparent samples like biological cells and tissues. These methods operate on the mechanism of wave interference to obtain thorough data related to variations in optical path length, which are linked to their physical and chemical properties. There are several interferometric techniques as follows:

5.2.1.1 Digital Holographic Microscopy (DHM)

It is an enhanced optical technique to offer quantitative, label-free imaging of live cells and tissue. DHM works on the principle of interferometric digital holography, in which both phase and amplitude variations of the light beam reflected or transmitted through the sample are recorded to reconstruct it mathematically [25]. As compared to conventional microscopy, which captures only intensity information, DHM provides phase information, which is directly related to the specimen's physical characteristics, making DHM more valuable in biomedical research as a non-invasive, high-contrast imaging method for transparent or turbid samples, such as biological specimens [26–29] (Figure 5.2).

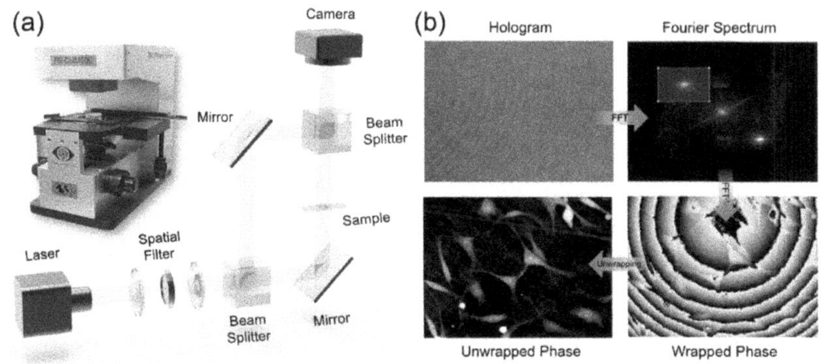

FIGURE 5.2 Digital holography microscopy. (a) Schematic of a DHM with a Mach-Zehnder configuration and (b) basic mechanism of digital recording and numerical reconstruction [33].

DHM uses a coherent light source divided into object and reference beams [30]. Initially, the object beam interacts with the sample, introducing phase shifts [31], while the reference beam provides a stable phase reference. Their interference forms a digital hologram containing both amplitude and phase information [32]. Using FFT and spatial filtering, the complex wavefront is reconstructed [33,30]. This allows analysis of cell morphology, dry mass and refractive index [31,32]. DHM also enables numerical refocusing for 3D imaging and supports real-time imaging with GPU acceleration, making it ideal for live-cell and biomedical applications [34].

5.2.1.2 Phase-Shifting Interferometry (PSI)

PSI is a precise method in QPI that retrieves phase information from transparent samples. In this technique, multiple interferograms are captured by using controlled phase shift, generally using a piezoelectric transducer. These shifts are partial sections of a wavelength, such as π/2 or π/4. The resultant intensity patterns are processed using algorithms for the reconstruction of a quantitative phase map. This phase map provides the length of the optical path of the sample along with a detailed morphological analysis of cellular elements [34,35].

The working principle of PSI involves recording several interferograms with increasing phase shifts. Mathematically, each intensity pattern $I_k\,(x, y)$ can be described as:

$$Ik\left(x,\ y\right) = I_0\left(x,\ y\right) + I_1\left(x,\ y\right)\ \cos\ [\phi\ \left(x,\ y\right) + \delta_k]$$

where I_0 is the background intensity, I_0 is the modulation, $\phi\ (x, y)$ is the unknown phase to be measured, δk is the known phase shift at kth exposure. PSI calculates phase $\phi\ (x, y)$ from multiple phase-shifted interferograms using trigonometric algorithms, which enables high-precision wavefront reconstruction [36,35]. The resulting phase is wrapped between $-\pi$ and $+\pi$, to obtain exact phase maps with a precision of nanometres. PSI is very sensitive to vibrations and is commonly executed in steady situations.

PSI provides high phase precision with nanometre resolution, which makes it superlative for measuring smooth surfaces, MEMS and thin films [37]. Also, it offers remote quantitative data to achieve higher signal-to-noise ratios from controlled phase shifts [34].

5.2.1.3 Spectral Modulation Interferometry (SMI)

SMI is a highly sensitive, speckle-free method for QPI that enables rapid acquisition processes. SMI combines spectral modulation and spectral-domain low-coherence interferometry, which encodes amplitude and phase data of a sample on the wider light spectrum. Thus, this technique can be demonstrated for the effective study of both static reflective samples and live biological samples.

This study describes a spectrally encoded interferometric imaging system using a superluminescent diode with a line illumination and a spectrometer. Light gets dispersed and focused on the sample and a reference mirror, producing an interference spectrum captured by a highly speed linescan camera. A 1D galvo scanner permits full-field imaging by sweeping the line across the sample. The system enables both surface metrology and imaging of thin and transparent samples such as live cells through a double-pass transmission process. As compared to conventional techniques, which analyse fringe patterns using short-time Fourier transforms, SMI analyses the whole interference spectrum, providing more precise and inclusive signal reconstruction and better understanding of modulation dynamics [38].

5.2.1.4 Common-Path Interferometry

Common path interferometry in QPI is a technique in which the reference and object beams pass in nearly undistinguishable optical paths inside the system. This arrangement is designed to obtain maximum phase stability and minimum noise, which can make it a highly appropriate method for sensitive biological measurements [39,40].

A coherent or moderately coherent light ray, like a laser or LED, is used as a light source. Generally, the illumination beam passes through a beam splitter or diffractive element, which generates both the reference and sample beams in the same optical axis [41,42]. The incident light interacts while passing through the biological specimen. This phenomenon creates a phase shift due to variations in the optical path length produced by dissimilarities in refractive index and thickness. Later, it results in a modulated wavefront [43]. In this technique, a partial light beam performs as the reference beam, and the remaining beam, which is modified after passing through or reflecting from the sample, acts as a sample beam [44,45]. These sample and reference beams get recombined to produce an interferogram, which encodes the phase variations between the two beams to reveal the optical properties of the sample [46].

Common path interferometry is highly stable against mechanical and environmental variations.

Also, it has a compact and simple optical setup with minimised alignment sensitivity [47].

5.2.2 Non-Interferometric Techniques

Non-interferometric techniques in QPI allow for phase retrieval without the requirement of separate reference beams or intricate interferometric setups. These techniques depend on computational algorithms for the reconstruction of phase information from intensity data. This principle makes them simpler and more robust against environmental instabilities compared to interferometric techniques [48–50].

5.2.2.1 Transport of Intensity Equation (TIE)

We consider traditional photodetectors, which can only calculate light intensity but fail to measure phase. Classical interferometry accomplishes phase information by overlaying coherent beams to produce fringe patterns, which can be used to extract phase variation using computational algorithms. However, naturally, phase gets converted into intensity through light propagation, which is known as the Transport of Intensity Equation (TIE) [33]. TIE-based phase retrieval techniques capture the variations in the intensity at defocused planes at a small distance to reconstruct phase information. However, TIE is a non-interferometric and label-free method in QPI; it involves a partially coherent light source. Thus, TIE proves that even the invisible phase can also affect intensity patterns during light propagation, allowing for reconstruction of the quantitative phase without the requirement of an interferometric setup [33].

In 1982, Teague [51,52] established Transport of Intensity Equation (TIE), depending on the axial change $\frac{\partial I}{\partial Z}$ in the intensity of a monochromatic beam to the transverse phase gradient of that beam. For a complex field, $U=(x, y)\sqrt{I(x,y)}e^{i\phi(x,y)}$, the TIE can be calculated as

$$-k\frac{\partial I(x,y)}{\partial Z} = \nabla \cdot [I\ (x,y)\nabla\phi\ (x)]$$

where $k=2\pi/\lambda$ and ∇ is the gradient in the transverse coordinate $x=(x, y)$. Thus, this approach shows that the quantitative phase measurements can be retrieved by using measured axial intensity variations without interferometry data.

5.2.2.2 Wavefront Sensing

In wavefront sensing, the Shack-Hartmann wavefront sensor (SHWFS) is a main configuration that is used to obtain the phase shift of the sample using QPI. The SHWFS is an extensively used phase measurement tool based on geometric optics. It includes an array of lenslets, focusing the light beam on the CCD or CMOS sensors. When a planar wavefront passes, the focal spots produce a regular grid. If the wavefront is distorted, the shifts get observed from their ideal positions. The shift of every spot is linked with the gradient of the wavefront through the equivalent lenslets. By measuring these spot shifts, the single wavefront tilts are gained, and later, the overall phase distribution is recreated through numerical integration. This method is being adapted for different fields like biomedical imaging, ophthalmology and optical system alignment [53,10]. The Hartmann screen was enhanced in 1971 by Shack and Platt using lenslets, improving precision and effectiveness. Recent SHWFSs are compact, broadband-compliant and suitable for non-interferometric wavefront sensing [54]. However, Shack–Hartmann sensors offer restricted spatial resolution because of their microlens size, which makes them inappropriate for high-resolution imaging like QPI and optical microscopy [53,10]. Considering the current improvements, Meta surface-enhanced SHWFS offers enhanced spatial resolution and enables single-shot phase imaging, but conventional SHWFS is restricted to QPI and high-resolution microscopy applications [53,10].

5.2.2.3 Phase Retrieval

It is a non-interferometric technique that evaluates the phase of light by using intensity measurements. The Gerchberg–Saxton (GS) algorithm was the first phase retrieval method to reconstruct phase by implementing consistency between intensity images and Fourier planes. This technique has been extensively applied to X-ray diffraction imaging, adaptive optics and quantitative phase microscopy [11,12]. The amendments to the GS algorithm, like Misell's method, using several diffraction patterns and Fienup's hybrid input-output algorithm, have enhanced conjunction and flexibility by overcoming problems such as stagnation and local minima [12]. Ptychography, a latest phase retrieval technique, uses spatially overlaid diffraction patterns for phase reconstruction along with enhanced stability and dependability. Now, this technique is vital in high-resolution X-ray imaging, electron microscopy and QPI [12]. Fourier ptychographic microscopy outspreads ptychography to the frequency domain, adding angle-based illuminations to enlarge the aperture and enhance resolution, aiding large field-of-view, high-resolution imaging. Despite their influence, iterative phase retrieval methods need large datasets and many iterations for stable convergence, restraining their speed and depending on high coherence, which is challenging for some light illumination sources [12,13].

5.3 APPLICATIONS OF QPI IN BIOMEDICAL RESEARCH

QPI provides non-invasive and label-free examination, making it a vital technique in several areas of biomedical research. The following content elaborates on the applications of QPI in various fields of biomedical research.

5.3.1 Cell Biology

QPI provides a label-free, high-resolution method to study biological samples, making it invaluable in biomedical research and cell biology. Most importantly, QPI is used to study cellular growth [14], cell division [15] and morphology. It empowers the quantitative parameters of shape, size, area and volume [16] of cellular components without the requirement of fluorescent probes. This method delivers a better understanding of cellular morphology and processes, along with diminishing phototoxicity. Also, QPI can be used to precisely measure the dry mass of neural cells, offering the segmentation of somas and neurites, along with inclusive mass information throughout the whole region. Especially, a research work was demonstrated to determine the ability of QPI to perform continuously for 120 hours [17]. In this study, the cell samples were not harmed due to the incident light beam. Also, QPI can be used to study the metabolism-induced atrophy, to detect structural damage, which confirms the variations in the membrane along with decelerated dynamics.

Additionally, it is possible to quantitatively track muscle wasting in vitro, studying variations in the cell thickness, volume, dry mass loss and intracellular density shifts [18]. Thus, QPI can be used for long-term and label-free examination to study cell growth, providing innovative perceptions about the circulation and dynamics of cellular mass through cellular growth and development.

5.3.2 Cancer Diagnosis

QPI has established the capability to detect nanoscale-level structural variations in tissue that cannot be recorded using conventional microscopy. In a previous study for the diagnosis of prostate cancer, QPI-derived light scattering anisotropy was observed around the stromal areas, which are adjacent to cancerous glands. Similarly, quantitative image examination using QPI was applied to pancreatic adenocarcinoma, by discovering important prognostic implications dependent on the structural arrangement of the extracellular matrix. This study highlights the ability of QPI to improve cancer detection in early stages [19]. In one more research study, QPI was applied to non-labelled ThinPrep urine cytological slides with urothelial carcinoma. By calculating parameters like nuclear mass, entropy, cell dry mass and nucleus-to-cell mass ratio, the researchers observed that QPI can easily differentiate between atypical, negative, suspicious and positive cytological diagnoses. This suggests QPI can enhance diagnostic precision, especially in complex cytology cases [55]. The molecular biology techniques with efficient authentication through QPI can become a powerful model to study nanoparticle drug development. QPI can be helpful to link cellular mechanisms and morphology, which adds depth and clarity to the study [56]. Also, QPI has been used to study fixed breast tissue samples, measuring refractive index variations. Another study observed that disorder strength was remarkably higher in malignant compared to benign breast tissue samples. As this marker is obtained from non-stained tissue, it is not affected by staining variability, which makes it a strong and objective diagnostic device. The study confirms that QPI can be effortlessly used in current pathology systems for assistance in the differentiation of benign and malignant breast lesions [57].

5.3.3 Haematology

QPI is widely applied to examine the blood and RBCs to diagnose diseases such as sickle cell disease [58], reticulocytosis [59], diabetes [60] and leukaemia [61]. By quantification of variations in the refractive index and dry mass of RBCs, QPI allows for automatic, rapid detection of abnormal cells. Such an approach raises diagnostic accuracy and speed. Another study by Kim et al. presents a rapid, label-free method for screening of hematologic disease using QPI combined with machine learning. By analysing refractive index from a single RBC, this method precisely distinguishes conditions like iron-deficiency anaemia, reticulocytosis, diabetes mellitus and hereditary spherocytosis with 98% precision. This method overpowers the confines of conventional blood tests and establishes the powerful

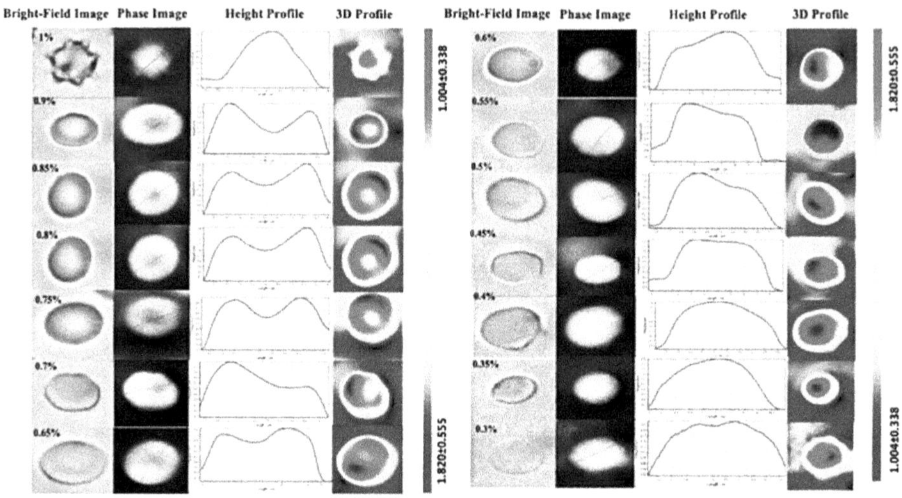

FIGURE 5.3 Variations in the morphology of RBCs under different concentrations of saline.

collaboration between machine learning and QPI in medical diagnosis [62]. Pillai et al. verified 3D QPI to analyse the response of human RBCs to a saline gradient, by using their sensitivity to osmotic variations. By detecting the phase profile, the range of 1%–0% saline, researchers measured the morphological changes in RBC, membrane integrity and haemolysis. This study provides a better understanding of RBC behaviour under hypotonic stress and supports wider applications in haematology and biomedicine [63] (Figure 5.3).

5.3.4 Neuroscience

In the field of neuroscience, QPI offers a non-invasive study of neuronal networks by detecting phase shifts in light passing through cells. QPI allows for real-time observation of neurons as they detect stimuli and establish networks without the need for fluorescent probes, which may cause damage to biological processes. This label-free technique provides a breakdown of dynamic neuronal relations inside the developing neural networks by protecting cell viability for long-term examination. Its ability to integrate with live-cell imaging makes it a vital tool in biomedical research. One study carried out by Kandel et al. determined the use of label-free spatial light interference microscopy (SLIM) to record images of tiny-sized microtubules. Also, SLIM prevents photobleaching and phototoxicity of biological samples. SLIM allows for wide-field imaging of microtubules. This work confirms SLIM as a powerful technique to study the molecular activity of neurons [64]. Also, according to some previous studies at the cellular level, QPI permits a quantitative study of morphology and biomolecule modifications in the cells. Yang et al. scrutinised the single neuron under 3D optical diffraction tomography to examine the initial neurotoxic consequence in *in-vitro* Parkinson's disease (PD) model. In this study, apoptotic symptoms were observed in the cells after 5 h of reaction with the chemicals, which cause PD similar symptoms [17] (Figure 5.4).

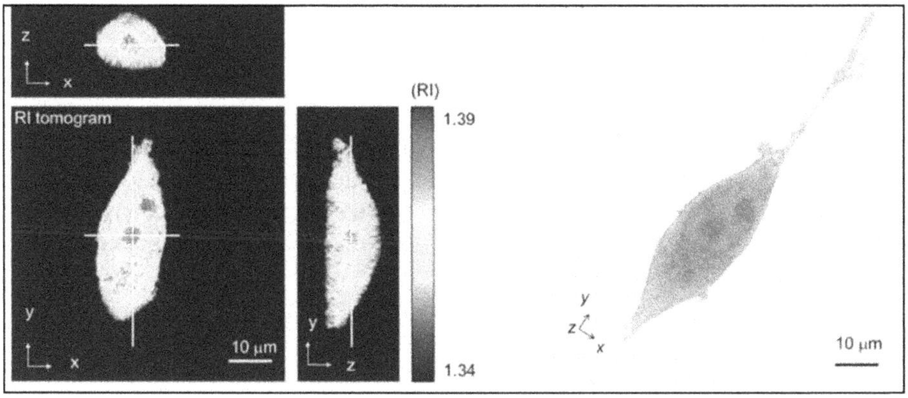

FIGURE 5.4 Left shows the cross-sectional view of the SH-SY5Y cell through RI tomogram, which is reconstructed using multiple 2D amplitude and phase information from the optical diffraction tomography algorithm. The right illustrates the 3D rendered image with a customised transfer function [17].

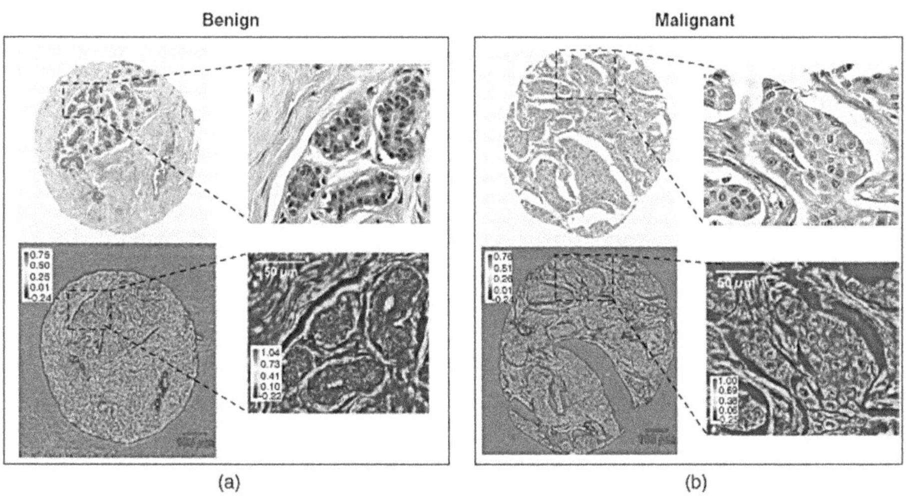

FIGURE 5.5 Comparison between H&E-labelled brightfield microscopy and SLIM images to study tissue morphology of (a) benign and (b) malignant tissues [65].

5.3.5 Histopathology

The research work by Majeed et al. explores SLIM as a label-free, quantitative technique and successfully compares it with conventional H&E staining for the diagnosis of breast cancer. SLIM produces phase images of non-labelled tissues, which can be classified as healthy and malignant tissues. These outcomes recommend SLIM as a durable tool for precise and effective screening and diagnosis of breast cancer [65] (Figure 5.5).

Kandel et al. introduced a non-labelled and quantitative method for surgical pathology by imaging of tissue biopsies. This method offers advanced software that enables computerised diagnosis and data archiving. The setup identified essential markers in colorectal diseases by highlighting their potential to enhance accuracy and effectiveness in a large number of pathology screenings [66]. One more study in which QPI combined with SLIM has been used as a rapid and label-free data about tissue scattering parameters such as mean free path l_s) and anisotropy factor (*g*), which offers morphological insights at the nanoscale level. These features are helpful to build an inclusive dataset of normal and abnormal tissues. The demonstrated method provides precise diagnosis and improves in vivo optical techniques by gaining analytical tissue information [67].

5.4 MULTIMODAL IMAGING APPROACHES INTEGRATED WITH QPI

QPI is being integrated with various multimodal techniques to improve molecular specificity, spatial resolution and morphological analysis of biological samples. These integrations can overcome characteristic limitations of QPI. Some of the important multimodal techniques are as follows:

5.4.1 QPI + Fluorescence Imaging

The combination of refractive index-dependent QPI and fluorescence imaging can be a vital tool to study morphological data and molecular examination of several biological samples. This allows for real-time tracking of dynamic cellular processes using fluorescence imaging and biomechanical parameters such as dry mass or membrane variations by using QPI. Also, another multimodal technique in which QPI is combined with chip-on nanoscopy offers the nanoscale study of the lateral size and the thickness of cellular compartments, which can become a widely interesting study in bioimaging and cell biology [68].

As we consider biomedical research, several imaging approaches, including wide-field microscopy [69], confocal microscopy [70], two-photon microscopy [71], fluorescence lifetime imaging microscopy [72] and light-sheet fluorescence microscopy [73], are established to study the morphology and mechanisms of cellular processes. The integration of these fluorescence imaging techniques with QPI can offer a better understanding of the morphology and mechanisms of cellular processes. Additionally, some of the research groups are working on the cellular studies of cancer cells for the early diagnosis of cancer. The study carried out by Li et al. combined QPI and fluorescence microscopy to record 3D images of fluorescently labelled mouse brain tissues using a conventional bright field microscope. Gu et al. obtained structural and molecular evidence, which will offer a comprehensive understanding of tissue morphology along with neuroscience and oncology [74]. One more research work observed and maximised the accuracy of diagnosis by studying the optimum number of cervical epithelial cells using FLIM. The upper half of the epithelium layers exhibited a high sensitivity and specificity of around 94% and 84%. This FLIM approach can enhance the diagnosis of cervical cancer [75].

5.4.2 QPI + Tomography

QPI combined with the tomography shows a beneficial application for measuring subcellular structures, along with the examination of membrane-less organelles called biomolecular condensates. Such a combined technique can be easily applied to study cell behaviours, such as bone metastasis [76]. Choi et al. presented a multimodal method for 3D mapping of refractive index using a phase-shifting laser interferometric microscopy. This technique permits tomographic images of cells and dynamic mechanical variations, which supports quantitative investigation of microscopy aberrations and tissue light scattering [77].

5.4.3 QPI + Artificial Intelligence (AI)

Recent improvements in AI, particularly in deep learning, significantly improve the capabilities of QPI in biomedical research. AI approaches offer quick and precise tasks like image processing, enhancement, classification, segmentation and data conversion. The high-resolution quantitative information obtained by using QPI supports the training of deep learning models to make it a vital tool for image analysis by simplifying innovative biological information.

This study, carried out by Choi et al., developed a label-free method for 3D segmentation of cellular compartments in living cells using a 3D U-Net-based deep learning model, which empowers detailed segmentation of the cell membrane, nucleoli, nucleus membrane and lipid droplets without the requirement of fluorescent tagging. The same approach can also be demonstrated for time-lapse imaging for the analysis of the active behaviour of immune cells, permitting an enduring observation in their natural state [78]. The figure shows the AI-based segmentation of recorded images using QPI (Figure 5.6).

Shu et al. described an artificial-intelligence-allowed, staining imaging haematology analyser, AIRFIHA, for the precise classification of leukocyte subtypes using label-free images. After training a two-step residual neural network on the images, AIRFIHA can differentiate T and B lymphocytes and CD8 and CD4 T cells with a higher precision than traditional haematology analysers [79].

5.5 CONCLUSION

In this chapter, we discussed the evolutionary applications of QPI in optical imaging for accurate, comprehensive and innovative developments in biomedical research. QPI is a vital, label-free imaging technique to provide high-resolution, quantitative information about live biological samples by calculating the phase variations. By using different interferometric and non-interferometric techniques, it offers both rapid 2D surface analysis and thorough and precise 3D analysis of the specimens. Current improvements have enhanced QPI's sensitivity and capability, making it a critical tool for biomedical research and diagnosis.

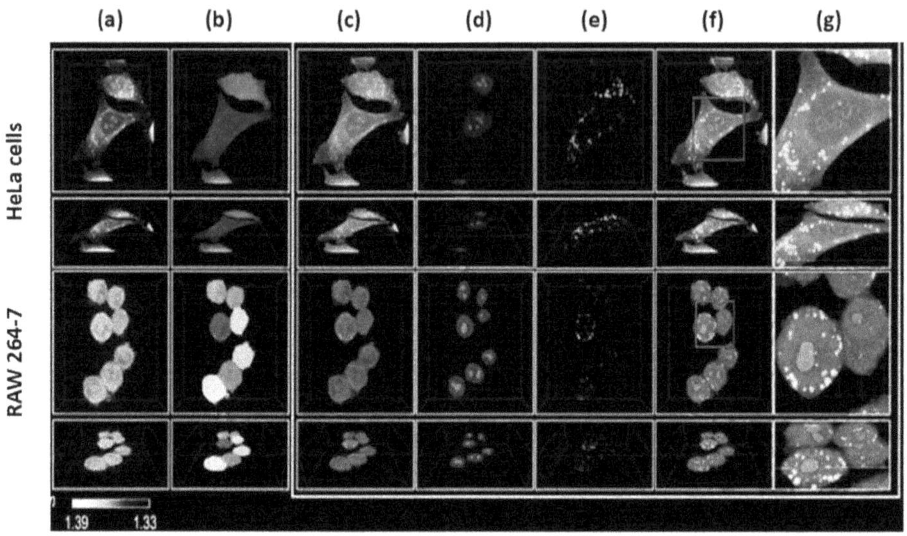

FIGURE 5.6 (a) 3D tomogram, (b) segmentation of the complete cellular region, (c) segmentation of the cell membrane, (d) segmentation of the nucleus and nucleolus, (e) segmentation of lipid droplets, (f) merged image after segmentation and (g) enlarged image with combined segmentation [78].

ACKNOWLEDGEMENT

The author of this chapter, Mr. Varad Mahesh Phadke, is grateful to the Manipal Academy of Higher Education, Manipal, for offering the Dr. T.M.A. Pai Scholarship for the research work.

REFERENCES

1. "Milestones in light microscopy," *Nat Cell Biol*, 11(10), 1165, 2009, doi:10.1038/NCB1009-1165.
2. Y. K. Park, C. Depeursinge, and G. Popescu, "Quantitative phase imaging in biomedicine," *Nat Photon*, 12(10), 578–589, 2018, doi:10.1038/s41566-018-0253-x.
3. R. Kasprowicz, R. Suman, and P. O'Toole, "Characterising live cell behaviour: traditional label-free and quantitative phase imaging approaches," *Int J Biochem Cell Biol*, 84, 89–95, 2017, doi:10.1016/j.biocel.2017.01.004.
4. S. S. Butt et al., "Quantitative phase imaging for characterization of single cell growth dynamics," *Lasers Med Sci*, 38(1), 2023, doi:10.1007/S10103-023-03902-2.
5. S. Kwon et al., "Mitochondria-targeting indolizino 3,2-c. quinolines as novel class of photosensitizers for photodynamic anticancer activity," *Eur J Med Chem*, 148, 116–127, 2018, doi:10.1016/j.ejmech.2018.02.016.
6. J. Sun et al., "Quantitative phase imaging through an ultra-thin lensless fibre endoscope," *Light Sci Appl*, 11(1), 1–10, 2022, doi:10.1038/s41377-022-00898-2.
7. A. Shanker et al., "Quantitative phase imaging endoscopy with a metalens," *Light Sci Appl*, 13(1), 1–14, 2024, doi:10.1038/s41377-024-01587-y.

8. P. C. Chaumet, P. Bon, G. Maire, A. Sentenac, and G. Baffou, "Quantitative phase microscopies: accuracy comparison," *Light Sci Appl*, 13(1), 1–29, 2024, doi:10.1038/s41377-024-01619-7.
9. T. L. Nguyen, S. Pradeep, R. L. Judson-Torres, J. Reed, M. A. Teitell, and T. A. Zangle, "Quantitative phase imaging: recent advances and expanding potential in biomedicine," *ACS Nano*, 16(8), 11516, 2022, doi:10.1021/ACSNANO.1C11507.
10. S. Aknoun et al., "Quantitative phase microscopy for non-invasive live cell population monitoring," *Sci Rep*, 11(1), 1–11, 2021, doi:10.1038/S41598-021-83537-X.
11. C.-Y. Shen et al., "Multiplane quantitative phase imaging using a wavelength-multiplexed diffractive optical processor," *Adv Photon* 6(5), 056003, 2024, doi:10.1117/1.AP.6.5.056003.
12. D. Pirone, V. Bianco, L. Miccio, P. Memmolo, D. Psaltis, and P. Ferraro, "Beyond fluorescence: advances in computational label-free full specificity in 3D quantitative phase microscopy," *Curr Opin Biotechnol*, 85, 03054, 2024, doi:10.1016/J.COPBIO.2023.103054.
13. J. Huang, A. Pan, H. Jin, G. Meng, and Q. Ye, "Non-interferometric accurate phase imaging via linear-convergence iterative optimization," *Opt Lasers Eng*, 144, 106630, 2021, doi:10.1016/J.OPTLASENG.2021.106630
14. A. Pillai, A. K. Asundi, and A. Patil, "Quantitative optical phase imaging techniques: interference-based basics and working principle," *Biophys Tech Biosci From Fundam Adv Appl,* 8, 134–151, 2025, doi:10.1201/9781003545712
15. Z. Huang and L. Cao, "Quantitative phase imaging based on holography: trends and new perspectives," *Light Sci Appl*, 13(1), 1–43, 2024, doi:10.1038/s41377-024-01453-x.
16. W. Choi, R. Dasari, M. S. Feld, T. Yamauchi, and Y. Park, "Spectroscopic phase microscopy for quantifying hemoglobin concentrations in intact red blood cells," *Opt Lett,* 34(23), 3668–3670, 2009, doi:10.1364/OL.34.003668.
17. S. A. Yang, J. Yoon, K. Kim, and Y. K. Park, "Measurements of morphological and biophysical alterations in individual neuron cells associated with early neurotoxic effects in Parkinson's disease," *Cytometry* A, 91,(5), 510–518, 2017, doi:10.1002/CYTO.A.23110.
18. B. Kemper et al., "Investigation of living pancreas tumor cells by digital holographic microscopy," *J Biomed Opt,* 11(3), 034005, 2006, doi:10.1117/1.2204609.
19. C. J. Mann, L. Yu, and M. K. Kim, "Movies of cellular and sub-cellular motion by digital holographic microscopy," *Biomed Eng*, 5(1), 1–10, 2006, doi:10.1186/1475-925X-5-21/FIGURES/4.
20. C. Zuo et al., "Transport of intensity equation: a tutorial," *Opt Lasers Eng*, 135, 106187, 2020, doi:10.1016/j.optlaseng.2020.106187.
21. J. Gao, J. A. Lyon, D. P. Szeto, and J. Chen, "In vivo imaging and quantitative analysis of zebrafish embryos by digital holographic microscopy," *Biomed Opt Express*, 3(10), 2623, 2012, doi:10.1364/BOE.3.002623.
22. G. O. Dwapanyin et al., "Investigation of refractive index dynamics during embryo development using digital holographic microscopy," *Biomed Opt Express, 14*(7), 3327–3342, 2023, doi:10.1101/2023.04.17.537152.
23. X. Zheng, S. Fang, B. Guo, B. Sia, and L. Yang, "Development of simultaneous dual-resolution digital holography system," *Appl Sci*, 13, 2856, 2023, doi:10.3390/APP13052856.
24. K. Creath, "V phase-measurement interferometry techniques," *Prog Opt*, 26(C), 349–393, 1988, doi:10.1016/S0079-6638(08)70178-1.
25. K. Creath and J. Schmit, "N-point spatial phase-measurement techniques for non-destructive testing," *Opt Lasers Eng*, 24(5–6), 365–379, 1996, doi:10.1016/0143-8166(95)00096-8.

26. K. Yassien, M. El Bakary, K. Yassien, and M. El Bakary, "Opto-mechanical properties of PTFE fibres using phase shifting interferometry," *Opt Photon J*, 4(6), 153–163, 2014, doi:10.4236/OPJ.2014.46016.
27. J. C. Wyant, "White light interferometry," *Proc SPIE*, 4737, 98–107, 2002, doi:10.1117/12.474947.
28. R. Shang, S. Chen, C. Li, and Y. Zhu, "Spectral modulation interferometry for quantitative phase imaging," Biomed Opt Express, 6(2), 473, 2015, doi:10.1364/BOE.6.000473.
29. A. Ahmad, V. Dubey, A. Butola, B. Singh Ahluwalia, and D. S. Mehta, "Highly stable common-path quantitative phase microscope for biomedical imaging", 2018, doi: 10.48550/arXiv.1812.01057
30. M. Mir, B. Bhaduri, R. Wang, R. Zhu, and G. Popescu, "Quantitative phase imaging chapter three," *Prog Opt*, 57, 133–217, 2012, doi:10.1016/B978-0-44-459422-8.00003-5.
31. G. Popescu, T. Ikeda, R. R. Dasari, and M. S. Feld, "Diffraction phase microscopy for quantifying cell structure and dynamics," *Opt Lett*, 31(6), 775, 2006, doi:10.1364/OL.31.000775.
32. T. Kim et al., "White-light diffraction tomography of unlabelled live cells," *Nat* Photon, 8(3), 256–263, 2014, doi:10.1038/NPHOTON.2013.350,
33. *Quantitative Phase Imaging of Cells and Tissues*. McGraw-Hill Education. .https://www.accessengineeringlibrary.com/content/book/9780071663427?implicit-login=true (accessed June 9, 2025).
34. P. Marquet et al., "Digital holographic microscopy: a noninvasive contrast imaging technique allowing quantitative visualization of living cells with subwavelength axial accuracy," *Opt Lett*, 30(5), 468, 2005, doi:10.1364/OL.30.000468.
35. M. Mir, B. Bhaduri, H. Pham, and G. Popescu, "Diffraction phase microscopy with white light," *Opt Lett*, 37(6), 1094–1096, 2012, doi:10.1364/OL.37.001094.
36. M. Takeda, H. Ina, and S. Kobayashi, "Fourier-transform method of fringe-pattern analysis for computer-based topography and interferometry," *J Opt Soc Am*, 72(1), 156–160, 1982, doi:10.1364/JOSA.72.000156.
37. K. R. Lee et al., "Quantitative phase imaging techniques for the study of cell pathophysiology: from principles to applications," *Sensors*, 13(4), 4170–4191, 2013, doi:10.3390/S130404170.
38. Y. Liu and S. Uttam, "Perspective on quantitative phase imaging to improve precision cancer medicine," *J Biomed Opt*, 29(2), cS22705, 2024, doi:10.1117/1.JBO.29.S2.S22705.
39. J. Park et al., "nature methods Artificial intelligence-enabled quantitative phase imaging methods for life sciences," *Nat Methods*, 20, 1645–1660, 2023, doi:10.1038/s41592-023-02041-4.
40. A. Pillai, A. K. Asundi, and A. Patil, "Quantitative optical phase imaging techniques: non-interference-based basics and working principle," *Biophys Tech Biosci*, 152–166, 2025, doi:10.1201/9781003545712-9.
41. M. R. Teague, "Deterministic phase retrieval: a Green's function solution," *JOSA*, 73(11), 1434–1441, 1983, doi:10.1364/JOSA.73.001434.
42. M. R. Teague, "Irradiance moments: their propagation and use for unique retrieval of phase," *JOSA*, 72(9), 1199–1209, 1982, doi:10.1364/JOSA.72.001199.
43. G. H. Go, D. G. Lee, J. Oh, G. Song, D. Lee, and M. Jang, "Meta Shack–Hartmann wavefront sensor with large sampling density and large angular field of view: phase imaging of complex objects," *Light Sci Appl,* 13(1), 1–11, 2024, doi:10.1038/s41377-024-01528-9.
44. A. Mikhaylov et al., "Shack-Hartmann wavefront sensors based on 2D refractive lens arrays and super-resolution multi-contrast X-ray imaging," *J Synchrotron Radiat*, 27(3), 788–795, 2020, doi:10.1107/S1600577520002830/VE5120SUP1.PDF.
45. D. R. Neal and J. Mansell, "Application of shack-hartmann wavefront sensors to optical system calibration and alignment," *Proc SPIE*, 3779, 234–240, 1999, doi:10.1142/9789812817815_0037.

46. T. Zhao and Y. Chi, "Modified Gerchberg–Saxton (G-S) algorithm and its application," *Entropy*, 22, 1354, 2020, doi:10.3390/E22121354.
47. M. A. Fiddy and U. Shahid, "Legacies of the Gerchberg–Saxton algorithm," *Ultramicroscopy*, 134, 48–54, 2013, doi:10.1016/J.ULTRAMIC.2013.05.009.
48. V. Chambouleyron et al., "Using the Gerchberg–Saxton algorithm to reconstruct non-modulated pyramid wavefront sensor measurements," *Astron Astrophys*, 681, A48, 2024, doi:10.1051/0004-6361/202347220.
49. E. R. Polanco et al., "Multiparametric quantitative phase imaging for real-time, single cell, drug screening in breast cancer," *Commun Biol*, 5(1), 1–12, 2022, doi:10.1038/s42003-022-03759-1.
50. V. P. Pandiyan and R. John, "Imaging cell division in E-coli using white light diffraction phase microscopy," In: *Workshop on Recent Advances in Photonics, WRAP 2015*, Bangalore, India. doi:10.1109/WRAP.2015.7805958.
51. A. Pillai, U. S. Kamath, S. Belurkar, A. K. Asundi, and A. Patil, "3D quantitative phase imaging via the transport of intensity equation: applications for studying red blood cells," *RSC Adv*, 15(15), 11655–11661, 2025, doi:10.1039/D5RA01071C.
52. S. Pradeep, T. Tasnim, H. Zhang, and T. A. Zangle, "Simultaneous measurement of neurite and neural body mass accumulation via quantitative phase imaging," *Analyst*, 146(4), 1361–1368, 2021, doi:10.1039/D0AN01961E.
53. T. Fukawa et al., "Excessive fatty acid oxidation induces muscle atrophy in cancer cachexia," *Nat Med*, 22(6), 666–671, 2016, doi:10.1038/nm.4093.
54. Y. Liu and S. Uttam, "Perspective on quantitative phase imaging to improve precision cancer medicine," *J Biomed Opt*, 29(S2), S22705, 2024, doi:10.1117/1.JBO.29.S2.S22705.
55. H. V. Pham, L. Pantanowitz, and Y. Liu, "Quantitative phase imaging to improve diagnostic accuracy of urine cytology," *Cancer Cytopathol*, 124(9), 641, 2016, doi:10.1002/CNCY.21734.
56. S. Sarangapani, A. Patil, Y. K. Ngeow, R. Elsa Mohan, A. Asundi, and M. J. Lang, "Chitosan nanoparticles' functionality as redox active drugs through cytotoxicity, radical scavenging and cellular behaviour," *Integr Biology*, 10(5), 313–324, 2018, doi:10.1039/C8IB00038G.
57. M. Takabayashi, H. Majeed, A. Kajdacsy-Balla, and G. Popescu, "Disorder strength measured by quantitative phase imaging as intrinsic cancer marker in fixed tissue biopsies," *PLoS One*, 13(3), e0194320, 2018, doi:10.1371/JOURNAL.PONE.0194320.
58. Y. O. N. Ansong-Ansongton and T. D. Adamson, "Computing Sickle Erythrocyte Health Index on quantitative phase imaging and machine learning," *Exp Hematol*, 131, 2024, doi:10.1016/j.exphem.2024.104166.
59. G. Kim et al., "Learning-based screening of hematologic disorders using quantitative phase imaging of individual red blood cells," *bioRxiv*, 091983, 2018, doi:10.1101/091983.
60. A. Doblas, J. Garcia-Sucerquia, G. Saavedra, and M. Martinez-Corral, "Digital holographic microscopy as a screening technology for diabetes," *Proc SPIE,* 10997, 100–106, 2019, doi:10.1117/12.2523416.
61. S. K. Paidi et al., "Raman and quantitative phase imaging allow morpho-molecular recognition of malignancy and stages of B-cell acute lymphoblastic leukemia," *Biosens Bioelectron*, 190(113403), 2021, doi:10.1016/J.BIOS.2021.113403.
62. G. Kim, Y. J. Jo, H. Cho, H. seok Min, and Y. K. Park, "Learning-based screening of hematologic disorders using quantitative phase imaging of individual red blood cells," *Biosens Bioelectron*, 123, 69–76, 2019, doi:10.1016/J.BIOS.2018.09.068.
63. A. Pillai, S. K. U, S. Belurkar, A. K. Asundi, and A. Patil, "3D quantitative phase imaging of human red blood cells," *Proc SPIE,* 13069, 172–178, 2024, doi:10.1117/12.3023411.
64. M. E. Kandel, K. W. Teng, P. R. Selvin, and G. Popescu, "Label-free imaging of single microtubule dynamics using spatial light interference microscopy," *ACS Nano*, 11(1), 647–655, 2017, doi:10.1021/ACSNANO.6B06945.

65. H. Majeed et al., "Breast cancer diagnosis using spatial light interference microscopy," *J Biomed Opt*, 20(11), 111210, 2015, doi:10.1117/1.JBO.20.11.111210.
66. M. E. Kandel et al., "Label-free tissue scanner for colorectal cancer screening," *J Biomed Opt* 22(6), 066016, 2017, doi:10.1117/1.JBO.22.6.066016.
67. H. Ding, X. Liang, S. A. Boppart, Z. Wang, K. Tangella, and G. Popescu, "Measuring the scattering parameters of tissues from quantitative phase imaging of thin slices," *Opt Lett*, 36(12), 2281–2283, 2011, doi:10.1364/OL.36.002281.
68. A. Butola et al., "Multimodal on-chip nanoscopy and quantitative phase imaging reveals the nanoscale morphology of liver sinusoidal endothelial cells," *Proc Natl Acad Sci USA*, 118(47), e2115323118, 2021, doi:10.1073/PNAS.2115323118/ASSET/77D9A7CE-E043-4F08-8D3D.
69. A. Kapur et al., "Enhanced uptake of luminescent quantum dots by live cells mediated by a membrane-active peptide," *ACS Omega*, 3(12), 17164–17172, 2018, doi:10.1021/acsomega.8b02918.
70. Y. Cao, H. Dong, S. Pu, and X. Zhang, "Photoluminescent two-dimensional SiC quantum dots for cellular imaging and transport," *Nano Res*, 11(8), 4074–4081, 2018, doi:10.1007/s12274-018-1990-3.
71. J. Kwon et al., "FeSe quantum dots for in vivo multiphoton biomedical imaging," *Appl Sci Eng* 5(12), eaay0044, 2019. doi:10.1126/sciadv.aay0044.
72. N. Mazumder et al., "Fluorescence lifetime imaging of alterations to cellular metabolism by domain 2 of the hepatitis c virus core protein," *PLoS One*, 8(6), 2013, doi:10.1371/journal.pone.0066738.
73. G. Raju et al., "Light sheet fluorescence microscopy for monitoring drug delivery: unlocking the developmental phases of embryos," *Adv Drug Deliv Rev*, 218, 115520, 2025, doi:10.1016/J.ADDR.2025.115520.
74. Z. Li, Z. Guang, P. C. Costa, C. E. Serafini, S. Bharadwaj, and F. E. Robles, "Multimodal fluorescence microscopy and 3D quantitative phase imaging/refractive index tomography in thick tissue samples,"Proc SPIE, 12852, 83–88, 2024, doi:10.1117/12.3003242.
75. J. Gu, C. Y. Fu, B. K. Ng, L. B. Liu, S. K. Lim-Tan, and C. G. L. Lee, "Enhancement of early cervical cancer diagnosis with epithelial layer analysis of fluorescence lifetime images," *PLoS One*, 10(5), e0125706, 2015, doi:10.1371/JOURNAL.PONE.0125706.
76. M. Esposito et al., "TGF-β-induced DACT1 biomolecular condensates repress Wnt signalling to promote bone metastasis," *Nature Cell Biol*, 23(3), 257–267, 2021, doi:10.1038/s41556-021-00641-w.
77. W. Choi et al., "Tomographic phase microscopy," *Nat Methods*, 4(9), 717–719, 2007, doi:10.1038/NMETH1078.
78. J. Choi et al., "Label-free three-dimensional analyses of live cells with deep-learning-based segmentation exploiting refractive index distributions," *bioRxiv,* 2021, doi:10.1101/2021.05.23.445351.
79. X. Shu et al., "Artificial-intelligence-enabled reagent-free imaging hematology analyzer," *Adv Intel Syst*, 3(8), 2000277, 2021, doi:10.1002/AISY.202000277.

6 Advances in Photoacoustic Imaging for Regenerative Medicine and Stem Cell Therapy

Vyasraj G. Bhat, Vibha Kamath, Gagan Raju, and Manikanth Karnati
Manipal Academy of Higher Education

Guan-Yu Zhuo
National Yang Ming Chiao Tung University

6.1 INTRODUCTION

Photoacoustic imaging (PAI), also called optoacoustic imaging (OAI), is an imaging method based on the principle of ultrasonic (US) emission of the energy absorbed, which is then converted to heat, allowing for in-depth optical spectroscopic imaging. The use of near-infrared pulses of light and the absence of background US signals result in increased depth resolution. The sample is irradiated by a pulsed source light, which expands and instantly generates US waves [1]. This technique stands out from traditional imaging methods because of its superior ability to visualize and analyse biological tissues with enhanced efficiency, high contrast, and excellent spatial resolution. The contrast in PAI arises from endogenous biomolecules, engineered contrast agents, or nanomaterials such as gold nanoparticles and nanodots. Depending on the focal mechanism, PAIs can be broadly classified into two major types: optical focus-based and acoustic focus-based. Optical focus-based PAI uses tightly focused light for high-resolution imaging at shallow depths, whereas the latter uses focused ultrasound detection for deeper tissue imaging with moderate resolution. Strong photoacoustic signals are generated where these two focal points intersect, ensuring optimal sensitivity. The incorporation of a transducer into the system enhances the central frequency and bandwidth, resulting in increased axial resolution and spatial precision below 100 μm. Recently, the development of three-dimensional PAI has enabled detailed investigations of biological tissues tagged with contrast agents, facilitating the assessment of gene expression, cellular behaviour, and tumor progression in xenograft mouse models. Moreover, integrating PAI with tomographic

DOI: 10.1201/9781003570011-6

approaches such as photoacoustic computed tomography (PACT) and photoacoustic microscopy (PAM) has expanded the imaging capabilities beyond those of fluorescence-based techniques, opening new avenues for exploring cell-based therapies [2]. Despite significant research in the field of stem cell therapy, only a few have been clinically implemented, which is due to the lack of imaging techniques to study the underlying mechanisms precisely. Several imaging techniques, such as magnetic resonance imaging, positron emission tomography, optical imaging, and ultrasound, have been used to obtain images with high resolution, sensitivity, penetration depth, safety, and cost efficiency. However, all these techniques have the major drawbacks of obtaining images with poor contrast, low sensitivity and penetration depth, side effects of radiation, and cytotoxicity due to nanoparticles. On the other hand, PAI addresses these challenges, as it provides high-contrast images by exploiting the different absorption levels in various tissue types. In addition, it provides real-time imaging feedback during therapy and monitors the challenges that obstruct the achievement of total therapeutic output [3]. Since the PAI technique has high potential for imaging biological samples even at greater depth, it stands out as one of the most convenient and accessible methods for imaging. The ease with which the imaging system works in tandem to obtain images with better resolution makes its field of application very wide. Owing to its ability to utilize an endogenous contrast of specimens with high spectral and lateral resolution, this imaging modality is rapidly emerging in the field of regenerative medicine. Additionally, the integration of PAI with tomographic methods such as PACT and PAM has significantly enhanced imaging capabilities, surpassing those of traditional fluorescence-based techniques and enabling new possibilities in the study of cell-based therapies. These increasing numbers of applications will, in turn, increase the availability of commercial photoacoustic systems for further research and development in regenerative therapies [4].

6.2 FUNDAMENTALS OF PAI

The fundamental principle of PAI is based on the photoacoustic effect. When a short laser pulse is applied to tissues, different chromophores, such as melanin and haemoglobin, absorb energy [1]. This subsequently leads to a localized increase in temperature as well as rapid thermoelastic expansion, producing pressure waves called ultrasound. An ultrasound detector placed on the surface of a tissue can detect these acoustic or ultrasound waves [5]. The arrival time and wave intensity help in reconstructing the location as well as the intensity of the original absorption of light. Various algorithms are used to reconstruct the images of the detected signals. A typical PAI system is designed to convert light energy into acoustic signals, which are then used to form high-resolution biological images. The detailed experimental setup of the PAI is schematically represented in Figure 6.1a. Figure 6.1b depicts a flowchart of PAI, which shows the sequential process from laser excitation to reconstruction of the image [6]. At the core, a pulsed laser source such as Nd:YAG is used, which emits nanosecond pulses of light, usually in the near-infrared region (680–1,064 nm), resulting in the excitation of tissue chromophores [7]. These short pulses are delivered to the biological sample through an optical delivery system that comprises optical fibers, lenses, and mirrors, ensuring accurate targeting of the region of interest. To effectively capture these waves, a sample is usually placed in contact with an

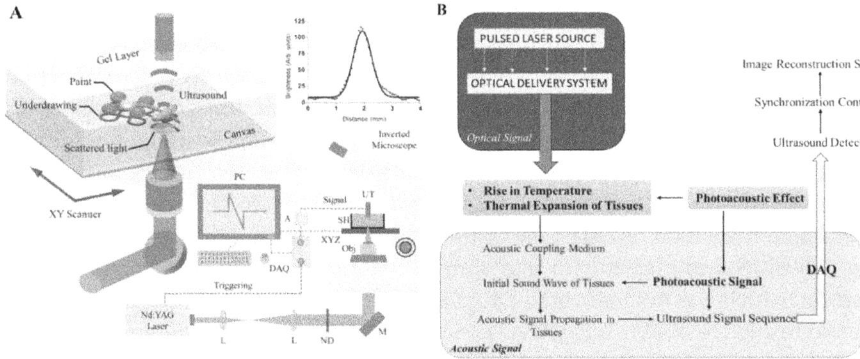

FIGURE 6.1 Schematic representation of a PAI system. (a) Three-dimensional setup showing laser excitation, tissue sample, ultrasound detection, and data processing (Reproduced from [11]). (b) Flowchart illustrating the sequential process from laser excitation to reconstruction of the image.

acoustic coupling medium, such as water or ultrasound gel, which helps transmit the generated ultrasound signals to an ultrasound transducer [8]. This transducer, which can be a single-element device or a multielement array, detects the acoustic signals and converts them into electrical signals. These signals are then sent to a data acquisition system (DAQ) that digitizes the information for processing [9]. A synchronization controller is used to coordinate the timing between the laser pulses and the ultrasound signal detection to ensure accuracy. Finally, specialized image reconstruction software processes the digitized data via algorithms such as back-projection or delay-and-sum, converting it into two-dimensional or three-dimensional images. The entire system may include a sample holder or imaging stage—especially in in vivo imaging setups—which can incorporate features such as temperature control, anaesthesia delivery, or motorized scanning stages for volumetric imaging. This integrated setup enables PAI to provide deep tissue imaging with optical contrast and ultrasound resolution, making it highly valuable for applications in oncology, vascular biology, and functional imaging [10].

6.3 PAI MODALITIES IN REGENERATIVE MEDICINE

PACT is well-positioned for imaging deep tissues, with a maximum depth capacity of 7 cm [12–14]. One of the benefits of the PACT technique is the use of the inherent optical absorbance properties of biological molecules such as haemoglobin, melanin, lipids, and nucleic acids. For example, haemoglobin enables functional imaging because of its strong absorption between 500 and 650 nm (in both oxygenated and deoxygenated forms), whereas melanin provides information for melanoma imaging, and lipids can be used to understand tissue structure [15]. PACT employs a pulsed laser to illuminate a wide area, and an array of ultrasound detectors arranged in a circular or hemispherical shape captures the result of sound waves from every direction. These systems are fast, with some operating at 50 frames per second, and more recent models are more reasonably priced by using LED arrays [16]. PACT is particularly helpful for measuring regenerative processes such as blood flow or stem cell activity, as well as for imaging entire organs [17,18]. Optical-resolution PAM (OR-PAM) focuses on laser light and uses small, high-powered lenses to focus on laser light. Imaging takes place only at shallow depths, which makes it perfect for imaging structures such as skin capillaries or small clusters of stem cells [19]. OR-PAM systems typically utilize a green laser (~532 nm) and scan the laser across the tissue with tiny mirrors as the light is delivered by pulses. It can only image features on a tissue surface and not deep into a tissue sample because of its limited concentrated resolution [20,21]. One of the studies presented a high-speed OR-PAM using a MEMS scanning mirror and simple calibration methods to correct spatial distortion, which was demonstrated on artificial and human fingerprint samples [22]. Acoustic-resolution PAM (AR-PAM) penetrates tissue more deeply, up to 7–8 mm, via ultrasound instead of optical focusing [23]. Although the resolution is not as high as that of OR-PAM, the system can still visualize small vessels and be able to image deeper structures [24]. Recent advances have used artificial intelligence tools such as GAGs to sharpen images even further. Some systems combine OR-PAM and AR-PAM to provide both deep and detailed imaging for scaffold vacuolation in tissue engineering [25]. The transverse resolution depends on the numerical aperture of the transducer and the center frequency of the ultrasound transducer. Using a transducer allows for some flexible imaging capability. For example, we can choose a 5 MHz transducer so that the imaging depth is 10.3 mm with a resolution of 590 μm or a 40 MHz transducer for a fine resolution of 85 μm at a depth of 3.1 mm [26]. Its nonionizing property, cost-effectiveness, and translatability make it a vital imaging modality in preclinical and clinical settings [27]. US-guided PAI (US/PA) is a technology that integrates photoacoustics and conventional ultrasound imaging [28]. It detects photoacoustic signals and emits ultrasonic pulses via the same probe. Using US/PA allows a physician or researcher to assign specific features and visualize the structure and function of tissues simultaneously. It is also a convenient way to guide stem cell injections into very specific areas, such as the spine, and to follow labelled cells within tumors [29]. Multispectral OR-PAM adds another layer of capability by using lasers that can change wavelengths. It allows for distinguishing between different molecules on the basis of their specific absorption patterns. For example, it can identify lipid-rich membranes in stem cells or distinguish between blood that is oxygen-rich and blood that is oxygen-poor. These systems are valuable for monitoring oxygen delivery and inflammation during regeneration [30,31].

6.4 APPLICATIONS IN STEM CELL THERAPY AND REGENERATIVE MEDICINE

6.4.1 Stem Cell Tracking and Viability Monitoring

Tracking and monitoring stem cell viability are essential for assessing their utility in treating a variety of diseases. Stem cells cannot be visualized via PAI [32]. As a result, various exogenous contrast agents have been developed to enhance the detection of viable stem cells in the body and enable real-time, non-invasive imaging. An example includes the development of a reactive oxygen species (ROS)-sensitive photoacoustic nanoprobe composed of gold nanorods conjugated with the IR775c dye. This probe can detect oxidative stress as an indicator of cell death in mesenchymal stem cells (MSCs).

This probe demonstrated low toxicity, efficient uptake by MSCs, and facilitated real-time monitoring of viability both in vitro and in vivo after transplantation. As shown in Figure 6.2, a ROS-sensitive photoacoustic nanoprobe enabled real-time, non-invasive imaging of MSC viability over 10 days, with ratiometric photoacoustic signals decreasing in response to oxidative stress following transplantation [33]. Similarly, Yoo et al. employed indocyanine green (ICG), a dye with strong absorption at 800 nm, for viability tracking; the dye diffused out of apoptotic cells, allowing for real-time discrimination between live and dying cells [34]. To avoid false-positive signals from dead cells or uptake by macrophages, Ricles et al., in 2014, developed a dual nanoparticle system including PEGylated gold nanospheres for macrophages and poly-L-lysine-coated silica gold nanorods for MSCs, thus enabling simultaneous tracking and monitoring of immune–stem cell interactions during tissue regeneration [35]. Long-term viability monitoring is critical for assessing stem cell biodistribution and engraftment. OAI, a form of PAI, is particularly well-suited for this purpose because of the stability of signals from contrast agents, which are not dependent on radioactive decay. While nanoparticles smaller than 5.5 nm are easily cleared renally, they are less suited for long-term tracking [36]. To address this, surface modification strategies—such as the use of citrate-coated Prussian blue particles (PBPs)—have been implemented. In 2017, Li et al. demonstrated effective longitudinal tracking

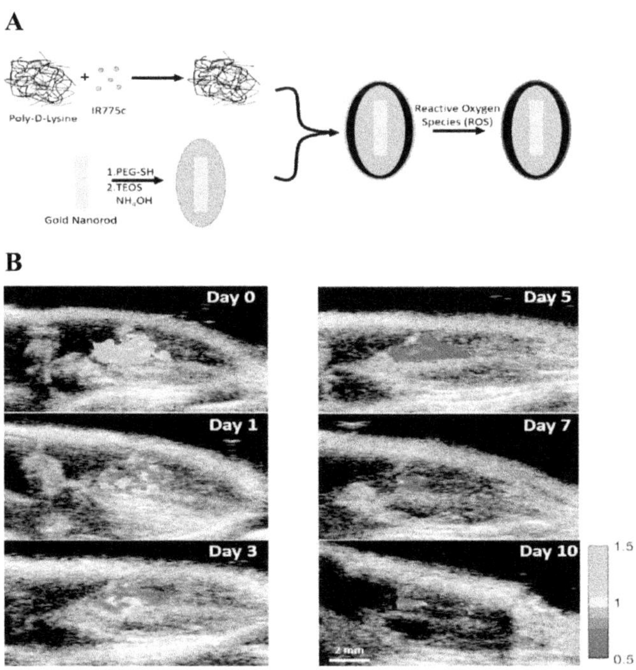

FIGURE 6.2 Longitudinal PAI of MSC viability using a ROS-sensitive IR775c–AuNR nanoprobe. (a) Schematic of nanoprobe synthesis and the ROS-triggered response. (b) Ratiometric photoacoustic images from Day 0–10 post transplantation show a progressive decrease in signal intensity, indicating stem cell death due to oxidative stress. The color scale represents the PA ratio (green = high viability, red = low viability). (Reproduced from [33].)

of MSCs in a traumatic brain injury model using PBPs, with less than 10% leakage photolabeling [37].

The incorporation of stem cells into 3D extracellular matrix (ECM)-based systems such as hydrogels has further enhanced their in vivo relevance. Encapsulation in PEGylated fibrin gels combined with gold nanotracer (Au NT) labelling allowed for imaging of MSCs for up to 10 days at tissue depths of 10–50 mm [38]. However, as Chung et al. (2013) reported, issues such as tracer excretion and gel interaction may cause false positives over extended monitoring periods [39]. Hence, optimizing nanoparticle retention, colloidal stability, and surface chemistry is crucial for reliable long-term imaging. Preclinical studies have successfully reported stem cell tracking for up to two weeks via tailored nanoparticles [40], whereas Zhang et al., in 2019, achieved up to 35 days of tracking via a dual-modal contrast agent [41]. In 2019, Kubelick et al. demonstrated the use of a US/PA imaging platform for longitudinal tracking of AuNS-labelled stem cells in *ex vivo* porcine eyes. This system enables real-time monitoring and successful differentiation of labelled stem cells from background tissues via spectroscopic analysis, while also highlighting the inefficiencies in cell delivery to target tissues [42]. These preclinical imaging strategies are essential for assessing the safety and efficacy of regenerative medicine therapies. Understanding each imaging modality's detection limitations and integrating multimodal approaches—such as combining PAI with fluorescence or magnetic imaging—can enhance data quality and reduce animal use, supporting ethical practices and improving risk–benefit evaluations of stem cell-based therapies [43,44], demonstrating the way to visualize human embryonic stem cell-derived cardiomyocytes transplanted into mouse hearts by semiconducting polymer-based photoacoustic nanoparticles as contrast agents. With a spatial resolution of 100 μm, the robust and precise photoacoustic signals allowed for the sensitive detection of up to 2,000 labelled cells, allowing them to be distinguished from background tissues [44]. Overall, PAI combined with engineered contrast agents is a powerful, minimally invasive tool for monitoring stem cell viability and tracking biodistribution. Advancements in biocompatible, multifunctional probes with strong NIR absorption and long-term stability are crucial for the successful clinical translation of stem cell therapies.

6.4.2 Tissue Engineering and Scaffold Characterization

PAI is increasing as a viable non-invasive imaging modality for tissue engineering, as it provides molecule-specific, high-resolution imaging of vascular structures and functional parameters, such as blood oxygenation. PAI enables detailed measurements of vascular structure and tissue oxygenation (noninvasively) through the combination of optical excitation of tissue and ultrasound detection of the resultant photoacoustic signal. For tissue engineering applications, this imaging process allows for superior mapping of the microvasculature of the tissue-engineered vascular graft and even functional measures of tissue oxygenation, particularly in preclinical studies [25]. In 2016, Avigo et al. used PAI and US imaging for 2D and 3D imaging of the microvasculature and tracking of gold nanorod CAs, as well as mapping of oxygen saturation in engineered tissues. This application of image fusion using PA and US also overcomes some of the limitations of traditional imaging by providing advanced depths of imaging as well as molecule specificity [45]. In another experiment, US/PA imaging was used to noninvasively longitudinally monitor poly(lactic-co-glycolic acid) (PLGA) scaffolds containing SWCNTs. The PLGA scaffolds were implanted subcutaneously in rats. The study was able to produce high-resolution images at a depth of 6 mm, allowing them to track the scaffold architecture and degradation, shape, size, and surrounding oxygenation levels over time [46]. In addition, PAM has been used for tracking the performance of scaffolds on the basis of scaffold degradation levels, tissue regeneration, and in vivo situations. In one study, MTT formazan-doped PLGA scaffolds were implanted in mice, which were subsequently cultured for six weeks. Using hybrid PAM systems, it was possible to obtain high-resolution images of cell behaviour and scaffolds and highlight the advantages of PAM in tissue engineering [47]. More advanced uses of PAM include multiscale PAM, OR-PAM, and AR-PAM. These findings helped reveal and quantify the porosity and pore sizes of the PLGA scaffolds containing SWNTs under physiological conditions. OR-PAM performs extremely well in terms of lateral resolution, whereas AR-PAM allows deeper imaging capabilities to provide comprehensive scaffold characterization [48]. To improve safety and accessibility, a recently reported study utilized LED-based PA imaging, a safe, affordable, and portable approach, as an alternative to class IV laser systems. In this LED-based PA-US image study, a resolution of approximately 200 μm and an imaging depth of up to 2 cm were demonstrated to noninvasively assess angiogenesis in fibrin scaffolds only, thus showing strong potential for clinical translation to regenerative medicine [49].

In another important advancement, PA imaging was used to monitor neovascularization in decellularized human tissue scaffolds over time (15-week study). This label-free, high-resolution imaging study achieved a maximum penetration depth of 6.8 mm and was able to identify and monitor progressive vascular growth significantly over time; histological examination showed similar findings. This study was the first to use PA imaging to examine and track angiogenesis in biological scaffolds, which is important for supporting clinical graft evaluation [50]. In addition to these imaging advancements, PEGDA hydrogel scaffolds were also produced from a stereolithography-based 3D printer (using UV laser curing) customized specifically for the fabrication of scaffolds. CAD-based printing allows control over square or hexagonal pore geometries; the final products provide very high porosity, and the surface area increases even more (700% for square pores and 350% for hexagonal pores). Scanning electron microscopy (SEM) analyses confirmed that the integrity of the structures was still intact and usable for acoustic and imaging-based characterization [51]. Most recently, Ferreira et al. developed a photoacoustic processing technique that integrates ultrasound and visible-light photo crosslinking to rapidly fabricate enzyme-free decellularized ECM (dECM) hydrogels from porcine mammary tissue. This approach preserved the native fibrillar architecture and enabled the production of SdECM hydrogels in various formats, including cast scaffolds, microbeads, and 3D-bioprinted constructs. These scaffolds supported fibroblast and cancer cell viability and exhibited tissue-like mechanical properties. The full workflow—from decellularization and photoacoustic processing to scaffold validation through structural, mechanical, and biological assays—is illustrated in Figure 6.3 [52].

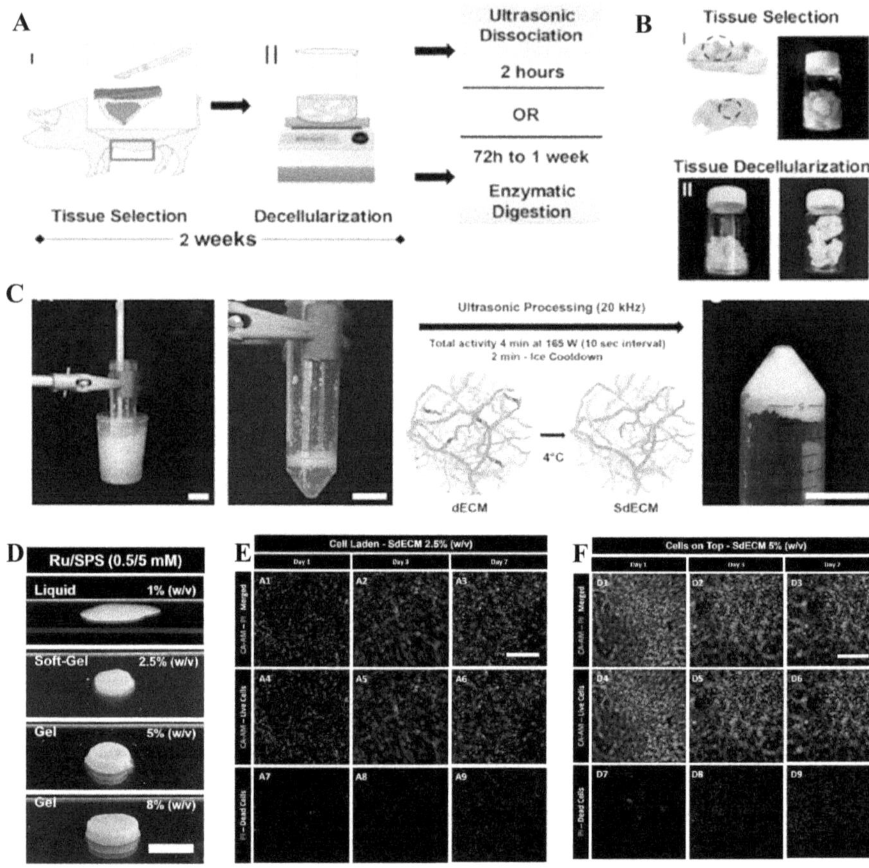

FIGURE 6.3 Photoacoustic processing and characterization of dECM scaffolds. (a) Schematic overview comparing traditional enzymatic digestion and photoacoustic (ultrasound-based) dECM processing. (b) Tissue selection and decellularization stages. (c) Ultrasonication setup and schematic of the molecular disruption leading to SdECM formation. (d) Photo crosslinking of SdECM at various concentrations showing liquid-to-gel transitions. (e, f) Live/dead staining of 3T3-L1 fibroblasts in SdECM hydrogels: (e) cells embedded within 2.5% SdECM, (f) cells seeded on a 5% SdECM surface for 7 days, showing high viability and distribution. Scale bars: 1 cm (d), 500 μm (e, f). (Reproduced from [52].)

6.4.3 Angiogenesis and Vascular Imaging

PAI has emerged as a valuable, label-free technology for visualizing and measuring vascular networks because of its inherent sensitivity to oxyhaemoglobin (HbO_2) and deoxyhaemoglobin. PAI enables non-invasive, high-resolution imaging of both vascular structure and functional oxygenation; therefore, PAI is well-suited for studying changes in vascularization (e.g., through angiogenesis) or vascular health across a range of clinical and research applications [53]. Many PAI configurations, from microscopic to tomographic, have been used to image superficial vasculature, particularly in human skin [11]. Microscopic and mesoscopic PAI systems have excellent spatial resolution, approaching the level of individual capillaries, which is necessary to measure angiogenic activity. These systems have generated detailed images of the cutaneous microvascular network and show potential as systems to identify early signs of peripheral vascular disease, with superficial vascular changes acting as surrogate biomarkers. Many conventional imaging modalities, including magnetic resonance angiography or magnetic resonance imaging (MRI), which entail exogenous agents and an extensive preparation process, are generally not comparable to PAI. Although laser speckle imaging has been used to assess blood flow, its inability to resolve vascular structure and measure oxygen saturation limits its use as a comprehensive vascular assessment technique [54]. Multispectral PAI systems allow for simultaneous structural and functional analysis of blood vessels, with very high specificity for haemoglobin oxygenation, which expands vascular imaging [55–58]. Tomographic PAI in handheld probe formats has indicated great promise for in vivo imaging of large and deep vessels. Several studies have documented the use of handheld 2D and 3D volumetric probes to visualize the human vasculature, supporting their utility for clinical applications [59]. An example of such a study utilized a multispectral optoacoustic tomography (MSOT) system with a 135° curved handheld detector [60]. The MSOT system was able to visualize human vasculature at depths of 10 mm or more; major arteries, such as the tibialis posterior and dorsalis pedis arteries, were resolved, and the microvasculature was resolved to 90–110 μm in the toe. The MSOT system was also able to dynamically track the pulsatile nature of the arterial flow in real time [60]. Yamaga et al., in the context of breast cancer, utilized a 3D full-breast MSOT system with a hemispherical detection aperture to detect vascular branching points (VBPs) in a malignant breast compared with a healthy contralateral breast, while finding VBPs in greater quantities in malignant breasts. They suggested that VBPs may be vascular biomarkers for breast cancer [61]. Another study by Oraevsky et al. in 2018 reported that a handheld optoacoustic-ultrasound total haemoglobin concentration and blood oxygen saturation of breast tumors using ultrawideband detectors and conventional ultrasound improved diagnostic confidence [62]. In addition to cancer, in 2018, Jo et al. reviewed the potential of PAI for musculoskeletal inflammation, particularly for the imaging of rheumatoid arthritis (RA), since it is non-invasive and has soft tissue contrast and hemoglobin sensitivity [63]. In support of this, van den Berg et al. also used a portable US/PA system to image the interphalangeal joints in RA patients. They reported that the PAI signal intensity of inflamed joints was, on average, 4–10 times greater than that of controls, providing evidence for inflammation assessment [64]. To demonstrate further potential for deeper vascular imaging, Merčep et al., in 2018, reported on a hybrid imaging system that integrated PAI (into ultrasound) with Doppler imaging. Their handheld multisegment detector array was able to visualize deep tissue oxygenation status in healthy volunteers from blood vessels such as the carotid artery and jugular vein at depths of 1–2 cm [65]. Most recently, Miao et al., in 2024, used PAI to visualize the angiogenesis process during wound healing in a murine oral ulcer model. The authors provided label-free, real-time monitoring of several vascular parameters, including vessel density, vessel diameter, and perfusion, over a 7-day observation period. The mice treated with phycocyanin or compound chamomile and lidocaine hydrochloride gel (CCLH) presented a significantly improved angiogenic response and healing time, which was confirmed by CD34 immunostaining and clinical recovery. The PAI setup permitted detection of the macro- and microvascular changes that occurred throughout the ulcer site, as depicted in Figure 6.4. Overall, the series of three-dimensional (3D) reconstructions provided visualization of haemorrhage, vascular recovery, and injury at three stages of the wound-healing process [66].

6.4.4 Therapeutic Applications

A PLGA-graphene oxide (GO) scaffold with PA stimulation was investigated in this study for bone defect repair. Bone marrow-derived stem cells (BMSCs) were seeded on the scaffold, and marker expression and bone regeneration analysis were used to assess the osteogenic differentiation of the cells. PA stimulation was applied via pulsed laser parameters (200 ns, 10 mJ, 10 Hz), and the authors reported enhanced osteogenesis and neovascularization. Bone healing was promoted the most in the PA group. The authors demonstrated the therapeutic potential of PA-activated scaffolds and proposed PA stimulation as an innovative strategy to grow bone tissue. As shown in Figure 6.5, GO–PLGA scaffolds were fabricated, seeded with BMSCs, stimulated with a pulsed laser to induce photoacoustic effects, and implanted into a rat femoral bone defect model to evaluate their osteogenic potential [67]. Imaging techniques facilitate the therapeutic application of 3D-bioprinted organoids through non-invasive assessments of their structure, cell viability, and function. Imaging ensures quality

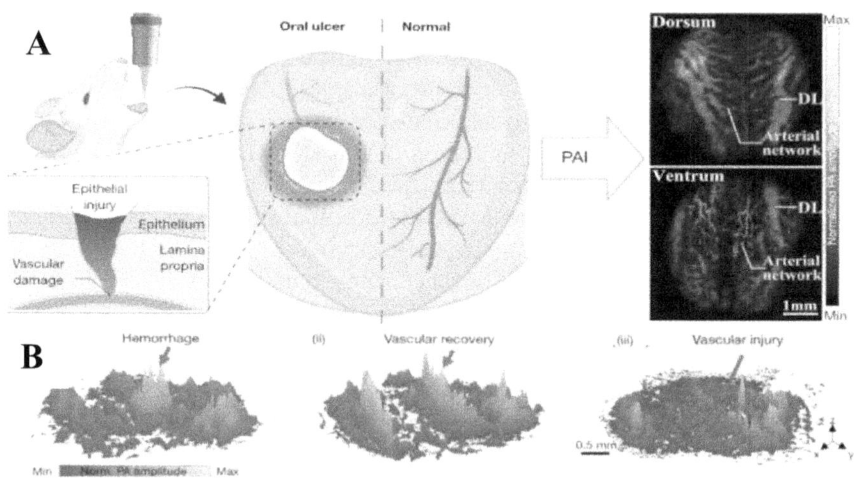

FIGURE 6.4 PAI of oral ulcer healing and vascular remodelling. (a) Schematic of the PAI setup in a mouse oral ulcer model, showing a comparison between normal and ulcerated tissue and corresponding PA vascular images of the dorsum and ventrum of the tongue. (b) 3D photoacoustic reconstructions showing dynamic vascular changes during different stages: (i) haemorrhage, (ii) vascular recovery, and (iii) vascular injury. Red arrows indicate localized vascular alterations. The colour scale denotes the normalized PA amplitude. (Reproduced from [66].)

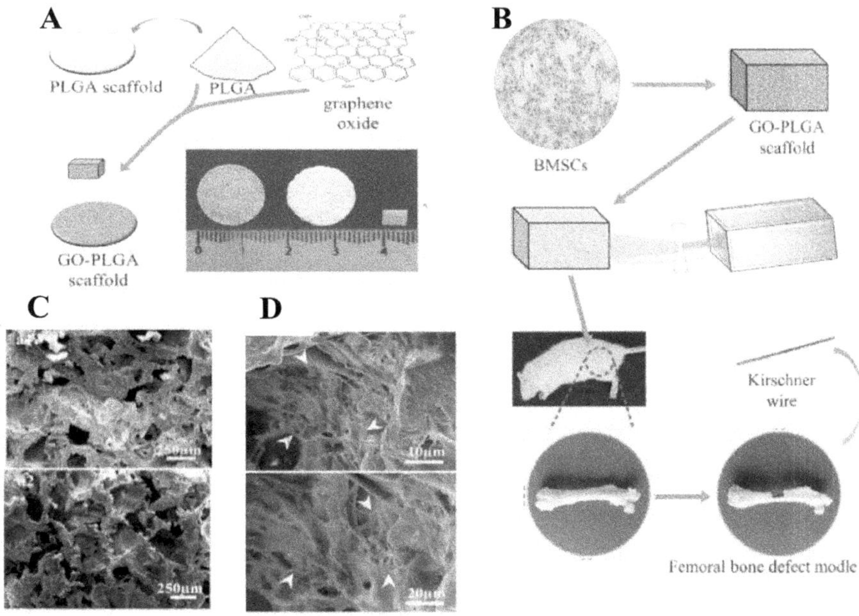

FIGURE 6.5 Fabrication and evaluation of PLGA-GO scaffolds for bone repair. (a) Schematic of PLGA-GO scaffold preparation. (b) Bone defect model established in the rat femoral midshaft. (c) SEM images showing the porous microstructures of the PLGA-GO (a1) and PLGA (a2) scaffolds. (d) SEM confirming BMSC attachment and spreading on the PLGA-GO scaffold after 72 h (white arrows). (Reproduced from [67]).

control when bioprinted organoids are produced and can be used to ensure that the organoids are ready for transplantation or disease modelling. High-resolution, multimodal imaging, such as fluorescence lifetime imaging (FLIM) and PAI, can provide far greater insights into the tissue responses and development of organoids. Imaging works effectively with functional assays to assess therapeutic effects and physiological relevance. Thus, imaging provides essential features that facilitate the growth of organoids in regenerative medicine and personalized therapy [68].

6.5 SYNERGISTIC MULTIMODAL IMAGING

Multimodal photoacoustic microscopy (MPAM) combines PAI with other techniques, such as optical coherence tomography (OCT) to increase image quality by providing both functional and anatomical details. This synergy improves the resolution, contrast, and tissue visualization beyond optical limits. Although alignment and system complexity pose challenges, the MPAM has shown promise in fields such as regenerative medicine, neuroscience, and ophthalmology, especially for visualizing retinal vessels and monitoring cell transplantation. A recent study demonstrated multimodal imaging synergies in regenerative medicine by developing a functional C@M/GLU scaffold integrating 3D printing and imaging technologies.

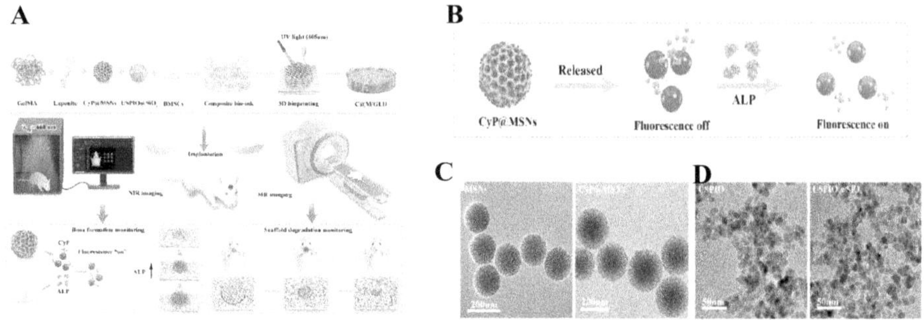

FIGURE 6.6 Multimodal C@M/GLU scaffold for simultaneous monitoring of bone regeneration and scaffold degradation. (a) Schematic illustration of scaffold fabrication via 3D bioprinting and its application in NIR fluorescence and MRI monitoring. (b) ALP-triggered fluorescence activation mechanism of CyP@MSNs. (c) TEM images of MSNs and CyP@MSNs. (d) TEM images of USPIO and USPIO@SiO_2 nanoparticles showing successful synthesis and coating. (Reproduced from [69].)

The scaffold incorporated CyP@MSNs for fluorescence-based monitoring of ALP expression and USPIO@SiO2 for MRI-based tracking of scaffold degradation. The release of osteoinductive ions (Si^{4+} and Mg^{2+}) enhances bone regeneration, whereas imaging enables real-time, non-invasive monitoring of tissue repair, as shown in Figure 6.6 [69]. Another study presented a multimodal PAOFM imaging system—combining PAM, OCT, and fluorescence microscopy (FM)—for real-time, non-invasive tracking of transplanted hiPSC-RPE cells. By utilizing gold nanoclusters with a red/NIR-shifted absorption peak (~620 nm), high spatial resolution can be achieved while avoiding haemoglobin interference [70]. Ou et al. 2025 developed a bio targeted nanoparticle system, *Bifidobacterium bifidum*@PC-CLs, by electrostatically adsorbing negatively charged *B. bifidum and multifunctional ROS producing nanoparticles* (abbreviated as PC-CLs) to achieve target specificity. The nanoparticle system incorporated perfluoro hexane for ultrasound imaging and cavitation and chlorin e6 (Ce6) for reactive oxygen species production under pulsed focused US (PFUS) treatment. Dual imaging modalities provided real-time visualization of therapy trials in 4T1 tumor models. PFUS treatment was shown to activate the nanoparticulates and release ROS, resulting in son sensitization, and sonodynamic therapy resulted in the death of tumor cells in a B. bifidum@PC-CL/ROS-dependent manner. The authors of the investigated B. bifidum@PC-CL group demonstrated enhanced targeting, imaging, and synergy of therapy over the control groups for antitumor efficacy [71].

6.6 CONCLUSION AND FUTURE DIRECTIONS

PAI is rapidly emerging as a transformative innovation in regenerative medicine by providing high-resolution and real-time imaging of biological tissues. It allows for real-time in vivo monitoring of stem cell behaviours and tissue repair and can direct the delivery of therapeutic procedures without causing ionizing radiation, thus affecting both research and clinicians. Despite these advantages, difficulties remain for the PAI to gain acceptance by the clinical community. The primary problem that exists is the depth of imaging, which, although it can image deeper than pure optical imaging modalities, is limited compared to MRI or ultrasound imaging of deep human tissue. Additionally, PAI is limited by the lack of standardization in how a system is designed, the ability to replicate and standardize how signal processing is completed, and the lack of clinically approved biocompatible contrast agents. In preclinical models for PAI, contrast agents such as gold nanoparticles and semiconducting polymers require full assessment for toxicity, stability, and clearance in humans. Currently, PAI is primarily used in vascular and cancer imaging, and the use of PAI in regenerative therapies is much earlier in development. Future directions for PAI will likely include increased imaging depth and resolution, new contrast agents selected for the target site, and the integration of PAI with additional imaging modalities such as MRI or optical coherence tomography for multimodal imaging. Machine learning and image reconstruction techniques can enhance image quality, reduce artifacts, and enable real-time monitoring. Furthermore, to prove the safety, effectiveness, and long-term advantages of PAI in human regeneration applications, carefully planned clinical trials are necessary. PAI is anticipated to emerge as a key imaging modality in tissue engineering, stem cell monitoring, and precision medicine as technological and regulatory issues are gradually resolved.

REFERENCES

1. Beard, P. (2011). Biomedical photoacoustic imaging. *Interface Focus,* 1(4), 602–631.
2. Shrestha, B., Deluna, F., Anastasio, M. A., Yong Ye, J., & Brey, E. M. (2020). Photoacoustic imaging in tissue engineering and regenerative medicine. *Tissue Engineering Part B: Reviews,* 26(1), 79–102.
3. Jeon, S., Kim, J., Lee, D., Baik, J. W., & Kim, C. (2019). Review on practical photoacoustic microscopy. *Photoacoustics,* 15, 100141.
4. Sengupta, P., Van Engelenburg, S. B., & Lippincott-Schwartz, J. (2014). Superresolution imaging of biological systems using photoactivated localization microscopy. *Chemical Reviews,* 114(6), 3189–3202.
5. Manohar, S., & Razansky, D. (2016). Photoacoustics: A historical review. *Advances in Optics and Photonics,* 8(4), 586–617.
6. Wu, Y., Wu, D., Wen, Y., Yang, Y., Zhang, J., Chi, Z., & Jiang, H. (2025, April). Photoacoustic imaging in visualization of acupuncture mechanisms. *Photonics,* 12(4), 365.
7. Chen, Z., Xing, X., & Yang, Y. (2024). Laser processing. In: S. H. Teoh (Ed.), *Designing Materials for Medical Devices: Fundamentals* (p. 443). World Scientific, Singapore.
8. Ren, D., Li, C., Shi, J., & Chen, R. (2021). A review of high-frequency ultrasonic transducers for photoacoustic imaging applications. *IEEE Transactions on Ultrasonics, Ferroelectrics, and Frequency Control,* 69(6), 1848–1858.
9. Tang, Z., He, D., Yu, B., Zhang, H., Yu, B., Li, W., … & Wu, S. (2024, November). Design and implementation of an advanced high-speed data acquisition system for photoacoustic tomography. *Proc. SPIE, 13242*, 108–117).
10. Jiang, D., Zhu, L., Tong, S., Shen, Y., Gao, F., & Gao, F. (2024). Photoacoustic imaging plus X: A review. *Journal of Biomedical Optics,* 29(S1), S11513.
11. Tserevelakis, G. J., Vrouvaki, I., Siozos, P., Melessanaki, K., Hatzigiannakis, K., Fotakis, C., & Zacharakis, G. (2017). Photoacoustic imaging reveals hidden underdrawings in paintings. *Scientific Reports,* 7, 747.

12. Gu, Y., Sun, Y., Wang, X., Li, H., Qiu, J., & Lu, W. (2023). Application of photoacoustic computed tomography in biomedical imaging: A literature review. *Bioengineering & Translational Medicine,* 8(2), e10419.
13. Riksen, J. J. M., Nikolaev, A. V., & van Soest, G. (2023). Photoacoustic imaging on its way toward clinical utility: A tutorial review focusing on practical application in medicine. *Journal of Biomedical Optics,* 28(12), 121205.
14. Lin, Y., Zhou, H. C., Chen, N., Ren, Y., Gao, R., Li, Q., & Ren, J. (2022). Unveiling the improved targeting migration of mesenchymal stem cells with CXC chemokine receptor 3-modification using intravital NIR-II photoacoustic imaging. *Journal of Nanobiotechnology,* 20(1), 307.
15. Vij, I., Sunder, M., Shetty, B. S. K., Zhuo, G. Y., & Mazumder, N. (2025). An insight into photoacoustic imaging in regenerative medicine. *Discover Imaging,* 2(1), 1–13.
16. Wray, P., Lin, L., Hu, P., & Wang, L. V. (2019). Photoacoustic computed tomography of human extremities. *Journal of Biomedical Optics,* 24(2), 026003–026003.
17. Kubelick, K. P., & Emelianov, S. Y. (2020). In vivo photoacoustic guidance of stem cell injection and delivery for regenerative spinal cord therapies. *Neurophotonics,* 7(3), 030501.
18. Zhang, Y. S., Wang, Y., Wang, L., Wang, Y., Cai, X., Zhang, C., Wang, L. V., & Xia, Y. (2013). Labeling human mesenchymal stem cells with gold nanocages for in vitro and in vivo tracking by two-photon microscopy and photoacoustic microscopy. *Theranostics,* 3(8), 532–543.
19. Rebling, J., Ben-Yehuda Greenwald, M., Wietecha, M., Werner, S., & Razansky, D. (2021). Long-term imaging of wound angiogenesis with large scale optoacoustic microscopy. *Advanced Science,* 8(13), 2004226.
20. Liu, C., Chen, J., Zhang, Y., Zhu, J., & Wang, L. (2021). Five-wavelength optical-resolution photoacoustic microscopy of blood and lymphatic vessels. *Advanced Photonics,* 3(1), 016002–016002.
21. Cao, R., Zhao, J., Li, L., Du, L., Zhang, Y., Luo, Y., … & Wang, L. V. (2023). Optical-resolution photoacoustic microscopy with a needle-shaped beam. *Nature Photonics,* 17(1), 89–95.
22. Shintate, R., Ishii, T., Ahn, J. et al. (2022). High-speed optical resolution photoacoustic microscopy with MEMS scanner using a novel and simple distortion correction method. *Scientific Reports,* 12, 9221.
23. Zhang, J., Shi, Y., Zhang, Y., Liu, H., Li, S., & Liu, L. (2024). Resolution enhancement strategies in photoacoustic microscopy: A comprehensive review. *Micromachines,* 15(12), 1463.
24. Durmuş, H. O. (2024). Photoacoustic imaging: Basic principles, techniques and biomedical applications. In: S. Yalin & M. Eraslan (Eds.), *Current Research in Medicine and Health Sciences–2024.* Livre de Lyon, France. doi:10.5281/zenodo.13943422.
25. Kim, C., Favazza, C., & Wang, L. V. (2010). In vivo photoacoustic tomography of chemicals: High-resolution functional and molecular optical imaging at new depths. *Chemical Reviews,* 110(5), 2756–2782.
26. Park, S., Lee, C., Kim, J., & Kim, C. (2014). Acoustic resolution photoacoustic microscopy. *Biomedical Engineering Letters,* 4, 213–222.
27. Gao, R., Xue, Q., Ren, Y., Zhang, H., Song, L., & Liu, C. (2022). Achieving depth-independent lateral resolution in AR-PAM using the synthetic-aperture focusing technique. *Photoacoustics,* 26, 100328.
28. Lee, H., Choi, W., Kim, C., Park, B., & Kim, J. (2023). Review on ultrasound-guided photoacoustic imaging for complementary analyses of biological systems in vivo. *Experimental Biology and Medicine,* 248(9), 762–774.
29. Klontzas, M. E., Kakkos, G. A., Papadakis, G. Z., Marias, K., & Karantanas, A. H. (2021). Advanced clinical imaging for the evaluation of stem cell-based therapies. *Expert Opinion on Biological Therapy,* 21(9), 1253–1264.

30. Tachi, K., Hirasawa, T., Okawa, S., Horiguchi, A., Ito, K., & Ishihara, M. (2021). Chromatic-aberration-free multispectral optical-resolution photoacoustic microscopy using reflective optics and a supercontinuum light source. *Applied Optics,* 60(31), 9651–9658.
31. Song, Y., Park, S. M., Jeong, Y., Kim, J., & Lee, H. (2025). Review on multispectral photoacoustic imaging using stimulated Raman scattering light sources. *Sensors,* 25(11), 3325.
32. James, S., Neuhaus, K., Murphy, M., & Leahy, M, (2021). Contrast agents for photoacoustic imaging: A review of stem cell tracking. *Stem Cell Research & Therapy,* 12, 511.
33. Dhada, K. S., Hernandez, D. S., & Suggs, L. J. (2019). In vivo photoacoustic tracking of mesenchymal stem cell viability. *ACS Nano,* 13(7), 7791–7799.
34. Yoo, J. M., Yun, C., Bui, N. Q., Oh, J., & Nam, S. Y. (2019). Photoacoustic monitoring of the viability of mesenchymal stem cells labeled with indocyanine green. *IRBM,* 40(1), 45–50.
35. Ricles, L. M., Nam, S. Y., Treviño, E. A., Emelianov, S. Y., & Suggs, L. J. (2014). A dual gold nanoparticle system for mesenchymal stem cell tracking. *Journal of Materials Chemistry B,* 2(46), 8220–8230.
36. Yoon, S. J., Mallidi, S., Tam, J. M., Tam, J. O., Murthy, A., Johnston, K. P., Sokolov, K. V., & Emelianov, S. Y. (2010). Utility of biodegradable plasmonic nanoclusters in photoacoustic imaging. *Optics Letters,* 35(22), 3751–3753.
37. Li, W., Chen, R., Lv, J., Wang, H., Liu, Y., Peng, Y., Qian, Z., Fu, G., & Nie, L. (2017). In vivo photoacoustic imaging of brain injury and rehabilitation by high-efficient near-infrared dye-labeled mesenchymal stem cells with enhanced brain barrier permeability. *Advanced Science,* 5(2), 1700277.
38. Nam, S. Y., Ricles, L. M., Suggs, L. J., & Emelianov, S. Y. (2012). In vivo ultrasound and photoacoustic monitoring of mesenchymal stem cells labeled with gold nanotracers. *PLoS ONE,* 7(5), e37267.
39. Chung, E., Nam, S. Y., Ricles, L. M., Emelianov, S., & Suggs, L. (2013). Evaluation of gold nanotracers to track adipose-derived stem cells in a PEGylated fibrin gel for dermal tissue engineering applications. *International Journal of Nanomedicine,* 8(1), 325–336.
40. Comenge, J., Fragueiro, O., Sharkey, J., Taylor, A., Held, M., Burton, N. C., Park, B. K., Wilm, B., Murray, P., Brust, M., & Lévy, R. (2016). Preventing plasmon coupling between gold nanorods improves the sensitivity of photoacoustic detection of labeled stem cells in vivo. *ACS Nano,* 10(7), 7106–7116.
41. Zhang, H., Wang, Z.-J., Wang, L.-J., Li, T.-T., He, S., Li, L.-P., Li, X.-Y., Liu, S.-J., Li, J.-D., Li, S.-J., & Zhang, R.-P. (2019). A dual-mode nanoparticle based on natural biomaterials for photoacoustic and magnetic resonance imaging of bone mesenchymal stem cells in vivo. *RSC Advances,* 9(60), 35003–35010.
42. Kubelick, K. P., Snider, E. J., Ethier, C. R., & Emelianov, S. (2019). Development of a stem cell tracking platform for ophthalmic applications using ultrasound and photoacoustic imaging. *Theranostics,* 9(13), 3812.
43. Scarfe, L., Brillant, N., Kumar, J. D., Ali, N., Alrumayh, A., Amali, M., … & Park, B. K. (2017). Preclinical imaging methods for assessing the safety and efficacy of regenerative medicine therapies. *NPJ Regenerative Medicine,* 2, 28.
44. Qin, X., Chen, H., Yang, H., Wu, H., Zhao, X., Wang, H., Chour, T., Neofytou, E., Ding, D., Daldrup-Link, H., Heilshorn, S. C., Li, K., & Wu, J. C. (2018). Photoacoustic imaging of embryonic stem cell-derived cardiomyocytes in living hearts with ultrasensitive semiconducting polymer nanoparticles. *Advanced Functional Materials,* 28(1), 1–12.
45. Avigo, C., Flori, A., Armanetti, P., Di Lascio, N., Kusmic, C., Jose, J., … & Menichetti, L. (2016). Strategies for non-invasive imaging of polymeric biomaterial in vascular tissue engineering and regenerative medicine using ultrasound and photoacoustic techniques. *Polymer International,* 65(7), 734–740.

46. Talukdar, Y., Avti, P., Sun, J., & Sitharaman, B. (2014). Multimodal ultrasound-photoacoustic imaging of tissue engineering scaffolds and blood oxygen saturation in and around the scaffolds. *Tissue Engineering Part C: Methods,* 20(5), 440–449.
47. Zhang, Y. S., Wang, L. V., & Xia, Y. (2016). Seeing through the surface: Noninvasive characterization of biomaterial–tissue interactions using photoacoustic microscopy. *Annals of Biomedical Engineering,* 44, 649–666.
48. Cai, X., Paratala, B. S., Hu, S., Sitharaman, B., & Wang, L. V. (2012). Multiscale photoacoustic microscopy of single-walled carbon nanotube-incorporated tissue engineering scaffolds. *Tissue Engineering Part C: Methods,* 18(4), 310–317.
49. Zhu, Y., Lu, X., Dong, X., Yuan, J., Fabiilli, M. L., & Wang, X. (2019). LED-based photoacoustic imaging for monitoring angiogenesis in fibrin scaffolds. *Tissue Engineering Part C: Methods,* 25(9), 523–531.
50. Ogunlade, O., Ho, J. O., Kalber, T. L., Hynds, R. E., Zhang, E., Janes, S. M., … & Beard, P. (2019). Monitoring neovascularization and integration of decellularized human scaffolds using photoacoustic imaging. *Photoacoustics,* 13, 76–84.
51. Aliabouzar, M., Zhang, G. L., & Sarkar, K. (2018). Acoustic and mechanical characterization of 3D-printed scaffolds for tissue engineering applications. *Biomedical Materials,* 13(5), 055013.
52. Ferreira, L. P., Jorge, C., Lagarto, M. R., Monteiro, M. V., Duarte, I. F., Gaspar, V. M., & Mano, J. F. (2024). Photoacoustic processing of decellularized extracellular matrix for biofabricating living constructs. *Acta Biomaterialia,* 183, 74–88.
53. Attia, A. B. E., Balasundaram, G., Moothanchery, M., Dinish, U. S., Bi, R., Ntziachristos, V., & Olivo, M. (2019). A review of clinical photoacoustic imaging: Current and future trends. *Photoacoustics,* 16, 100144.
54. Deng, Z., Wang, Z., Yang, X., Luo, Q., & Gong, H. (2012). In vivo imaging of hemodynamics and oxygen metabolism in acute focal cerebral ischemic rats with laser speckle imaging and functional photoacoustic microscopy. *Journal of Biomedical Optics,* 17(8), 081415.
55. Deán-Ben, X. L., & Razansky, D. (2013). Portable spherical array probe for volumetric real-time optoacoustic imaging at centimeter-scale depths. *Optics Express,* 21(23), 28062–28071.
56. Deán-Ben, X. L., Ozbek, A., & Razansky, D. (2013). Volumetric real-time tracking of peripheral human vasculature with GPU-accelerated three-dimensional optoacoustic tomography. *IEEE Transactions on Medical Imaging,* 32(11), 2050–2055.
57. Deán-Ben, X. L., & Razansky, D. (2013). Functional optoacoustic human angiography with handheld video rate three-dimensional scanner. *Photoacoustics,* 1(3–4), 68–73.
58. Manohar, S., & Gambhir, S. S. (2020). Clinical photoacoustic imaging. *Photoacoustics,* 19, 100196.
59. Buehler, A., Kacprowicz, M., Taruttis, A., & Ntziachristos, V. (2013). Real-time handheld multispectral optoacoustic imaging. *Optics Letters,* 38(9), 1404–1406.
60. Taruttis, A., Timmermans, A. C., Wouters, P. C., Kacprowicz, M., van Dam, G. M., & Ntziachristos, V. (2016). Optoacoustic imaging of human vasculature: Feasibility by using a handheld probe. *Radiology,* 281(1), 256–263.
61. Yamaga, I., Kawaguchi-Sakita, N., Asao, Y., Matsumoto, Y., Yoshikawa, A., Fukui, T., … & Toi, M. (2018). Vascular branching point counts using photoacoustic imaging in the superficial layer of the breast: A potential biomarker for breast cancer. *Photoacoustics,* 11, 6–13.
62. Oraevsky, A. A., Clingman, B., Zalev, J., Stavros, A. T., Yang, W. T., & Parikh, J. R. (2018). Clinical optoacoustic imaging combined with ultrasound for coregistered functional and anatomical mapping of breast tumors. *Photoacoustics,* 12, 30–45.

63. Jo, J., Tian, C., Xu, G., Sarazin, J., Schiopu, E., Gandikota, G., & Wang, X. (2018). Photoacoustic tomography for human musculoskeletal imaging and inflammatory arthritis detection. *Photoacoustics,* 12, 82–89.
64. van den Berg, P. J., Daoudi, K., Moens, H. J. B., & Steenbergen, W. (2017). Feasibility of photoacoustic/ultrasound imaging of synovitis in finger joints using a point-of-care system. *Photoacoustics,* 8, 8–14.
65. Merčep, E., Deán-Ben, X. L., & Razansky, D. (2018). Imaging of blood flow and oxygen state with a multisegment optoacoustic ultrasound array. *Photoacoustics,* 10, 48–53.
66. Miao, X., Ma, R., Li, J., You, W., He, K., Meng, F., ... & Wang, X. (2024). Dynamic characterization of vascular response and treatment in oral traumatic ulcer in mice via photoacoustic imaging. *Quantitative Imaging in Medicine and Surgery,* 14(7), 4333.
67. Huang, Z., Xu, J., Chen, J., Chen, H., Wang, H., Huang, Z., ... & Hu, J. (2017). Photoacoustic stimulation promotes the osteogenic differentiation of bone mesenchymal stem cells to enhance the repair of bone defect. *Scientific Reports,* 7(1), 15842.
68. Maharjan, S., Ma, C., Singh, B., Kang, H., Orive, G., Yao, J., & Zhang, Y. S. (2024). Advanced 3D imaging and organoid bioprinting for biomedical research and therapeutic applications. *Advanced Drug Delivery Reviews, 208,* 115237.
69. Feng, Q., Fatima, K., Yang, A., Li, C., Chen, S., Yang, G., ... & He, C. (2024). Multimodal imaging for dynamic visualization of osteogenesis and implant degradation in 3D bioprinted scaffolds. *Bioactive Materials,* 37, 119–131.
70. Nguyen, C. (2024). *Combined Photoacoustic and Acoustic Radiation Force Impulse Imaging for Compositional and Biomechanical Assessment of Biologically Relevant Samples* (Doctoral dissertation, Tufts University).
71. Ou, X., Tang, Y., Yang, H., Du, Y., Zhang, Z., Lin, L., ... & Zou, J. (2025). Evaluation of multimodal imaging guided multifunctional biotargeted reactive oxygen species production synergists: A preliminary study. *Frontiers in Materials,* 12, 1550323.

7 Leveraging Quantum Dot Probes for High-Resolution and Multiplexed Microscopy

Gitanjal Deka
Prospexis Research PVT. Ltd

Vyasraj G. Bhat, Vibha Kamath, Nirmal Mazumder
Manipal Academy of Higher Education

Smritimala Sarmah
B. Barooah College

Ankur Gogoi
Jagannath Barooah University

7.1 INTRODUCTION

High-resolution microscopy encompasses a suite of advanced techniques that allow scientists to visualize structures at the cellular and molecular levels with exceptional clarity and detail by surpassing the diffraction limit. The diffraction-limited resolution is a fundamental constraint of conventional optical microscopy, which constrains the resolution to approximately 200–300 nm laterally and 500–700 nm axially (Heilemann, 2010). It is determined by the wavelength of light and the numerical aperture of the imaging system (Heilemann, 2010; Faust & House, 2018). This limit imposes restrictions on the level of detail that can be observed during an imaging event. However, advancements in high-resolution microscopy have revolutionized various fields, including biology, medicine, and materials science, by providing insights into the intricate architecture and dynamic processes of living organisms and complex materials (Garini et al., 2005; Leung & Chou, 2011).

Notably, many high-resolution microscopy techniques rely on fluorescence for imaging contrast. However, traditional organic fluorophores, though widely used, face challenges such as photobleaching, lack of robustness, and broad emission spectra, which can compromise imaging performance (Terai & Nagano, 2013). On the contrary, quantum dots (QDs) are semiconductor nanocrystals with unique optical properties, such as narrow emission spectrum, high quantum yield, robustness, and

 DOI: 10.1201/9781003570011-7

exceptional photostability, making them ideal for advanced imaging applications (Gould et al., 2008; Resch-Genger et al., 2008).

These nanostructures are typically composed of elements from groups II-VI or III-V of the periodic table, including materials such as cadmium selenide (CdSe), cadmium telluride (CdTe) (Donegan & Rakovich, 2013; da Costa et al., 2024), or indium arsenide (InAs) (Jalali et al., 2022), Zinc oxide (ZnO) (Fonoberov & Balandin, 2006), and zinc sulfide (ZnS) (Li et al., 2010). Their distinctive physical and chemical characteristics position them as versatile, application-driven materials for bio-imaging and bio-sensing (Jamieson et al., 2007). QDs can be conjugated either directly or indirectly to biological molecules, enabling detailed studies of various biological processes (Soldado et al., 2022) with unparalleled precision and accuracy.

Notably, in high-resolution imaging, fluorophores are often exposed to prolonged illumination by excitation light, leading to photobleaching and a decline in their quantum yield (Demchenko, 2020; Sanderson et al., 2014). In contrast, QDs address these challenges by offering bright, stable, and tunable fluorescence signals, enabling precise localization and imaging of structures at the nanoscale (Jamieson et al., 2007; Hyldahl et al., 2009; Bharadwaj & Novotny, 2011). In imaging techniques, where fine structural details are critical, the use of QDs as contrast agents has the potential to provide a significant advantage over conventional organic fluorophores. Undoubtedly, the integration of QDs with advanced imaging techniques expands the capabilities of microscopy, paving the way for groundbreaking discoveries in biological and materials science research (Keseroglu, 2017). This book chapter delves into the applications of QDs as contrast agents in high-resolution imaging, highlighting their transformative impact on advancing the field.

7.2 PRINCIPLE OF DIFFRACTION-LIMITED RESOLUTION

Spatial resolution in the context of imaging and optics refers to the ability of an imaging system to distinguish between closely spaced objects or details in an image. In other words, it refers to the minimum distance between two closely situated points on a specimen that can be resolved and observed as separate entities by the system. Thus, it determines how fine the details in an image can be and is a critical factor in the performance of cameras, microscopes, telescopes, and other optical instruments.

However, diffraction fundamentally limits the resolution of an imaging system. When light passes through an aperture (e.g., the lens of a camera or a microscope) or from a point source enters an optical system through an aperture, instead of traveling in perfectly straight lines, it spreads out or diffracts due to its wave nature, preventing light rays from focusing into an infinitely small point, even with an ideal optical system free of aberrations or imperfections. This spreading is again described by the point spread function (PSF), which describes the way an optical system responds to a point source of light and applies both to the imaging of a point source and to the collection of light from it. The shape of the PSF for a circular aperture is typically a pattern of concentric rings known as the Airy pattern. At the center of this pattern lies the Airy disk, which represents the brightest region where most of the light is concentrated. The size of this disk determines the resolution limit of the system. Figure 7.1a schematically demonstrates the Rayleigh criterion of partially and fully resolved point objects.

The resolution limit is described by two key formulas (Willingale, 2024; Sanderson, 2020). According to the Rayleigh criterion, two-point sources are resolvable when the peak of one Airy disk aligns with the first dark ring of the other, with the angular resolution given by $\theta = 1.22\frac{\lambda}{D}$, where λ is the wavelength of light and D is the diameter of the aperture. For lateral resolution, Rayleigh's criterion can be written as (Soliman, 2017; Sanderson, 2020):

$$\left(d_{x,y}\right)_{\text{Rayleigh}} = \frac{0.61\lambda}{NA}$$

where λ is the wavelength of light and NA is the numerical aperture of the optical system.

On the other hand, Abbe's formula provides a complementary definition of the spatial resolution in terms of physical distances between objects in the specimen in microscopy and other near-field optical systems. The formula relates the smallest resolvable detail to λ and NA of the optical system as (Sanderson, 2020):

$$\left(d_{x,y}\right)_{\text{Abbe}} = \frac{\lambda}{2NA}$$

or equivalently,

$$\left(d_{x,y}\right)_{\text{Abbe}} = \frac{\lambda}{2n \cdot \sin\theta}$$

where n is the refractive index of the medium and θ is the half-angle of the maximum cone of light that can enter or exit the lens. The NA is a dimensionless parameter that defines the ability of an optical system to gather or emit light over a range of angles. It is defined as:

$$NA = n \cdot \sin\theta$$

The lateral resolution of an imaging system is thus influenced by λ and NA. Shorter wavelengths provide better resolution (smaller $d_{x,y}$), allowing finer details to be resolved. The aperture size influences the NA indirectly through the collection angle θ. Larger apertures gather more diffracted light from the sample at wider angles (θ), increasing NA, reducing $d_{x,y}$ and improving resolution. Figure 7.1b schematically shows the concept of NA.

For axial resolution, which refers to the minimum distance along the optical axis where two points can be distinguished, the equation is given by:

$$d_z = \frac{2n\lambda}{NA^2}$$

This shows that axial resolution is also influenced by λ and NA, but with a stronger dependence on NA, as it is inversely proportional to its square.

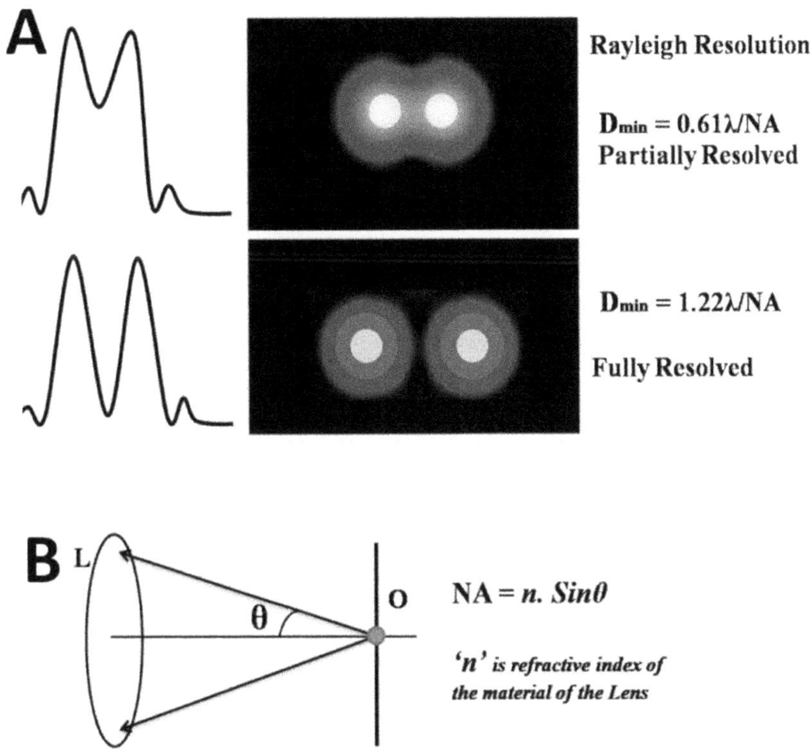

FIGURE 7.1 (a) Schematic of Rayleigh's criterion for partially resolved and fully resolved point objects. (b) Schematic of the concept of numerical aperture (NA), where L is the lens and O is the object.

7.3 SURPASSING THE DIFFRACTION LIMIT IN MICROSCOPY

Achieving resolution beyond the diffraction limit requires innovative approaches that either manipulate light in unconventional ways or extract additional information from the light collected by the system. Several groundbreaking microscopy techniques, such as confocal microscopy, multiphoton microscopy (MPM), structured illumination microscopy (SIM), stimulated emission depletion (STED) microscopy, single-molecule localization microscopy (SMLM), and so forth, have been developed to significantly enhance resolution, each leveraging unique principles of light interaction and detection.

Confocal microscopy uses point illumination and a spatial pinhole to eliminate out-of-focus light, resulting in enhanced image resolution and contrast (Fellers & Davidson, 2014). This method effectively reduces background noise and enhances image sharpness, providing better optical sectioning capabilities compared to wide-field microscopy (Figure 7.2a). This technique is particularly advantageous for creating three-dimensional images of thick specimens, such as tissues and large cells, by collecting images from multiple focal planes (Fujita et al., 2007; Garini et al., 2005; Fellers & Davidson, 2014; Wilson, 1990).

SIM utilizes patterned light illumination to capture high-frequency spatial details. By computationally reconstructing multiple images with different illumination patterns, SIM can extract information beyond the diffraction limit, doubling the effective resolution in both lateral and axial dimensions. SIM is particularly useful for live-cell imaging, allowing researchers to observe dynamic cellular processes in real time with enhanced clarity.

STED microscopy improves resolution by selectively depleting fluorescence in the periphery of the focal spot, leaving a smaller central region to emit light (Gogoi et al., 2019; Golm et al., 2023; Hell & Wichmann, 1994; Wildanger et al., 2012). This technique achieves nanoscale resolution, making it possible to visualize sub-cellular structures and protein interactions at an unprecedented level of detail, which again requires fluorescent markers of the order of nanometers (Figure 7.2b).

SMLM techniques such as photoactivated localization microscopy (PALM) and stochastic optical reconstruction microscopy (STORM) rely on the precise localization of individual fluorescent markers to reconstruct high-resolution images. By fitting the fluorescence emission of single molecules to a point spread function, these methods achieve resolutions of 10–20 nm, enabling the detailed visualization of molecular arrangements and interactions within cells.

Fluorescence saturation excitation microscopy (SAX or saturated excitation microscopy) is another cutting-edge optical imaging technique that enhances resolution by leveraging the nonlinear properties of fluorescence excitation under high-intensity illumination (Kumar Singh et al., 2021; Yamanaka et al., 2011; Yamanaka et al., 2013). In conventional fluorescence microscopy, excitation light intensities are maintained below the saturation point of fluorophores. In contrast, SAX microscopy intentionally increases the excitation intensity beyond the saturation threshold, where the fluorescence response of the fluorophore becomes nonlinear relative to the excitation light intensity. This nonlinear behavior is central to achieving super-resolution imaging, as it encodes spatial frequencies beyond the diffraction limit, enabling the resolution of finer details. By selectively analyzing the nonlinear fluorescence responses, SAX microscopy overcomes conventional optical limitations to achieve higher resolution (Yamanaka et al., 2011, 2013) (Figure 7.2c).

These techniques use fluorescent markers and controlled illumination to bypass the diffraction limit and have revolutionized optical microscopy, enabling the

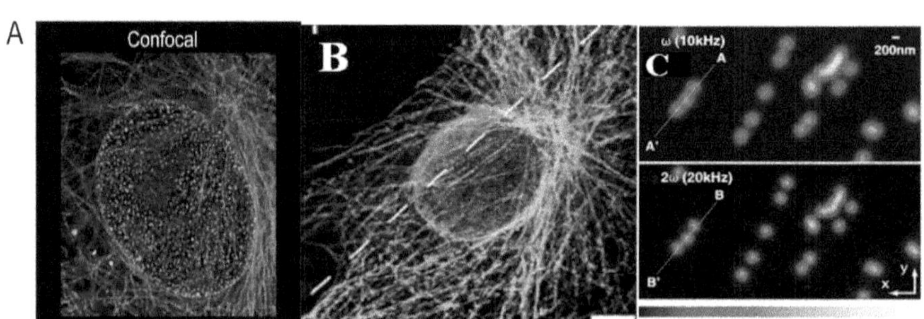

FIGURE 7.2 The resolution enhancement in microscopy using fluorephores as markers. (a) Confocal microscopy (Clausen et al., 2013). (b) STED microscopy (Bianchini et al., 2015). (c) SAX microscopy (Fujita et al., 2007).

visualization of biological structures and processes at scales previously unattainable, and have opened new frontiers in fields such as cell biology, materials science, and nanotechnology. Table 7.1 summarizes the resolution, PSF of each of the microscopies.

TABLE 7.1
Summary of the Resolution, PSF of the Microscopic Technique (Gogoi et al., 2019)

Microscopy Technique	Resolution		PSF	Remarks
	Lateral	Axial		
Widefield microscopy	$\frac{0.61\lambda}{NA}$	$\frac{2n\lambda}{NA^2}$	Diffraction-limited Airy disk with broad axial spread.	Suitable for general imaging; lacks optical sectioning and suffers from out-of-focus blur.
Confocal microscopy	~200 nm	~500 nm	Sharpened PSF by rejecting out-of-focus light using a pinhole.	Improved resolution and contrast; effective optical sectioning.
MPM (2PEM, 3PEM)	~200 nm	~500 nm	A smaller PSF is achieved due to the lack of fluorescence emission from the out-of-focus region.	Since it uses IR excitation light, it finds huge application in tissue and cell imaging due to less photo damage because of
SIM	100/50 nm	100 nm	PSF modulated by patterned illumination, enhancing spatial frequency capture.	Doubles resolution via computational reconstruction.
STED	40 nm	40–45 nm	PSF narrowed by depletion beam, reducing the effective excitation spot size.	Achieves super-resolution using high-intensity lasers; requires specific fluorophores.
SMLM (PALM, STORM)	30 nm	20–30 nm (3D STORM)	Effective PSF is localized fluorescence from individual molecules.	Nanometer precision achieved through sequential single-molecule activation and localization.
SAX	10 nm	30 nm	PSF is narrowed by selectively extracting the higher-order nonlinear signals.	Resolution enhancement is achieved by using an intensity-modulated excited laser and collecting the nonlinear signals of the modulated frequency (Ding et al., 2017).
Near-field optical microscopy			Highly localized evanescent field; PSF determined by probe geometry.	Bypasses diffraction limit; limited to near-surface imaging (Umakoshi, 2023; Walker et al., 2012).

7.4 CONVENTIONAL FLUORESCENT MARKERS IN HIGH-RESOLUTION MICROSCOPY

Fluorescent markers are indispensable in high-resolution microscopy, providing the specificity and sensitivity required to visualize biological structures and processes at the nanometer scale. Till date, a number of fluorescent markers have been developed, such as Alexa Fluor, Cy dyes, Rhodamine, GFP, mCherry, YFP, RFP, and so forth, which have become a traditional choice for microscopy. A representative list of traditional dyes used in microscopy techniques like confocal microscopy, MPM, SIM, STED, PALM, STORM, SAX, and NSOM is tabulated below in Table 7.2.

7.4.1 Limitations of Conventional Fluorescent Markers

Conventional fluorophores, such as organic dyes and fluorescent proteins, are widely used in super-resolution microscopy due to their ease of conjugation and availability.

TABLE 7.2
Summary of the Traditional Dyes Used in the Microscopy Technique

Microscopy Technique	Traditional Dyes Commonly Used	Key Features	References
Confocal microscopy	Fluorescein (FITC), rhodamine (TRITC), Alexa Fluors	Bright fluorescence, high photostability, used for general labeling.	Caprifico et al. (2021) and Entwistle and Noble (1992)
MPM	Fluorescein (FITC), Alexa Fluors, auto fluorescence, green fluorescence protein (GFP)	Deep tissue imaging, less tissue damage.	McNamara (2011) and Mao et al. (2023)
SIM	Cy3, Cy5, DAPI, Alexa Fluor series	High quantum yield and compatibility with structured illumination patterns.	McDonald et al. (2015) and Golm et al. (2023)
STED	Atto 647N, Atto 594, Alexa Fluor 594, Oregon Green	High photostability and compatibility with depletion lasers for precise control over emission.	Müller et al. (2012) and Lee (2020)
PALM	PA-GFP, mEos2, Dendra2	Photoactivatable proteins, enabling precise single-molecule localization.	Armes (2017) and Lelek et al. (2021)
STORM	Cy3, Cy5, Alexa Fluor 647, Alexa Fluor 568	Photo-switchable dyes with stable fluorescence under stochastic activation conditions.	Abdelsayed et al. (2022), Schirripa, Spagnolo, and Luin (2022)
SAX	Fluorescein, rhodamine derivatives, Nile red	High-intensity tolerance and efficient fluorescence yield for nonlinear excitation studies.	Ding et al. (2017)
NSOM	Rhodamine B, dii, coumarin, crystal violet	Small molecules with efficient fluorescence in near-field excitation for nanoscale resolution.	Klapetek et al. (2010)

However, as mentioned in section 1, these fluorophores exhibit significant drawbacks that hinder their performance in high-resolution imaging (Terai & Nagano, 2013). One of the primary limitations is photobleaching, where fluorophores lose their ability to fluoresce after prolonged or repeated excitation (Kwon et al., 2022). This is particularly problematic in techniques requiring extended imaging, such as SMLM, as it reduces signal intensity and compromises data quality over time. Another critical issue is their limited photostability under the high-intensity illumination often required for super-resolution methods like STED or SIM (Hanne et al., 2015). This instability can degrade fluorophores, resulting in inconsistent signals and reduced resolution reliability during prolonged imaging sessions. Additionally, fluorophores typically exhibit broad emission spectra, which leads to spectral overlap, complicating the resolution of multiple targets in multicolor imaging (McLeod & Ozcan, 2016). Fluorophores also suffer from short fluorescence lifetimes, which restrict their application in time-resolved imaging techniques and limit the dynamic range of fluorescence detection. Furthermore, many fluorophores generate reactive oxygen species under illumination, causing cytotoxicity that can damage living cells, thereby restricting their use in live-cell imaging (Winterbourn, 2014). Lastly, photoblinking, the random switching of fluorophores between fluorescent and non-fluorescent states, disrupts signal consistency, making quantitative analyses in techniques like PALM and STORM challenging (Urban et al., 2021).

7.5 SCOPE AND APPLICATIONS OF QDS IN HIGH-RESOLUTION MICROSCOPY

QDs, semiconductor nanocrystals with size-tunable optical properties, provide a robust alternative to conventional fluorophores, addressing many of their limitations. A key advantage of QDs is their high photostability, as they resist photobleaching and degradation under intense or prolonged illumination (Zhao & Zeng, 2015; Chacko, 2020; Kargozar et al., 2020). This enables long-term imaging, essential for time-lapse studies and single-molecule tracking. QDs also exhibit narrow and tunable emission spectra, with sharp size-dependent peaks that can be excited by a single wavelength. This feature minimizes spectral overlap, significantly enhancing multicolor imaging by allowing precise differentiation of multiple fluorophores in complex biological samples. Their high brightness and quantum yield further improve the signal-to-noise ratio, providing superior detection sensitivity and enabling finer resolution in super-resolution imaging (Yang et al., 2016). A QD-RGD probe serves as a highly specific fluorescence marker for pancreatic cancer cell imaging, leveraging quantum dots' superior photostability and multiplexing capabilities for high-resolution microscopy, also having the ability to bind integrin $\alpha\nu\beta3$ receptors, enabling targeted imaging and potential therapeutic applications (Figure 7.3) (Li et al., 2016).

Figure 7.4 presents 4Pi microscopy imaging of quantum dot-labeled PtK2 cells, highlighting the mitochondrial network with ATP synthase immunolabeled using Qdot 655. The axial projection in Figure 7.4a provides an overview of the 3D data stack, while Figure 7.4b and c showcase *xz*-slices, demonstrating high signal intensity with minimal background. The raw data in the upper frames retain some side-maxima (~30%), which are effectively removed through a simple 5-point deconvolution, as

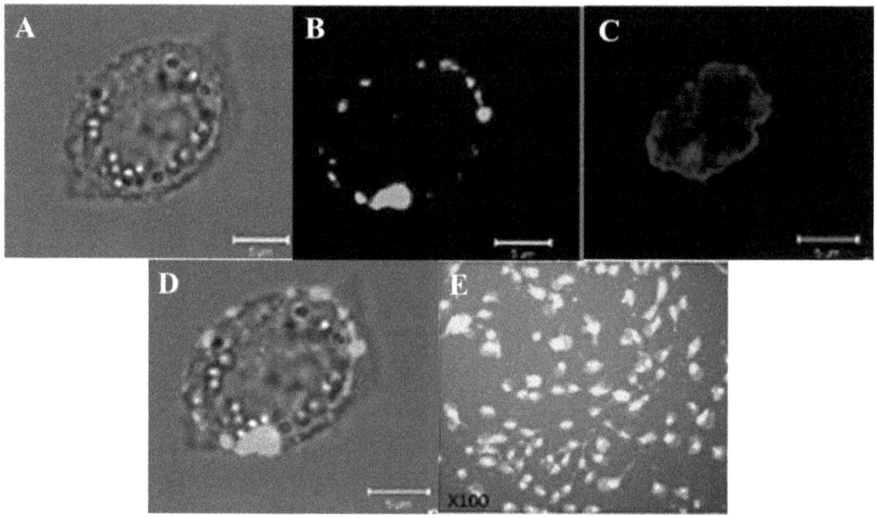

FIGURE 7.3 Confocal images of SW1990 pancreatic cells marked with the QD-RGD (Type II CdTe/CdS QDs) probe. (a) A single SW1990 cell. (b) The fluorescence image of the QD-RGD probe with laser excitation at 405 nm. (c) DAPI staining of an SW1990 cell nucleus. (d) SW1990 cells marked with the QD-RGD probe. (e) Fluorescence image of SW1990 cells stained with the QD-RGD probe (Li et al., 2016)

seen in the lower frames. Figure 7.4d and e further illustrate the improved clarity of mitochondrial structures post-deconvolution. Notably, photobleaching remained negligible even after prolonged imaging, confirming that 4Pi microscopy of quantum dot-labeled samples allows for extended recording times without significant signal loss (Figure 7.4) (Medda et al., 2006).

Multiphoton microscopy enabled deep-tissue imaging at a depth of 480 μm, revealing GFP-expressing MCF7 tumor cells (green), fibrillar collagens via second harmonic generation (SHG) (blue), and FeSe QDs (magenta, converted from cyan for better distinction). The successful targeting of FeSe QDs to the tumor was confirmed by their colocalization with GFP signals. The blue SHG signals highlighted fibrillar collagens in the dermis, while the magenta FeSe QDs demonstrated efficient tumor labeling. Imaging before and after anti-HER2–QD injection further emphasized the potential of FeSe QDs for tumor imaging and drug delivery (Figure 7.5) (Kwon et al., 2019).

Figure 7.6a–c illustrates the enhancement in mitochondrial resolution using SOFI-MSIM imaging. The comparison of mitochondrial images processed through widefield imaging, SOFI, and SOFI-MSIM demonstrates a significant improvement in resolution, with SOFI-MSIM achieving an approximate resolution of 55 nm. Decorrelation analysis provided parameter-free resolution estimates, showing values of 317 nm for widefield imaging, 141 nm for SOFI imaging, and 59 nm for SOFI-MSIM, indicating a nearly fourfold resolution increase. The incorporation of multifocal arrays improved the signal-to-noise ratio in widefield imaging, reducing inter-array spacing and enhancing imaging quality. The number of acquired frames also played a role in resolution enhancement, with 120 multifocal arrays using frame accumulations of 50, 30, and 20 per array for SOFI-MSIM reconstruction.

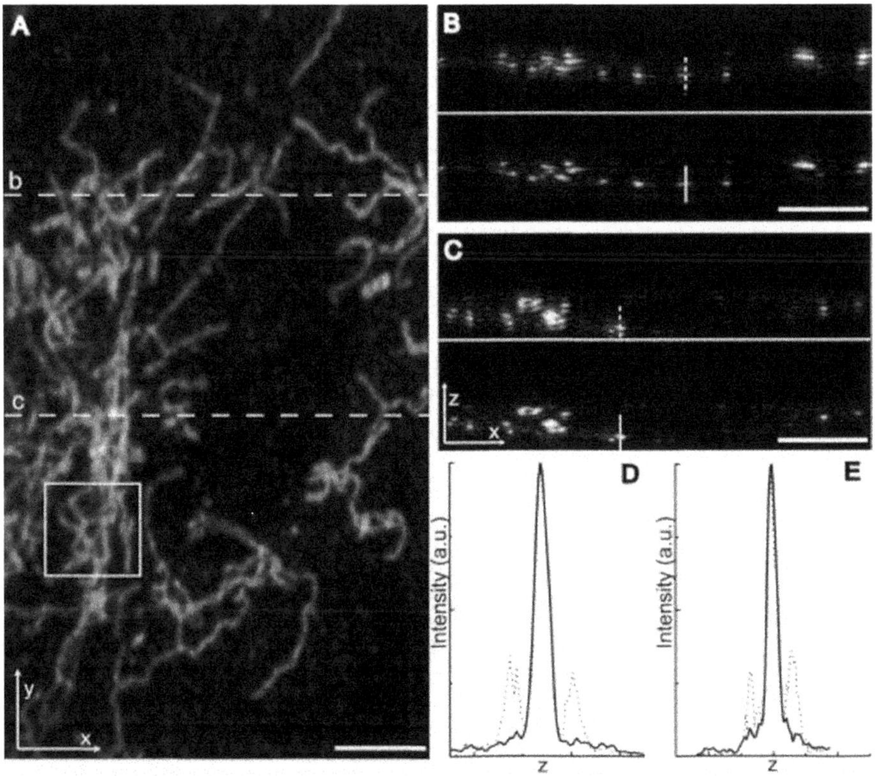

FIGURE 7.4 4Pi imaging of the mitochondrial structure in PtK2 cells labeled for the α-subunit of F1F0-ATP synthase using quantum dots. (a) Axial average projection of the 3D data stack. (b, c) xz sections along dashed lines in (a), with raw (top) and deconvolved (bottom) images. Intensity profiles along white lines in (b, c) are shown in (d, e) (Medda et al., 2006).

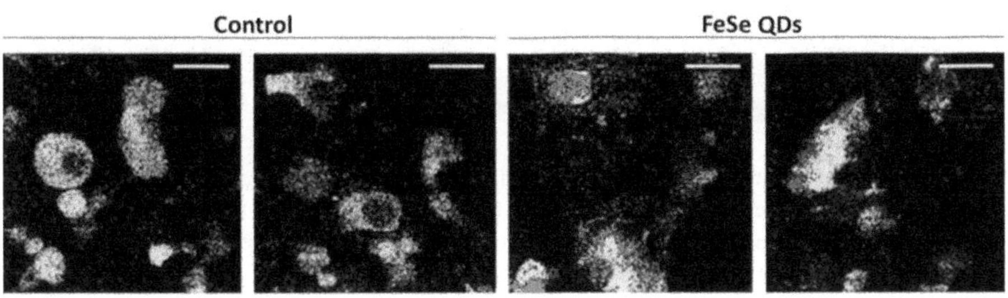

FIGURE 7.5 *In vivo* multiphoton microscopy images before and after anti-HER2–QD injection. The green signal indicates GFP-expressing MCF7 cells, blue represents second harmonic generation from fibrillar collagens, and magenta denotes FeSe QDs, converted from cyan for better visibility (Kwon et al., 2019).

The imaging improvements were further validated for both microtubules and mitochondria, confirming that SOFI-MSIM provides superior resolution compared to SOFI or MSIM alone (Figure 7.6a–c) (Zhou et al., 2023). Figure 7.6d and e presents multicolour wide-field and STORM images, highlighting mitochondrial structures and their association with microtubules. STORM imaging provides higher resolution than wide-field imaging, revealing a heterogeneous mitochondrial distribution, with a higher density near the nucleus where energy demand is greater. Mitochondria exhibit diverse morphologies, including elongated tubular and short globular forms, and most align along microtubules, though some show noncontiguous, inchworm-like interactions. Additionally, STORM imaging reveals thin, extended tubular connections between mitochondrial compartments, which are not clearly visible in wide-field images. These findings demonstrate the advantages of STORM microscopy in resolving mitochondrial structures and their dynamic interactions (Figure 7.6d and e) (Xu et al., 2015). SAX imaging of HeLa cells was performed using Qdot® 655 ITK™ CdSe/ZnS core/shell quantum dots. To specifically label actin fibers, HeLa cells were first treated with biotinylated phalloidin, which selectively binds to filamentous actin. Streptavidin-coated Qdot® 655 ITK™ CdSe/ZnS quantum dots were then introduced, allowing them to attach to the biotinylated phalloidin,

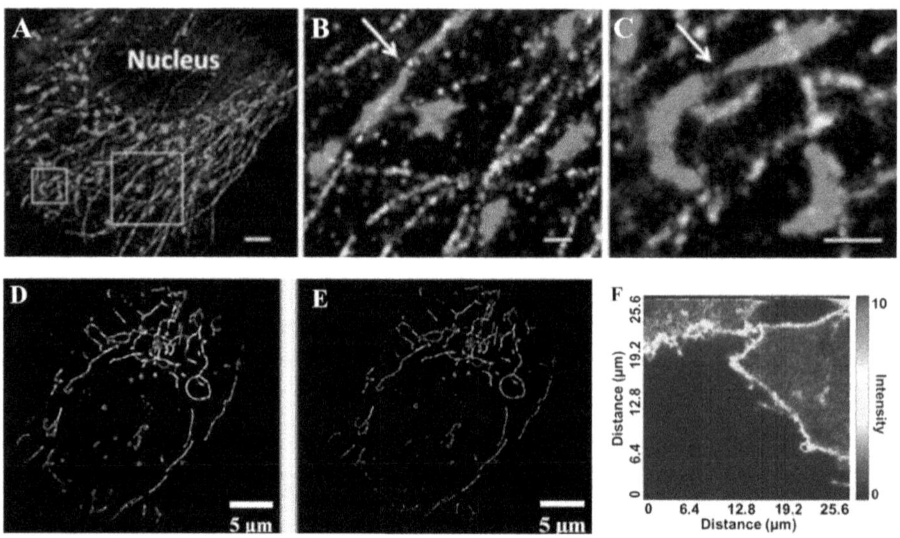

FIGURE 7.6 (a) Super-resolution STORM image of QD565-labeled microtubules (blue) and QD705-labeled microchondia (red) across the entire field of view (Xu et al., 2015). (b, c) Magnified zoomed-in STORM images of selected regions from (a), revealing detailed mitochondrial and microtubule structures (Scale bars: 500 nm) (Xu et al., 2015). (d) Super-resolution optical fluctuation imaging (SOFI) enhances mitochondrial resolution by utilizing temporal fluorescence fluctuations to improve spatial resolution and contrast (Zhou et al., 2023). (e) Combining SOFI with multiphton structured illumination microscopy (MSIM) further enhances resolution, achieving approximately 55 nm (red line), as demonstrated in mitochondrial imaging of fixed U2OS cells (Zhou et al., 2023). (f) SAX image of HeLa cells labeled with Qdot® 655 ITK™ CdSe/ZnS core/shell QDs. Actin fibers in HeLa cells were first labeled with biotinylated phalloidin. Streptavidin-coated Qdot® 655 ITK™ CdSe/ZnS core/shell QDs were bound to the biotinylated phalloidin (Gould et al., 2008; Kumar Singh et al., 2021).

enabling high-resolution visualization of the actin cytoskeleton (Figure 7.6f) (Gould et al., 2008; Kumar Singh et al., 2021).

Another notable advantage is the long fluorescence lifetime of QDs, which surpasses that of organic dyes. This property makes them highly suitable for time-resolved fluorescence microscopy and reduces background autofluorescence, particularly in dense biological specimens (Vu et al., 2015). Advances in QD surface chemistry have also mitigated photo blinking, providing consistent fluorescence essential for high-precision quantitative analyses in techniques such as STORM and PALM (Li et al., 2022). Additionally, QDs can be engineered for biocompatibility through surface modifications with biocompatible coatings. This reduces their cytotoxicity, allowing their use in live-cell imaging and dynamic studies (Sarkar et al., 2020). Collectively, these features make QDs a transformative tool for advanced microscopy, enabling researchers to achieve higher resolutions, improved multicolor imaging, and reliable long-term observations of biological and material systems at the nanoscale (Shivaji et al., 2018). Table 7.3 summarizes the applications of QDs as fluorophores in high-resolution microscopy techniques, including their advantages and limitations.

TABLE 7.3
Applications of Quantum Dots in High-resolution Microscopy

Microscopy Technique	Type of QD	Key Features	References
Confocal microscopy	CdSe/ZnS core-shell QDs	High brightness, narrow emission spectra	Vyshnava et al. (2022), Peuschel et al. (2016), Li et al. (2020), and Kumar et al. (2021)
	CdTe/CdS	Good light stability, imaging capabilities, and persistence	Li et al. (2016)
	Carboxylated QDs	Biocompatible	Kundrotas et al. (2019)
	Carbon dots	High fluorescence performance, low cytotoxicity, and biocompatible	Zhuo et al. (2015)
	Graphene QDs	Tunable emission, multifunctional, biocompatible	Qiu et al. (2015), Narasimhan et al. (2017), and Zhang et al. (2012)
	CdSe, CdSe/CdS, and CdSe/ZnSe, coated with l-arginine-modified β-cyclodextrin (β-CD-l-Arg)	Good dispersive, uniform, and excellent fluorescent properties as bio-probes	Zhao et al. (2010)
4Pi microscopy		Focused illumination and fluorescence detection from two objective lenses	Medda et al. (2006)

(Continued)

TABLE 7.3 (*Continued*)
Applications of Quantum Dots in High-resolution Microscopy

Microscopy Technique	Type of QD	Key Features	References
Multiphoton microscopy	CdSe/ZnS QDs	Greater depths than standard fluorophores due to their high two-photon absorption cross-section	Zeng et al. (2020) and Larson et al. (2003)
	FeSe QDs	Deep tissue QD imaging with two-photon excitation	Kwon et al. (2019)
	Graphene QDs		Singh et al. (2019), Wang et al. (2018), Feng et al. (2017), and Zhao et al. (2016)
SIM	PbSQDs	Near-infrared emission, tunable wavelengths	
	InP/ZnSe/ZnS	Controlled photoblinking	Zhou et al. (2023)
	Graphene QDs	Highly specific	Singh et al. (2019)
STED	InAs/ZnSe core-shell QDs	High photostability under depletion lasers	Jang and Chen (2010)
	CsPbBr3 perovskite QD	Super-resolution imaging, high fluorescence stability	Ye et al. (2018)
	CdTe	Blinking can be reduced under combined illumination	Hanne et al. (2015)
	ZnS-coated CdSe	High photostability, maximum resolution, time-lapse imaging	Hanne et al. (2015)
	Carbonized polymer dots	Highly photostability, highly efficient excitation feature, responsive fluorescence lifetime	Liu et al. (2021)
	Carbon dots	Spectrally distinguishable fluorescence, photostable	Han et al. (2019), Hua et al. (2019), and Li et al. (2019)
SMLM (PALM, STORM)	CdTeQDs	Sharp emission peaks, good photostability	Zhu et al. (2015)
	CdSe/CdS QDs	Fast blinking, extremely photostable	Urban et al. (2021)
	CdSe/ZnS QDs	Broad absorption, narrow emission spectra	Xu et al. (2015)
	$CsPbBr_3$ perovskite QDs	Higher photostability, larger absorption cross-section	Feld et al. (2021)
SOFI	Carbon dots	Super-resolution	Chizhik et al. (2016)
	CdSe/ZnS colloidal Qdots	Easily adaptable, highly reproducible	Dopf et al. (2014)
SAX	CdSeTeQDs	High-intensity tolerance	Marandi and Abadi (2020)
	CdSe/ZnS core-shell QDs	Distinct photoluminescence saturation	Gould et al. (2008) and Kumar Singh et al. (2021)
NSOM	CdS/ZnS quantum dots	Small size, high brightness	Holton (2007)
	Quantum dot-decorated nano-pyramid fiber tip	High scanning stability	Pan et al. (2019)

7.6 CONCLUSION AND FUTURE DIRECTIONS

QD high-resolution microscopy has significantly advanced our understanding of cellular and molecular biology by enabling more detailed imaging, multiplexed detection, and real-time monitoring of dynamic processes. These studies illustrate the versatility and transformative potential of QDs in biological research, ranging from fundamental studies of molecular interactions to applications in disease diagnostics and therapeutic development. Continued innovations in QD technology and imaging methodologies promise to further expand their impact in advancing biomedical sciences. Future advancements in QD technology for microscopy will likely improve biocompatibility, enhance stability, expand multiplexing capabilities, and integrate with advanced imaging modalities. Addressing these challenges will pave the way for broader adoption of QD-based microscopy in basic research, clinical diagnostics, and personalized medicine, thereby unlocking new insights into biological systems and disease mechanisms at the nanoscale.

ACKNOWLEDGMENTS

NM gratefully acknowledges the ASEAN-India Science & Technology Development Fund (AISTDF), Anusandhan National Research Foundation (ANRF), Government of India, India (Project Number- ANRF/F/84/2025–2026), and the Indian Council of Medical Research (ICMR) (Project Number-Dev/SG-00170/2024), Government of India, India, for financial support. AG gratefully acknowledges the financial support received from the Science and Engineering Research Board (SERB), Department of Science and Tech-nology (DST), India (Grant No. CRG/2019/004868). The authors thank the Manipal School of Life Sciences, Manipal Academy of Higher Education, Manipal, Karnataka, India, for providing the infrastructure.

REFERENCES

Abdelsayed, V., Boukhatem, H., & Olivier, N. (2022). An optimized buffer for repeatable multicolor STORM. *ACS Photonics*, 9(12), 3926–3934.

Armes, H. E. H. (2017). Implementing Super-Resolution Palm Microscopy in Fission Yeast. Doctoral dissertation, University of Sussex.

Bharadwaj, P., & Novotny, L. (2011). Robustness of quantum dot power-law blinking. *Nano Letters*, 11(5), 2137–2141.

Bianchini, P., Peres, C., Oneto, M., Galiani, S., Vicidomini, G., & Diaspro, A. (2015). STED nanoscopy: a glimpse into the future. *Cell and Tissue Research*, 360, 143–150.

Caprifico, A. E., Polycarpou, E., Foot, P. J., & Calabrese, G. (2021). Biomedical and pharmacological uses of fluorescein isothiocyanate chitosan-based nanocarriers. *Macromolecular Bioscience*, 21(1), 2000312.

Chacko, J. V. (2020). Nanobiophotonics and Fluorescence Nanoscopy in 2020. In: S. Thomas, Y. Grohens, G. Vignaud, N. Kalarikkal, & J. James (Eds.), *Nano-Optics: Fundamentals, Experimental Methods, and Application* (pp. 113–162). Elsevier, Amsterdam. doi: 10.1016/b978-0-12-818392-2.00006-8

Chizhik, A. M., Stein, S., Dekaliuk, M. O., Battle, C., Li, W., Huss, A., … & Chizhik, A. I. (2016). Super-resolution optical fluctuation bio-imaging with dual-color carbon nanodots. *Nano Letters*, 16(1), 237–242.

Clausen, M. P., Galiani, S., de la Serna, J. B., Fritzsche, M., Chojnaki, J., Gehmlich, K., … & Eggeling, C. (2013). Pathways to optical STED microscopy. *NanoBioImaging*, 1(1), 1–12.
da Costa, P. F. G. M., Merízio, L. G., Wolff, N., Terraschke, H., & de Camargo, A. S. S. (2024). Real-time monitoring of CdTe quantum dots growth in aqueous solution. *Scientific Reports*, 14(1), 7884.
Demchenko, A. P. (2020). Photobleaching of organic fluorophores: quantitative characterization, mechanisms, protection. *Methods and Applications in Fluorescence*, 8(2), 022001.
Ding, H. X., Li, K. Y., Deka, G., Su, I. C., & Chu, S. W. (2017, April). Resolution enhancement down to 10-nm based on saturated excitation (SAX) microscopy plus novel nonlinear response. *Proc. SPIE*, 10071, 15.
Donegan, J., & Rakovich, Y. (Eds.). (2013). *Cadmium Telluride Quantum Dots: Advances and Applications*. Jenny Stanford Publishing, Singapor. https://doi.org/10.1201/b16378
Dopf, K., Heunisch, S., Schwab, P., Moosmann, C., Habermehl, A., Lemmer, U., & Eisler, H. J. (2014, May). Superresolution optical fluctuation imaging (SOFI) aided nanomanipulation of quantum dots using AFM for novel artificial arrangements of chemically functionalized colloidal quantum dots and plasmonic structures. *Proc. SPIE*, 9126, 35–43).
Entwistle, A., & Noble, M. (1992). The use of Lucifer yellow, bodipy, FITC, TRITC, RITC and Texas red for dual immunofluorescence visualized with a confocal scanning laser microscope. *Journal of Microscopy*, 168(3), 219–238.
Faust, J. A., & House, J. E. (Eds.). (2018). *Physical Chemistry of Gas-Liquid Interfaces*. Elsevier, Amsterdam.
Feld, L. G., Shynkarenko, Y., Krieg, F., Rainò, G., & Kovalenko, M. V. (2021). Perovskite quantum dots for super-resolution optical microscopy: where strong photoluminescence blinking matters. *Advanced Optical Materials*, 9(18), 2100620.
Fellers, T. J., & Davidson, M. W. (2014). Confocal microscopy. S. W. Paddock (Ed.), *Confocal Microscopy* (pp. 9–47). Humana Press, New York.
Feng, L. L., Wu, Y. X., Zhang, D. L., Hu, X. X., Zhang, J., Wang, P., … & Tan, W. (2017). Near infrared graphene quantum dots-based two-photon nanoprobe for direct bioimaging of endogenous ascorbic acid in living cells. *Analytical Chemistry*, 89(7), 4077–4084.
Fonoberov, V. A., & Balandin, A. A. (2006). ZnO quantum dots: physical properties and optoelectronic applications. *Journal of Nanoelectronics and Optoelectronics*, 1(1), 19–38.
Fujita, K.; Kobayashi, M.; Kawano, S.; Yamanaka, M.; Kawata, S. High-resolution confocal microscopy by saturated excitation of fluorescence. *Physical Review Letters*, 2007, 99 (22), 228105.
Garini, Y., Vermolen, B. J., & Young, I. T. (2005). From micro to nano: recent advances in high-resolution microscopy. *Current Opinion in Biotechnology*, 16(1), 3–12.
Gogoi, A., Keiser, G., Kao, F.J. and Chiang, A.S., 2019. Multiscale and multimodal imaging for connectomics. In: F. J. Kao and A. S. Chiang (Eds.), *Advanced Optical Methods for Brain Imaging* (pp. 3–45) Springer, Singapore.
Golm, S. K., Hübner, W., & Müller, K. M. (2023). Fluorescence microscopy in adeno-associated virus research. *Viruses*, 15(5), 1174.
Gould, T. J., Bewersdorf, J., & Hess, S. T. (2008). A quantitative comparison of the photophysical properties of selected quantum dots and organic fluorophores. *Zeitschrift* für *Physikalische Chemie*, 222(5–6), 833–849.
Han, G., Zhao, J., Zhang, R., Tian, X., Liu, Z., Wang, A., … & Zhang, Z. (2019). Membrane-penetrating carbon quantum dots for imaging nucleic acid structures in live organisms. *Angewandte Chemie*, 131(21), 7161–7165.
Hanne, J., Falk, H. J., Görlitz, F., Hoyer, P., Engelhardt, J., Sahl, S. J., & Hell, S. W. (2015). STED nanoscopy with fluorescent quantum dots. *Nature Communications*, 6(1), 7127.
Heilemann, M. (2010). Fluorescence microscopy beyond the diffraction limit. *Journal of Biotechnology*, 149(4), 243–251.

Hell, S. W., & Wichmann, J. (1994). Breaking the diffraction resolution limit by stimulated emission: stimulated-emission-depletion fluorescence microscopy. *Optics Letters*, 19(11), 780–782.
Holton, M. D. (2007). *Near-Field Spectroscopy of Colloidal Quantum Dots and Vertical Cavity Surface Emitting Lasers*. Swansea University, United Kingdom.
Hua, X. W., Bao, Y. W., Zeng, J., & Wu, F. G. (2019). Nucleolus-targeted red emissive carbon dots with polarity-sensitive and excitation-independent fluorescence emission: high-resolution cell imaging and in vivo tracking. *ACS Applied Materials & Interfaces*, 11(36), 32647–32658.
Hyldahl, M. G., Bailey, S. T., & Wittmershaus, B. P. (2009). Photo-stability and performance of CdSe/ZnS quantum dots in luminescent solar concentrators. *Solar Energy*, 83(4), 566–573.
Jamieson, T., Bakhshi, R., Petrova, D., Pocock, R., Imani, M., & Seifalian, A. M. (2007). Biological applications of quantum dots. *Biomaterials*, 28(31), 4717–4732.
Jang, E. S., & Chen, X. (2010). Quantum dots for targeted tumor imaging. *Semiconductor Nanomaterials*, 63, 104.
Jalali, H. B., De Trizio, L., Manna, L., & Di Stasio, F. (2022). Indium arsenide quantum dots: an alternative to lead-based infrared emitting nanomaterials. *Chemical Society Reviews*, 51(24), 9861–9881.
Kargozar, S., Hoseini, S. J., Milan, P. B., Hooshmand, S., Kim, H. W., & Mozafari, M. (2020). Quantum dots: a review from concept to clinic. *Biotechnology Journal*, 15(12), 2000117.
Keseroglu, K. O. (2017). Super-Resolution Imaging via Spectral Separation of Quantum Dots. Doctoral dissertation, Queen Mary University of London.
Klapetek, P., Bujdák, J., & Buršík, J. (2010). Near-field scanning optical microscopy local luminescence studies of rhodamine dye. *Open Physics*, 8(3), 312–317.
Kumar Singh, A., Burkhow, S. J., Wijesooriya, C., Boote, B. W., Petrich, J. W., & Smith, E. A. (2021). Inorganic semiconductor quantum dots as a saturated excitation (sax) probe for sub-diffraction imaging. *ChemPhotoChem*, 5(3), 253–259.
Kumar, B. A., Kumar, P., Elangovan, T., Ramalingam, G., Ravi, G., Mohanapriya, P., & Natarajan, T. S. (2021). Surface functionalization of core-shell QDs for solar photovoltaic and anti-cancer applications. *Applied Surface Science Advances*, 5, 100122.
Kundrotas, G., Karabanovas, V., Pleckaitis, M., Juraleviciute, M., Steponkiene, S., Gudleviciene, Z., & Rotomskis, R. (2019). Uptake and distribution of carboxylated quantum dots in human mesenchymal stem cells: cell growing density matters. *Journal of Nanobiotechnology*, 17, 1–13.
Kwon, J., Elgawish, M. S., & Shim, S. H. (2022). Bleaching-resistant super-resolution fluorescence microscopy. *Advanced Science*, 9(9), 2101817.
Kwon, J., Jun, S. W., Choi, S. I., Mao, X., Kim, J., Koh, E. K., … & Lee, J. (2019). FeSe quantum dots for in vivo multiphoton biomedical imaging. *Science Advances*, 5(12), eaay0044.
Larson, D. R., Zipfel, W. R., Williams, R. M., Clark, S. W., Bruchez, M. P., Wise, F. W., & Webb, W. W. (2003). Water-soluble quantum dots for multiphoton fluorescence imaging in vivo. *Science*, 300(5624), 1434–1436.
Lee, M. C. (2020). A General Method for Three Color STED Microscopy with One Depletion Laser: Application to Primary Neuronal Culture. Doctoral Dissertation, Massachusetts Institute of Technology.
Lelek, M., Gyparaki, M. T., Beliu, G., Schueder, F., Griffié, J., Manley, S., … & Zimmer, C. (2021). Single-molecule localization microscopy. *Nature Reviews Methods Primers*, 1(1), 39.
Leung, B. O., & Chou, K. C. (2011). Review of super-resolution fluorescence microscopy for biology. *Applied Spectroscopy*, 65(9), 967–980.
Li, H., Shih, W. Y., & Shih, W. H. (2010). Highly photoluminescent and stable aqueous ZnS quantum dots. *Industrial & Engineering Chemistry Research*, 49(2), 578–582.

Li, H., Ye, S., Guo, J., Wang, H., Yan, W., Song, J., & Qu, J. (2019). Biocompatible carbon dots with low-saturation-intensity and high-photobleaching-resistance for STED nanoscopy imaging of the nucleolus and tunneling nanotubes in living cells. *Nano Research*, 12, 3075–3084.

Li, S. L., Yang, J., Lei, X. F., Zhang, J. N., Yang, H. L., Li, K., & Xu, C. Q. (2016). Peptide-conjugated quantum dots act as the target marker for human pancreatic carcinoma cells. *Cellular Physiology and Biochemistry*, 38(3), 1121–1128.

Li, X., Zhang, H., & Sun, F. (2020). CdSe/ZnS quantum dots exhibited nephrotoxicity through mediating oxidative damage and inflammatory response. *Aging*, 13(8), 12194.

Li, W., Kaminski Schierle, G. S., Lei, B., Liu, Y., & Kaminski, C. F. (2022). Fluorescent nanoparticles for super-resolution imaging. *Chemical Reviews*, 122(15), 12495–12543.

Liu, Y., Song, Y., Zhang, J., Yang, Z., Peng, X., Yan, W., & Qu, J. (2021). Responsive carbonized polymer dots for optical super-resolution and fluorescence lifetime imaging of nucleic acids in living cells. *ACS Applied Materials & Interfaces*, 13(43), 50733–50743.

MacDonald, L., Baldini, G., & Storrie, B. (2015). Does super-resolution fluorescence microscopy obsolete previous microscopic approaches to protein co-localization? In B. L. Tang (Ed.), *Membrane Trafficking* (pp. 255–275, 2nd edn.). Springer, New York. https://doi.org/10.1007/978-1-4939-2309-0_19.

Mao, W., Bui, H. T. D., Cho, W., & Yoo, H. S. (2023). Spectroscopic techniques for monitoring stem cell and organoid proliferation in 3D environments for therapeutic development. *Advanced Drug Delivery Reviews*, 201, 115074.

Marandi, M., & Abadi, S. H. (2020). Aqueous synthesis of colloidal $CdSe_xTe_{1-x}$–CdS core–shell nanocrystals and effect of shell formation parameters on the efficiency of corresponding quantum dot sensitized solar cells. *Solar Energy*, 209, 387–399.

McLeod, E., & Ozcan, A. (2016). Unconventional methods of imaging: computational microscopy and compact implementations. *Reports on Progress in Physics*, 79(7), 076001.

McNamara, G. (2011). McNamara 2011 MPMicro-multi-probe microscopy (10/31/2011). *Microscopy*, *10*, 31.

Medda, R., Jakobs, S., Hell, S. W., & Bewersdorf, J. (2006). 4Pi microscopy of quantum dot-labeled cellular structures. *Journal of Structural Biology*, 156(3), 517–523.

Müller, T., Schumann, C., & Kraegeloh, A. (2012). STED microscopy and its applications: new insights into cellular processes on the nanoscale. *ChemPhysChem*, 13(8), 1986–2000.

Narasimhan, A. K., Santra, T. S., Rao, M. R., & Krishnamurthi, G. (2017). Oxygenated graphene quantum dots (GQDs) synthesized using laser ablation for long-term real-time tracking and imaging. *RSC Advances*, 7(85), 53822–53829.

Pan, B., Yang, Y., Bian, J., Hu, X., & Zhang, W. (2019). Quantum dot decorated nano-pyramid fiber tip for scanning near-field optical microscopy. *Optics Communications*, 445, 273–276.

Peuschel, H., Ruckelshausen, T., Kiefer, S., Silina, Y., & Kraegeloh, A. (2016). Penetration of CdSe/ZnS quantum dots into differentiated vs undifferentiated Caco-2 cells. *Journal of Nanobiotechnology*, 14, 1–19.

Qiu, J., Zhang, R., Li, J., Sang, Y., Tang, W., Rivera Gil, P., & Liu, H. (2015). Fluorescent graphene quantum dots as traceable, pH-sensitive drug delivery systems. *International Journal of Nanomedicine*, *10*, 6709–6724.

Resch-Genger, U., Grabolle, M., Cavaliere-Jaricot, S., Nitschke, R., & Nann, T. (2008). Quantum dots versus organic dyes as fluorescent labels. *Nature Methods*, 5(9), 763–775.

Sanderson, J. (2020). Fundamentals of microscopy. *Current Protocols in Mouse Biology*, *10*(2), e76.

Sanderson, M. J., Smith, I., Parker, I., & Bootman, M. D. (2014). Fluorescence microscopy. *Cold Spring Harbor Protocols*, *2014*(10), pdb.top071795. doi:10.1101/pdb.top071795.

Sarkar, S., Le, P., Geng, J., Liu, Y., Han, Z., Zahid, M. U., … & Smith, A. M. (2020). Short-wave infrared quantum dots with compact sizes as molecular probes for fluorescence microscopy. *Journal of the American Chemical Society*, 142(7), 3449–3462.
Schirripa Spagnolo, C., & Luin, S. (2022). Choosing the probe for single-molecule fluorescence microscopy. *International Journal of Molecular Sciences*, 23(23), 14949.
Shivaji, K., Mani, S., Ponmurugan, P., De Castro, C. S., Lloyd Davies, M., Balasubramanian, M. G., & Pitchaimuthu, S. (2018). Green-synthesis-derived CdS quantum dots using tea leaf extract: antimicrobial, bioimaging, and therapeutic applications in lung cancer cells. *ACS Applied Nano Materials*, 1(4), 1683–1693.
Singh, H., Sreedharan, S., Tiwari, K., Green, N. H., Smythe, C., Pramanik, S. K., … & Das, A. (2019). Two photon excitable graphene quantum dots for structured illumination microscopy and imaging applications: lysosome specificity and tissue-dependent imaging. *Chemical Communications*, 55(4), 521–524.
Soldado, A., Barrio, L. C., Díaz-Gonzalez, M., de la Escosura-Muñiz, A., & Costa-Fernandez, J. M. (2022). Advances in quantum dots as diagnostic tools. *Advances in Clinical Chemistry*, 107, 1–40.
Soliman, D. M. (2017). Augmented Microscopy: Development and Application of High-Resolution Optoacoustic and Multimodal Imaging Techniques for Label-Free Biological Observation. Doctoral dissertation, Technische Universität München.
Terai, T., & Nagano, T. (2013). Small-molecule fluorophores and fluorescent probes for bioimaging. *Pflügers Archiv-European Journal of Physiology*, 465, 347–359.
Umakoshi, T. (2023). Near-field optical microscopy toward its applications for biological studies. *Biophysics and Physicobiology*, 20(1), e200011.
Urban, J. M., Chiang, W., Hammond, J. W., Cogan, N. M., Litzburg, A., Burke, R., … & Krauss, T. D. (2021). Quantum dots for improved single-molecule localization microscopy. *Journal of Physical Chemistry* B, 125(10), 2566–2576.
Vu, T. Q., Lam, W. Y., Hatch, E. W., & Lidke, D. S. (2015). Quantum dots for quantitative imaging: from single molecules to tissue. *Cell and Tissue Research*, 360(1), 71–86.
Vyshnava, S. S., Kanderi, D. K., & Dowlathabad, M. R. (2022). Confocal laser scanning microscopy study of intercellular events in filopodia using 3-mercaptopropoinc acid capped CdSe/ZnS quantum dots. *Micron*, 153, 103200.
Walker, K. A. D., Morgan, C., Doak, S. H., & Dunstan, P. R. (2012). Quantum dots for multiplexed detection and characterisation of prostate cancer cells using a scanning near-field optical microscope. *PLoS One*, 7(2), e31592.
Wang, L., Li, W., Li, M., Su, Q., Li, Z., Pan, D., & Wu, M. (2018). Ultrastable amine, sulfo cofunctionalized graphene quantum dots with high two-photon fluorescence for cellular imaging. *ACS Sustainable Chemistry & Engineering*, 6(4), 4711–4716.
Wildanger, D., Patton, B. R., Schill, H., Marseglia, L., Hadden, J. P., Knauer, S., … & Smith, J. M. (2012). Solid immersion facilitates fluorescence microscopy with nanometer resolution and sub-ångström emitter localization. *Advanced Materials (Deerfield Beach, Fla.)*, *24*(44), OP309.
Wilson, T. (1990). The role of the pinhole in confocal imaging systems. In: J. B. Pawley (Ed.), *Handbook of Biological Confocal Microscopy* (pp. 113–126). Springer US, Boston, MA.
Willingale, R. (2024). Diffraction-limited optics and techniques. In: C. Bambi & A. Santangelo (Eds.), *Handbook of X-ray and Gamma-ray Astrophysics* (pp. 289–326). Springer, Singapore.
Winterbourn, C. C. (2014). The challenges of using fluorescent probes to detect and quantify specific reactive oxygen species in living cells. *Biochimica et Biophysica Acta (BBA)-General Subjects*, *1840*(2), 730–738.
Xu, J., Tehrani, K. F., & Kner, P. (2015). Multicolor 3D super-resolution imaging by quantum dot stochastic optical reconstruction microscopy. *ACS Nano*, 9(3), 2917–2925.

Yamanaka, M., Tzeng, Y. K., Kawano, S., Smith, N. I., Kawata, S., Chang, H. C., & Fujita, K. (2011). SAX microscopy with fluorescent nanodiamond probes for high-resolution fluorescence imaging. *Biomedical Optics Express*, 2(7), 1946–1954.

Yamanaka, M., Yonemaru, Y., Kawano, S., Uegaki, K., Smith, N. I., Kawata, S., & Fujita, K. (2013). Saturated excitation microscopy for sub-diffraction-limited imaging of cell clusters. *Journal of Biomedical Optics*, 18(12), 126002–126002.

Yang, X., Zhanghao, K., Wang, H., Liu, Y., Wang, F., Zhang, X., … Xi, P. (2016). Versatile application of fluorescent quantum dot labels in super-resolution fluorescence microscopy. *ACS Photonics,* 3(9), 1611–1618. doi:10.1021/acsphotonics.6b00178

Ye, S., Yan, W., Zhao, M., Peng, X., Song, J., & Qu, J. (2018). Low-saturation-intensity, high-photostability, and high-resolution STED nanoscopy assisted by CsPbBr3 quantum dots. *Advanced Materials*, 30(23), 1800167.

Zeng, H., Yang, H., Liu, G., Zhang, S., Zhang, X., & Zhang, Y. (2020). Simultaneous multicolour imaging using quantum dot structured illumination microscopy. *Journal of Microscopy*, 277(1), 32–41.

Zhang, M., Bai, L., Shang, W., Xie, W., Ma, H., Fu, Y., … & Yang, S. (2012). Facile synthesis of water-soluble, highly fluorescent graphene quantum dots as a robust biological label for stem cells. *Journal of Materials Chemistry*, 22(15), 7461–7467.

Zhao, M. X., Xia, Q., Feng, X. D., Zhu, X. H., Mao, Z. W., Ji, L. N., & Wang, K. (2010). Synthesis, biocompatibility and cell labeling of L-arginine-functional β-cyclodextrin-modified quantum dot probes. *Biomaterials*, 31(15), 4401–4408.

Zhao, M. X., & Zeng, E. Z. (2015). Application of functional quantum dot nanoparticles as fluorescence probes in cell labeling and tumor diagnostic imaging. *Nanoscale Research Letters*, 10, 1–9.

Zhao, W., Li, Y., Yang, S., Chen, Y., Zheng, J., Liu, C., … & Yang, R. (2016). Target-activated modulation of dual-color and two-photon fluorescence of graphene quantum dots for in vivo imaging of hydrogen peroxide. *Analytical Chemistry*, 88(9), 4833–4840.

Zhou, L., Cao, H., Huang, L., Jing, Y., Wang, M., Lin, D., … & Qu, J. (2023). Narrowband photoblinking InP/ZnSe/ZnS quantum dots for super-resolution multifocal structured illumination microscopy enhanced by optical fluctuation. *Nanophotonics*, 12(9), 1777–1785.

Zhuo, Y., Miao, H., Zhong, D., Zhu, S., & Yang, X. (2015). One-step synthesis of high quantum-yield and excitation-independent emission carbon dots for cell imaging. *Materials Letters*, 139, 197–200.

8 Spectroscopic and Microscopic Characterization of Millet Starch Digestibility

Nandana Bijulal, Baishali Das, and Ishita Chakraborty
Manipal Academy of Higher Education

Md. Hafizur Rahman
Khulna Agricultural University

Norma Sammán
Universidad Nacional de Jujuy

Nirmal Mazumder
Manipal Academy of Higher Education

8.1 INTRODUCTION

Millets and their products are becoming increasingly popular because of their numerous health-promoting characteristics. Several foods and beverages, including beer, porridge, nonalcoholic drinks, and fermented or nonfermented bread, are made from millet. The primary component of millet is starch (approximately 70% of total millet grain), which has a significant effect on the quality of millet-based food products because of its composition, structure, properties, and interactions [1]. In addition, millet starch can play a vital role in the packaging industry in terms of developing bioplastics [2]. Moreover, millet also contains other components, such as proteins (6%–19.5%), dietary fiber (12%–20%), fats (1.3%–6.5%), vitamins and minerals (2%–4%), and compounds with antioxidant properties, depending upon their species or type [3]. Owing to its high fiber, protein, and antioxidant contents as well as gluten-free status, millet is becoming increasingly popular around the world.

Starch is the primary carbohydrate found in the seeds of millet plants [4]. It is a polysaccharide that is exclusively composed of D-glucose residues with $\alpha(1\rightarrow4)$ linkages in linear amylose and α-$(1\rightarrow4)$ linkages and $\alpha(1\rightarrow6)$ branch linkages in amylopectin. The ratio of amylose to amylopectin can affect its functional properties and digestibility [5].

DOI: 10.1201/9781003570011-8

However, the digestibility of starch can also be influenced by factors such as the amylose content, granule size, and presence of other components in the seed. For example, some millet varieties have a relatively high amylose content, making their starch less digestible than those with relatively high amylopectin contents [6]. In addition, starch is classified as rapidly digestible starch (RDS), slowly digestible starch (SDS), or resistant starch (RS) on the basis of the degree of α-amylase hydrolysis. Millets are highly preferred since they contain a significant amount of RS. RS resists digestion in the small intestine and behaves more like dietary fiber [7]. RS also has various health benefits, including promoting gut health. The glycemic index (GI) and RS have been established as two important indicators of starch digestibility [8].

Moreover, the GI is typically calculated by dividing the increase in blood glucose levels after a test food is consumed by the increase in blood glucose caused by the consumption of a reference food that contains the same amount of carbohydrates. This comparison helps us understand how a test food influences blood glucose levels compared to the reference food. The GI values of various millet-based food products are summarized in Table 8.1. Unfortunately, despite their increasing interest, millets are still underexplored and underutilized compared with other grains. To the best of our knowledge, information on the digestibility of millet starch is limited. Therefore,

TABLE 8.1
Glycemic Index Values of Various Millet-based Food Products

Food Item	Method of Analysis	Predicted GI	References
Millet-based food mix (roti)	*In vivo* using the glucometer (Precision IQD)	53	[9]
Millet-based food mix (dosa)	*In vivo* using the glucometer (Precision IQD)	37	
Millet-based food mix (dumplings)	*In vivo* using the glucometer (Precision IQD)	48	
Pearl millet, finger millet, and semolina-based *uttapam*	*In vivo* using the glucometer (Accu-Check/One Touch)	38.7	[10]
Pearl millet, finger millet, and *besan* (Bengal gram flour) made *cheela*	*In vivo* using the glucometer (Accu-Check/One Touch)	36.8	
Pearl millet and finger millet made cheela	*In vivo* using the glucometer (Accu-Check/One Touch)	36.8	
Fox tail millet made dosa	*In vivo* using the glucometer (Accu-Check Meter	59.25	[11]
Foxtail millet made bread	*In vivo* using a glucometer	49.53	[12]
Finger millet made muffins	*In vitro* using pancreatic α-amylase	<55	[13]
Little millet made RTC flakes	*In vivo* using glucometer (Precision IQD)	52.11	[14]
Foxtail millet made Biscuit	*In vivo* using the glucometer-Glucotrend 2	50.8	[15,16]
Finger millet made buns	*In vivo* using the glucometer (Accu-Check Meter	36.57	[17]

studies on the digestibility of millet starch to better understand and expand its applications have received much attention from the food industry.

Investigating the characteristics of starch through biophysical techniques could aid in understanding its structure and properties, which directly influence its nutritional value and digestibility. Biophysical analytical techniques can also help to evaluate the susceptibility of starch to enzymatic breakdown during digestion [18]. This information is crucial for determining the potential impact of starvation-based foods on blood glucose levels and overall nutritional benefits. Moreover, these analytical techniques provide insights into the behavior of starch during food processing. Understanding how starch interacts with other ingredients, undergoes gelatinization, and affects the texture and quality of food products can aid in optimizing food formulations and enhancing their overall acceptability. While the nutritional and functional properties of millet starch have been studied, detailed biophysical characterization using advanced techniques is underreported. There is limited information on how the biophysical properties of millet starch and starch from other grains correlate with their functional properties, such as digestibility. This review explores these correlations, offering insights into how specific structural features influence starch performance in food and industrial applications, along with the effects of different processing methods (e.g., milling and thermal treatment) on the biophysical properties of starch. The data and information created and compiled in this review with respect to millet starch will provide readers with the knowledge they need to use millet starch effectively when designing new products and how to use biophysical characterization for a better understanding of starch properties.

8.2 BIOPHYSICAL METHODS FOR EVALUATING THE DIGESTIBILITY OF MILLET STARCH

8.2.1 X-Ray Diffraction (XRD)

XRD is a nondestructive analytical tool that is applied to determine the type of starch and its degree of crystallinity [19]. This could provide detailed information about the alterations in the crystalline structure during starch digestion and deeper insights into the factors influencing starch digestibility, contributing to the development of more nutritious and digestible food products [20]. Starch is composed of both crystalline and amorphous regions. Crystalline regions are less susceptible to enzymatic hydrolysis, whereas amorphous regions are more accessible to digestive enzymes [21]. XRD was used to quantify the crystalline content of millet starch before and after digestion. By comparing the changes in crystallinity during digestion, the susceptibility of millet starch to enzymatic attack can be assessed. The arrangement and alterations in the properties of millet starches in their native (NA) and microwave-treated (MT) states were investigated by [22] using XRD. In its native state, NA displayed a relative crystallinity of 35.7% and exhibited distinct diffraction peaks at 2θ values of 15.3° and 23.1°, indicating singular peaks, along with a dual peak present at approximately 17°–18°, which indicated its classification as A-type starch (Figure 8.1a). The microwave treatment reduced the peak intensities and degree of crystallinity due to significant disruption of the starch structure, including damage to the crystalline domains, disappearance of the double helices, and disintegration of the starch granules. This subsequently led to

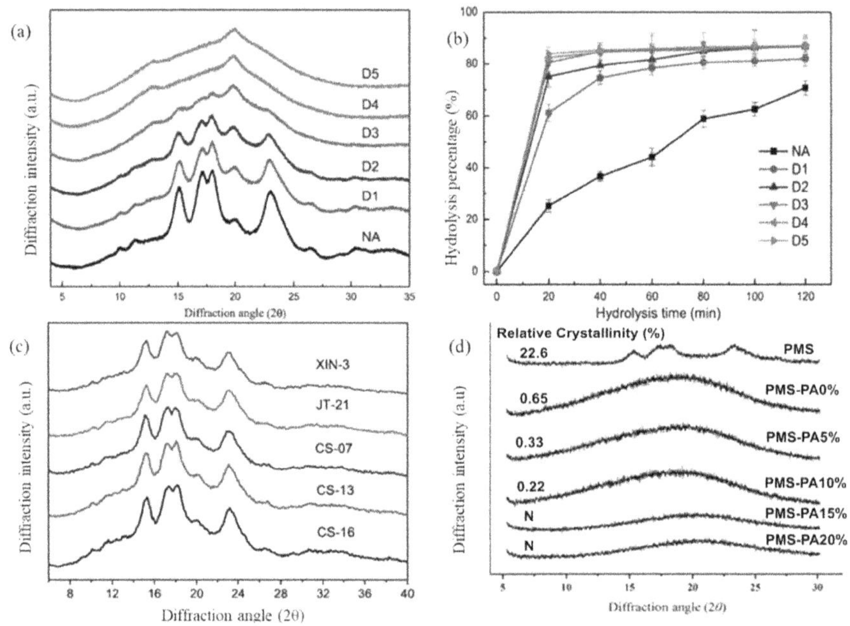

FIGURE 8.1 XRD patterns (a) and dynamic curves of *in vitro* enzymatic digestion (b) of NA and MT millets; (c) XRD patterns of millet starches with various amylose contents (CS-16, CS-13, CS-07, JT-21, and XIN-3). (d) XRD patterns of proso millet starch in the presence of proanthocyanidins. (Figures are reproduced from Refs. [22,25,26] with permission from Elsevier.)

increased digestibility (Figure 8.1b). Hence, XRD provides information about the interactions between millet starch and digestive enzymes. These interactions can alter the crystalline structure of starch, impacting its digestibility. By analyzing the diffraction patterns after enzymatic digestion, the enzyme–substrate interactions and their effects on starch digestion are better understood. Changes in the RS, SDS, and RDS contents and expected glycemic index (eGI) values of whole quinoa flour (WQ) in response to treatments such as heat–moisture treatment (HMT), HMT combined with pullulanase (HMT+P), HMT combined with microwaves (HMT+M), and HMT combined with citric acid (HMT+A) were investigated by Dong et al. [23]. The results revealed a significant increase in RS and SDS after the treatments and a reduction in the RDS content. Additionally, various treatments lowered the eGI value to different degrees. This outcome might be attributed to alterations in the starch profile that decreased starch digestibility. The increase in the RS content and decrease in the RDS content contributed to a reduced eGI value. These alterations may result from the reorganization of the crystalline structure within the granules and the improved interaction between the amylose and the amylose–amylopectin chains during HMT. Similar results may also be obtained from millet starches [24]. However, it is important to note that the relationship between the degree of hydrolysis and the GI is not always straightforward. The GI of millet starch is influenced by various factors in addition to the degree of hydrolysis and the RS content. Other factors, such as food composition, cooking methods, and individual differences in digestion, can also play a significant role in determining the glycemic response to a particular food. XRD can also be used to monitor the rate of crystalline-to-amorphous transition during the enzymatic digestion of millet starch. In a different investigation, five indigenous millet varieties (Changsheng 16: CS-16, Changsheng 13: CS-13, Changsheng 07: CS-07, Jintai 21: JT-21, and Xin 3: XIN-3) with amylose contents ranging from 12.3% to 27.4% were studied by Shi et al. [25]. They reported that a relatively high amylose content was correlated with reduced order in starch double-helical structures and decreased crystalline formations, as indicated by the XRD patterns. All varieties exhibited an A-type crystalline structure with prominent peaks at 15°, 17°, 18°, and 23° (2θ). CS-16 demonstrated the highest crystallinity (28.2%), followed by CS-13 (27.1%), CS-07 (25.8%), JT-21 (24.7%), and XIN-3 (23.8%) (Figure 8.1c). A higher amylose content in starch disrupts crystallite packing, reducing the ordered arrangement in granules and hindering enzymatic breakdown in the small intestine, leading to a lower glycemic response. Starch granules undergo gelatinization during digestion, involving crystalline structure disruption through water absorption and granule swelling. XRD detects crystalline pattern changes during gelatinization, offering insights into millet starch gelatinization temperature and enthalpy. The digestibility of proso millet starch (PMS) with various concentrations of proanthocyanidins (PAs) was investigated by Xiao et al. [26] who found an A-type diffraction pattern characterized by strong peaks at 2θ values of approximately 15° and 23°, along with a doublet at 17° and 18°. However, the starch is transformed into B-type crystallinity after cooking, as reflected by a distinct peak at 17°. This peak formation resulted from the disruption of the PMS crystal structure due to gelatinization. The intensity of the peak close to 17° in the PMS samples containing 5%–20% PA was notably absent in the typical X-ray diffractogram. This absence varied with PA concentration, leading to the disappearance of the B-type pattern. The reduction in relative crystallinity can be attributed to the combined effects of heat and moisture during treatment, which leads to the breakdown of the original hydrogen bonds between molecular chains within starch granules. This, in turn, causes changes in the molecular packing within the crystalline region (Figure 8.1d).

XRD analysis can compare the digestibility of millet starch with that of other starch sources, including rice, wheat, or corn. This helps in analyzing the unique features of millet starch that may impact its digestion and nutritional characteristics. XRD can also be used to investigate the impact of various processing methods (e.g., cooking and extrusion) on the crystalline structure of millet starch and its subsequent digestibility.

8.2.2 Differential Scanning Calorimetry (DSC)

DSC is a thermoanalytical method that is applied to investigate the gelatinization of starch [19]. This method can measure the heat flow associated with thermal transitions in starch, such as starch gelatinization, which is a critical process that affects starch digestibility [27]. By analyzing the endothermic peak in the DSC endotherm, the gelatinization temperature (T_o) of millet starch can be determined. The DSC thermogram also provides information about the enthalpy change associated with starch gelatinization. The enthalpy is a measure of the energy absorbed or released during a phase transition and is related to the degree of crystallinity and the amount of energy required to disrupt the starch granules [28]. Several studies have evaluated the impact of different processing techniques on the digestibility of millet starch. For example, DSC can be used to compare the gelatinization properties of raw millet starch with those of processed starches, such as cooked or extruded starch. Similarly, Rahman et al. [19] have investigated the comparative gelatinization properties of maize, potato, and sweet potato starches treated with high hydrostatic pressure and reported diverse results. This allows researchers to understand how processing affects starch structure and its subsequent digestibility. PMS was modified via HMT and annealing, and both modified starches presented an increased T_o and a decreased gelatinization enthalpy (ΔH) [29]. It was hypothesized that interactions involving amylose-amylose and amylose-lipids restricted the movement of the amorphous region during HMT. This constraint is reflected in the elevated T_o value and reduced ΔH value. Starches with higher T_o values presented more intricate and organized structures, which hindered their susceptibility to digestive enzymes. Finger millet-resistant starch (FRS) was prepared from various modifications, such as ultrasonication (UP-FRS), ultrasonication followed by autoclaving (UAP-FRS), enzymatic hydrolysis, and ozonation (OP-FRS), and the highest gelatinization peak temperatures (T_p) were for UP-FRS and UAP-FRS, indicating that both FRSs can withstand elevated temperatures during food processing [30]. However, UP-FRS required the most energy for gelatinization, as indicated by the ΔH value of 145 J/g. In contrast, both UP-FRS and OP-FRS showed strong resistance to *in vitro* digestion, with less than 30% digestion rates, suggesting increased RS content and enhanced anti-digestive properties (Figure 8.2a). Moreover, Navyashree et al. [31] investigated the thermal properties of native, roasted and germinated White Finger Milnavyalet (WFM) and reported that the onset temperature (T_o), peak temperature (T_p), conclusion temperature (T_c), and gelatinization temperature range (ΔT_r) of the roasted and germinated samples were significantly greater than those of the native WFM (Figure 8.2b).

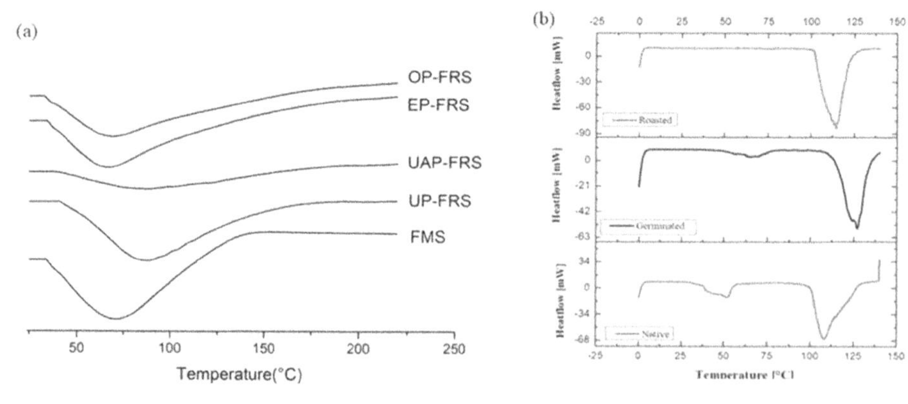

FIGURE 8.2 Differential scanning calorimetry (DSC) thermograms of (a) finger millet-resistant starches and (b) roasted, germinated, and native WFM flours. (Figures are reproduced from Refs. [30,31] with permission from John Wiley and Sons and Elsevier.)

This finding suggests that the properties of millet starch might be affected by different modification techniques as well as the type of millet starch, including its respective gelatinization properties, which are determined by DSC.

8.2.3 Rheometry

Rheometry is another analytical tool that is widely used for the evaluation of the rheological characteristics of different food ingredients, indicating their viscous and elastic properties [1,32]. This method can also play a significant role in analyzing millet starch digestion, which is influenced by the texture and consistency of foods, oral processing, enzymatic activity, gastric processing, intestinal transit, and nutrient absorption. The complex interplay between rheological properties and starch digestion highlights the importance of understanding how food structure and properties affect the physiological processes involved in nutrient digestion and absorption. For example, rotational viscometry measures the viscosity of a fluid under controlled shear rates. To study millet starch digestibility, starch samples can be subjected to varying shear rates, and the resulting changes in viscosity can indicate the breakdown of starch granules and the formation of digestible products. As starch is digested, its viscosity may decrease due to the breakdown of larger starch molecules into smaller, more soluble fragments. The rheological properties of the apparent viscosity of the doughs made from pearl, barnyard, little, sorghum, and kodo millet were examined via rotational viscometry, with the shear rate varying from 0 to $20\,s^{-1}$, and the findings revealed that the apparent viscosity of the dough made from nongluten millet varieties was notably greater than that of the dough made from regular wheat flour (RWF) [33]. This difference might be attributed to the fact that millet doughs are rich in water-insoluble dietary fibers, which disrupts the starch matrix, leading to a competitive hydration process among the insoluble dietary fibers. Consequently, this process causes the starch granules to break, releasing amylopectin, which ultimately results in an increase in apparent viscosity. This, in turn, is also responsible for slowing the digestion process. Oscillatory rheology involves subjecting a millet starch sample to oscillatory shear stress or strain. By measuring the complex modulus (G^*) and the loss tangent ($\tan \delta$) of the sample, information about the viscoelastic properties of the starch digesta can be obtained [1,32]. Changes in these parameters during digestion can provide insights into the gelatinization and retrogradation processes and the overall digestibility of starch. Small-amplitude oscillatory rheological measurements were performed for pearl millet starches from different cultivars (HC-10, HHB-67, HHB-223, HHB-226, and W-445) by Siroha et al. [34]. The rheological characteristics of starches exhibited elastic tendencies during the heating process. They also reported that the storage modulus (G') value consistently surpassed the loss modulus (G'') value across all frequency levels in the case of starch pastes, affecting the rheological properties. Moreover, the rheological characteristics of cross-linked modified starches also underwent significant changes and showed a reduction peak at G', which could be attributed to a decreased degree of swelling and subsequently diminished intergranular interactions. In the following study of Siroha et al. [35]. They revealed changes in the G' and G'' values of pearl millet starch with varying degrees of acetylation as a function of frequency. For example,

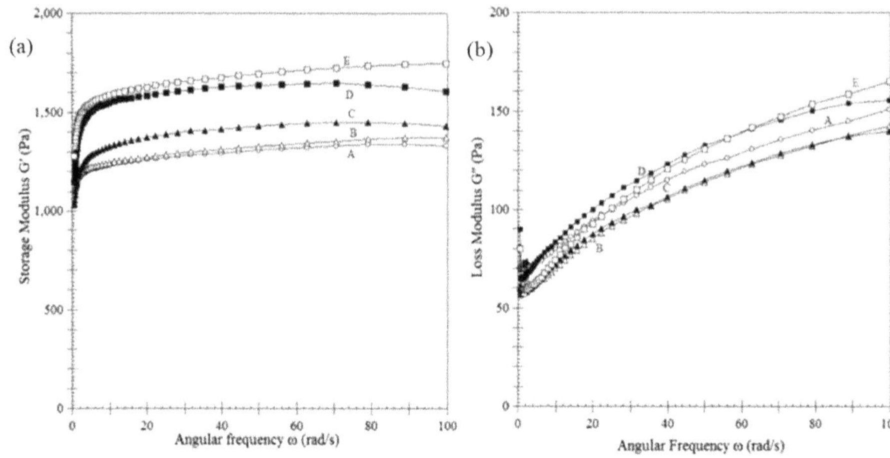

FIGURE 8.3 Storage modulus (a) value and loss modulus (b) value with respect to angular frequency dependence at 25°C for native and acetylated starches (the curves are as follows: A, native; B, 1.25% acetic anhydride; C, 2.5% acetic anhydride; D, 3.25% acetic anhydride; and E, 5.0% acetic anhydride). (The figure is reproduced from Ref. [35] with permission from Elsevier.)

values of G' ranging from 1,216 to 1,557 Pa were found to be much higher than G″ ranging from 64.7 to 76.0 Pa at all values of the testing frequency ω, reflecting the viscoelastic behavior of starch pastes (Figure 8.3a and b). However, by analyzing G' and G", the viscoelastic behavior and phase transitions occurring during digestion can be evaluated. These changes can indicate the digestibility and structural modifications of starch. It was investigated by Dangi et al. [36]. The impact of different concentrations (1% and 2% w/w) of guar gum and its hydrolysate on the rheological characteristics of pearl millet starch revealed that no intersection point between G' and G'' was identified, indicating relatively weak gel formation from a rheological perspective. However, the G' and G'' values are more pronounced in pearl millet starch-gum blends than in pure native pearl millet starch. The most substantial G' and G'' values were recorded in starch–gum blends containing 2% native guar gum, whereas the lowest values were noted in pure native starch samples. To differentiate materials on the basis of their ability to deform and recover, a frequency sweep test is applied to measure how $G'G''$ varies with ω, reflecting the viscoelastic properties of the materials [32].

The rheological properties of maize dough with varying levels of extruded finger millet (EF) were checked via a creep-recovery test at 25°C with consistent shear stress (50 Pa) for 300 s by Kumar et al. [37]. The dough with lower compliance and higher viscosity showed greater resistance to deformation than did the dough with higher compliance and lower viscosity. However, an increase in permanent deformation during the recovery phase indicated weakening of the dough structure and breaking of elastic bonds [1]. The proportion of relative elastic compliance (J_e/J_{max}) and relative viscous compliance (J_v/J_{max}) during the recovery phase demonstrated the ability of the dough to regain its original shape and stickiness, respectively [32]. However, a lower $J_e/_{Jmax}$ value in maize dough signifies reduced elasticity, which significantly increased with the incorporation of EF [37]. Similarly, the increase in J_e/J_{max} and decrease in J_v/J_{max} for the composite maize dough underscored the importance of extruded flour in enhancing the dough's elastic characteristics. In addition, steady shear flow experiments involve subjecting the millet starch digesta to a constant shear rate. The resulting stress–strain relationship can reveal the flow behavior and rheological properties of the digesta during digestion.

8.2.4 Scanning Electron Microscopy (SEM)

SEM is a tool that is used to investigate morphological changes in starch microstructures [19], including the breakdown of granules, the release of digestible products, and interactions with digestive enzymes, influencing the digestion of starchy foods. This information, which is usually combined with data from other biophysical techniques, can also provide a comprehensive understanding of millet starch digestibility, which is essential for developing millet-based food products with improved nutritional properties. In a previously mentioned study, the morphology of millet starch varieties with various amylose contents, namely, CS-16, CS-13, CS-07, JT-21, and XIN-3, was studied via SEM (Figure 8.4a(i–v)). All the samples displayed diverse shapes, including polygons, spheres, and irregular forms, and they featured pores and shallow indentations on their surfaces. These starch structural traits may result from the concentrated accumulation of endosperm or depressions caused by proteoplasts during development. In contrast to the other four types of millet, XIN-3 starch, which has a relatively high amylose content, presented fewer blemishes, fractures, and crevices on the granule surface (Figure 8.4a(i)). However, the changes in surface attributes indicated that an increased number of amylose molecule chains tended to encourage their interaction with amylopectin (for example, intertwining and framework reinforcement), consequently increasing the stiffness, density, and overall strength of starch granules [25]. Indications can

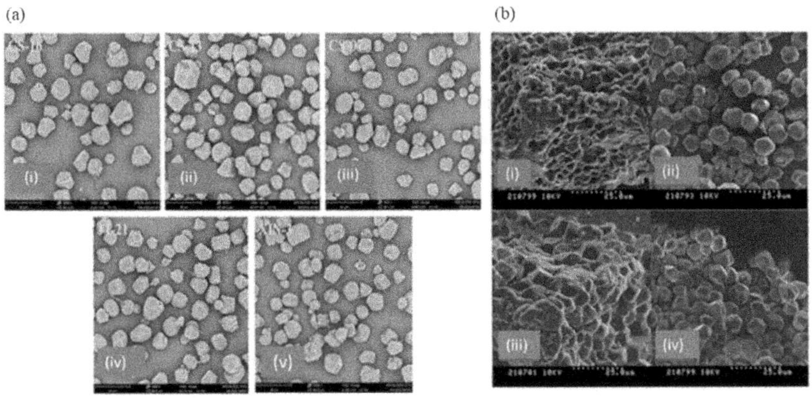

FIGURE 8.4 SEM images of (a) millet starch with various amylose contents: (i) CS-16, (ii) CS-13, (iii) CS-07, (iv) JT-21, and (v) XIN-3. (b) SEM images of various millet fraction samples; (i) MF, (ii) MF without proteins, (iii) MF without lipids, and (iv) MF without lipids and proteins. (Figures are reproduced from Refs. [25,41] with permission from Elsevier and John Wiley and Sons.)

also increase the overall surface area of starch granules. While this might seem beneficial for digestion, the accessibility of the surface within the indentations might be limited, potentially slowing down the enzymatic breakdown. However, if enzymes can reach indentations, the increased surface area could promote more efficient digestion. The shape of starch granules can also influence their digestibility. Compared with perfectly spherical granules, irregularly shaped granules may have more exposed areas, increasing their susceptibility to enzymatic attack and leading to faster digestion [38]. Starch granules with intact structures are more resistant to digestion than are damaged or disrupted granules. The semicrystalline structure of starch can protect the inner amorphous regions from enzymatic attack. Physical disruptions caused by processing or cooking can alter granule integrity and subsequently affect the digestion rate. The size of starch granules plays a significant role in their digestion [39]. The sizes of the granules in PMS, hybrid barnyard millet starch, Barnyard millet starch, and Finger millet starch were revealed by Mahajan et al. [40] and was found to be comparable, falling within the diameter range of 2.5–17 μm. The granules in foxtail millet starch and pearl millet had diameters ranging from 3 to 24 μm. The larger the starch granules are, the smaller their overall surface area-to-volume ratio. Digestive enzymes, such as amylases, work on the surface of starch granules to break down starch molecules into simpler sugars such as maltose and glucose. A larger surface area provides more sites for enzyme action, facilitating faster and more efficient digestion. Defatting and treating kodo millet flour (MF) with protease increased the rate of starch hydrolysis, with lipids having a more significant impact than proteins [41]. Proteins create a barrier around starch granules, hindering amylase access and thereby reducing starch digestibility. Lipids affect starch digestibility by forming complexes with amylose in starch. Under physiological conditions, fatty acids in lipids quickly form complexes with amyloses, creating resistant complexes. Compared with pure amylose, amylose complexed with lipids is more resistant to enzymatic breakdown (Figure 8.4b).

In Figure 8.4(b)(i), millet starch granules are clearly embedded within the protein matrix of the flour. The MF samples treated with protease displayed starch granules released from the protein matrix and exhibited large indentations, making the starch granules more accessible to hydrolytic enzymes. Removing proteins, lipids, or both significantly increased the RDS of the samples. This suggests that defatting and/or removing proteins from kodo millet would lead to a greater spike in blood glucose levels. The removal of lipids increased the RDS level of MF from 11.48% to 20.07%, whereas the removal of proteins increased it to 14.21%.

8.2.5 Fourier Transform Infrared (FTIR) Spectroscopy

FTIR spectroscopy is a valuable biophysical method used to explore the digestion of millet starch. This could provide information about the molecular structure and chemical bonds present in a sample by measuring the absorption and transmission of infrared radiation [19]. IR spectroscopy can also be used for quantitative analysis of starch digestion. By measuring the intensity of specific absorption bands related to starch or its breakdown products, researchers can estimate the extent of starch hydrolysis and determine the conversion of starch into digestible compounds [42]. By monitoring the changes in the IR spectra of starch in the presence of enzymes,

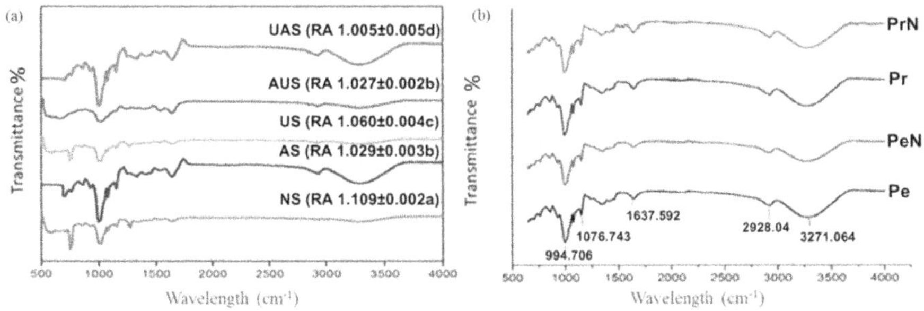

FIGURE 8.5 FT-IR spectra of (a) native and modified foxtail millet starches and (b) native pearl millet starch, nano pearl millet starch, native proso millet starch, and nano proso millet starch. (Figures are reproduced from Refs. [43,44] with permission from Elsevier.)

researchers can understand the enzymatic hydrolysis process and the specific interactions between the enzymes and starch molecules. Starch from foxtail millet was subjected to modifications by annealing (AS), ultrasonication (US), and a combination of both treatments, annealing and ultrasonication (AUS). AUS was remodified to obtain ultrasonicated-annealed starch (UAS) [43]. The FTIR spectra of both native starch (NS) and modified starch exhibited similar major peaks; however, there were discernible differences in peak amplitude (Figure 8.5a).

Amylose starches (AS and UAS) exhibit increased hydrogen bonding and α-1,4 glycosidic linkages, suggesting enhanced resistance to digestion, whereas modifications such as sonication (US and AUS) disrupt these bonds, potentially forming unstable intermediates. Jhan et al. [44] investigated the structural organization and changes in pearl millet (Pe) and proso (Pr) starch upon nanoreduction to form nanopearl millet starch (PeN) and nanoproso millet starch (PrN), respectively. All the samples presented absorption peaks in the range of 3,500–3,300 cm^{-1}, corresponding to the stretching of O–H bonds in starch (Figure 8.5b). Another absorption peak appeared in the 2,922–3,000 cm^{-1} range and was attributed to the stretching vibrations of C–H bonds in the starch glucose unit. The characteristic peak at approximately 1,600 cm^{-1} is due to tightly bound water in the starch, whereas the absorption bands at approximately 1,149 cm^{-1} and 1,076 cm^{-1} indicate the stretching of the anhydrous glucose ring's C–O and C–O–H bonds in the starch samples. The peak at approximately 990 cm^{-1} corresponds to the vibrations of the α-1,4 glycosidic linkages. There were no significant differences in the functional groups between the native and nanoreduced starch samples, suggesting that the basic chemical structure, which is primarily a physical method of nanoreduction, remained unchanged after ball milling. However, in PeN and PrN, the bands shifted to lower intensities around the characteristic α-1,4 glycosidic linkages than those in Pe and Pr, indicating the breakdown of intermolecular bonds (Figure 8.5b).

This alteration in the starch granule structure leads to decreased crystallinity and increased amorphous structure.

Hence, FTIR spectroscopy is advantageous for studying starch digestion because of its ability to analyze molecular structural changes, monitor enzymatic reactions, and track the breakdown of starch molecules into simpler components, providing insights into the digestion process in real time without the need for extensive sample preparation.

8.3 CONCLUSION

Millet starch has several unique characteristics that distinguish it from other types of starches found in various grains and cereals. These features contribute to its suitability for different food applications and potential health benefits. In the food industry, it may be used in the preparation of gluten-free bakery products, noodles, pasta, and food thickeners. However, the use of regular *in vitro* assays to predict the GI of food products has certain disadvantages. However, biophysical techniques provide valuable information about the structural and functional properties of millet starch molecules during digestion, which can shed light on various aspects of starch behavior. Some techniques, such as XRD, FTIR spectroscopy, and SEM, allow researchers to directly observe and quantify changes in starch structure after modification as well as during the digestion of millet starches. These findings provide insights into the breakdown of starch granules and the release of individual starch molecules. However, the insights gained from biophysical studies of starch digestion can have important dietary implications. Understanding the rate of starch digestion and the resulting glycemic response can help in designing healthier and more balanced diets, especially for individuals with diabetes or metabolic disorders. Despite these advantages, these methods have limitations. This may not fully replicate the complexity of *in vivo* digestion, and the transition from biophysical observations to physiological outcomes can require careful interpretation. Therefore, a combination of biophysical techniques, *in vitro* assays, and *in vivo* studies is often used to provide a comprehensive understanding of starch digestion.

STATEMENTS AND DECLARATIONS

Acknowledgments: The authors thank the Manipal Academy of Higher Education, Manipal, for providing the infrastructure and facilities. Ms. Ishita Chakraborty thanks Savitribai Jyotirao Phule Single Girl Child fellowship (Reg ID: UGCES-22-GE-KAR-F-SJSGC-943) from the University Grants Commission (UGC), Government of India, India. The authors thank the Manipal School of Life Sciences, Manipal Academy of Higher Education, Manipal, Karnataka, India, for providing the infrastructure.

Funding: We thank the ASEAN-India Science & Technology Development Fund (AISTDF), Anusandhan National Research Foundation (ANRF), Government of India, India (Project Number- ANRF/F/84/2025-2026), and the Indian Council of Medical Research (ICMR) (Project Number-Dev/SG-00170/2024), Government of India, India, for financial support.

Author Contributions: NB and BD: Conception, design, and preparation of the first draft of the manuscript. IC, MHR, and NS: Review. NM: Review and funding acquisition.

Competing Interests: The authors declare no financial or nonfinancial competing interests.

Availability of data and material: Data sharing is not applicable to this article, as no new data were created or analyzed in this review.

REFERENCES

1. Rahman, M. H.; Zhang, M.; Sun, H. N.; Mu, T. H. Comparative study of thermomechanical, rheological, and structural properties of gluten-free model doughs from high hydrostatic pressure treated maize, potato, and sweet potato starches. *Int. J. Biol. Macromol.* 2022, 204, 725–733. doi:10.1016/j.ijbiomac.2022.01.164.
2. Pooja, N.; Chakraborty, I.; Rahman, M. H.; Mazumder, N. An insight on sources and biodegradation of bioplastics: a review. *3 Biotech.* 2023, 13(7), 220.
3. Sharma, R.; Sharma, S.; Dar, B. N.; Singh, B. Millets as potential nutri-cereals: a review of nutrient composition, phytochemical profile and techno-functionality. *Int. J. Food Sci. Technol.* 2021, 56(8), 3703–3718. doi:10.1111/ijfs.15044.
4. Chauhan, M.; Sonawane, S. K.; Arya, S. S. Nutritional and nutraceutical properties of millets: a review. *Clin. J. Nutr. Diet.* 2018, 1(1), 1–10.
5. Qiao, J.; Jia, M.; Niu, J.; Zhang, Z.; Xing, B.; Liang, Y.; Li, H.; Zhang, Y.; Ren, G.; Qin, P., et al. Amylopectin chain length distributions and amylose content are determinants of viscoelasticity and digestibility differences in mung bean starch and proso millet starch. *Int. J. Biol. Macromol.* 2024, 267, 131488. doi:10.1016/j.ijbiomac.2024.131488.
6. Tyl, C.; Marti, A.; Hayek, J.; Anderson, J.; Ismail, B. P. Effect of growing location and variety on nutritional and functional properties of proso millet (*Panicum miliaceum*) grown as a double crop. *Cereal Chem.* 2018, 95(2), 288–301. doi:10.1002/cche.10028.
7. Arp, C. G.; Correa, M. J.; Ferrero, C. Resistant starches: a smart alternative for the development of functional bread and other starch-based foods. *Food Hydrocoll.* 2021, 121, 106949. doi:10.1016/j.foodhyd.2021.106949.
8. Lal, M. K.; Singh, B.; Sharma, S.; Singh. M. P.; Kumar, A. Glycemic index of starchy crops and factors affecting its digestibility: a review. *Trends Food Sci. Technol.* 2021, 111, 741–755. doi:10.1016/j.tifs.2021.02.067.
9. Geetha, K.; Yankanchi, G. M.; Hulamani, S.; Hiremath, N. Glycemic index of millet based food mix and its effect on pre diabetic subjects. *J. Food Sci. Technol.* 2020, 57, 2732–2738.
10. Nidhi, K.; Alka, G.; Saritha, S. Development of low glycemic foods with the use of pearl millet and finger millet. *Int. J. Sci. Res.* 2014, 3, 193–195.
11. Narayanan, J.; Sanjeevi, V.; Rohini, U.; Trueman, P.; Viswanathan, V. Postprandial glycemic response of foxtail millet dosa in comparison to a rice dosa in patients with type 2 diabetes. *Indian J. Med. Res.* 2016, 144(5), 712. doi:10.4103/ijmr.IJMR_551_15.
12. Chhavi, A.; Saritha, S. Evaluation of composite millet breads for sensory and nutritional qualities and glycemic response. *Mal. J. Nutr.* 2012, 18(1).
13. Ulfath, T. K.; Nikhitha, M.; Praditi, S.; Kavitha, G. S.; Myrene, R. D.; Sangeetha, P. Development of value added product to reduce the risk of hypoglycemia in diabetic population. *Eur. J. Pharm. Med. Res.* 2016, 3(3), 514–518.
14. Patil, K. B.; Chimmad, B. V.; Itagi, S. Glycemic index and quality evaluation of little millet (*Panicum miliare*) flakes with enhanced shelf life. *J. Food Sci. Technol.* 2015, 52, 6078–6082.
15. Anju, T.; Saritha, S. Suitability of foxtail millet (*Setaria italica*) and barnyard millet (*Echinochloa frumentacea*) for development of low glycemic index biscuits. *Mal. J. Nutr.* 2010, 16(3), 361–368.
16. Thathola, A.; Srivastava, S.; Singh, G. Effect of foxtail millet (*Setaria italica*) supplementation on serum glucose, serum lipids and glycosylated hemoglobin in type 2 diabetics. *Diabetol. Croat.* 2011, 40(1), 23–29.
17. Tiwari, N.; Srivasi, S. Effect of finger millet (*Eleusine coracana*) buns supplementation on serum glucose and serum lipids level in type 2 diabetics. *Asian J. Dairy Food Res.* 2017, 36(4), 337–340. doi:10.18805/ajdfr.DR-933.

18. Hayes, A. M.; Swackhamer, C.; Mennah-Govela, Y. A.; Martinez, M. M.; Diatta, A.; Bornhorst, G. M.; Hamaker, B. R. Pearl millet (*Pennisetum glaucum*) couscous breaks down faster than wheat couscous in the Human Gastric Simulator, although has slower starch hydrolysis. *Food Funct.* 2020, 11(1), 111–122. doi:10.1039/C9FO01461F.
19. Rahman, M. H.; Mu, T. H.; Zhang, M.; Ma, M.; Sun, H. N. Comparative study of the effects of high hydrostatic pressure on physicochemical, thermal, and structural properties of maize, potato, and sweet potato starches. *J. Food Process. Preserv.* 2020, 44(11). doi:10.1111/jfpp.14852.
20. Punia, S. Barley starch: structure, properties and *in vitro* digestibility: a review. *Int. J. Biol. Macromol.* 2020, 155, 868–875. doi:10.1016/j.ijbiomac.2019.11.219.
21. Wang, S.; Copeland, L. Effect of acid hydrolysis on starch structure and functionality: a review. *Crit. Rev. Food Sci. Nutr.* 2015, 55(8), 1081–1097. doi:10.1080/10408398.2012.684551.
22. Li, Y.; Hu, A.; Wang, X.; Zheng, J. Physicochemical and *in vitro* digestion of millet starch: effect of moisture content in microwave. *Int. J. Biol. Macromol.* 2019, 134, 308–315. doi:10.1016/j.ijbiomac.2019.05.046.
23. Dong, J.; Huang, L.; Chen, W.; Zhu, Y.; Dun, B.; Shen, R. Effect of heat-moisture treatments on digestibility and physicochemical property of whole quinoa flour. *Foods.* 2021, 10(12), 3042. doi:10.3390/foods10123042.
24. Mahajan, P.; Bera, M. B.; Panesar, P. S.; Chauhan, A. Millet starch: a review. *Int. J. Biol. Macromol.* 2021, 180, 61–79. doi:10.1016/j.ijbiomac.2021.03.063.
25. Shi, P.; Zhao, Y.; Qin, F.; Liu, K.; Wang, H. Understanding the multiscale structure and physicochemical properties of millet starch with varied amylose content. *Food Chem.* 2023, 410, 135422. doi:10.1016/j.foodchem.2023.135422.
26. Xiao, Y.; Zheng, M.; Yang, S.; Li, Z.; Liu, M.; Yang, X.; Lin, N.; Liu, J. Physicochemical properties and *in vitro* digestibility of proso millet starch after addition of Proanthocyanidins. *Int. J. Biol. Macromol.* 2021, 168, 784–791. doi:10.1016/j.ijbiomac.2020.11.136.
27. Chakraborty, I.; N, P.; Mal, S. S.; Paul, U. C.; Rahman, M. H.; Mazumder, N. An insight into the gelatinization properties influencing the modified starches used in food industry: a review. *Food Bioproc. Tech.* 2022, 15(6), 1195–1223.
28. Zhao, X.; Wang, Y.; Li, D.; Wang, L. Insight into the biphasic transition of heat-moisture treated waxy maize starch through controlled gelatinization. *Food Chem.* 2021, 341, 128214. doi:10.1016/j.foodchem.2020.128214.
29. Wang, J.; Ye, Y.; Chai, Z.; Liu, H.; Wei, X.; Ye, X.; Tian, J.; Fang, H. Physicochemical properties and *in vitro* digestibility of proso millet starch modified by heat-moisture treatment and annealing processing. *Int. J. Biol. Macromol.* 2023, 235, 123829. doi:10.1016/j.ijbiomac.2023.123829.
30. Wang, L.; Chen, J.; Lu, S.; Xiao, P.; Li, C.; Yi, C. Structural characterization, physicochemical properties and *in vitro* digestion of finger millet–resistant starch prepared by different methods. *Int. J. Food Sci. Technol.* 2022, 57(1), 416–425. doi:10.1111/ijfs.15432.
31. Navyashree, N.; Sengar, A. S.; Sunil, C. K.; Venkatachalapathy, N. White finger millet (KMR-340): a comparative study to determine the effect of processing and their characterization. *Food Chem.* 2022, 374, 131665. doi:10.1016/j.foodchem.2021.131665.
32. Rahman, M. H.; Sun, H. N.; Zhang, M.; Mu, T. H.; Khan, N. M. Effect of sucrose, trehalose, maltose and xylose on rheology, water mobility and microstructure of gluten-free model dough based on high hydrostatic pressure treated starches. *Int. J. Biol. Macromol.* 2023, 231, 123184. doi:10.1016/j.ijbiomac.2023.123184.
33. Chakraborty, S. K.; Kotwaliwale, N.; Navale, S. A. Rheological characterization of gluten free millet flour dough. *J. Food Meas. Charact.* 2018, 12, 1195–1202.
34. Siroha, A. K.; Sandhu, K. S. Physicochemical, rheological, morphological, and *in vitro* digestibility properties of cross-linked starch from pearl millet cultivars. *Int. J. Food Prop.* 2018, 21(1), 1371–1385. doi:10.1080/10942912.2018.1489841.

35. Siroha, A. K.; Sandhu, K. S.; Kaur, M.; Kaur, V. Physicochemical, rheological, morphological and *in vitro* digestibility properties of pearl millet starch modified at varying levels of acetylation. *Int. J. Biol. Macromol.* 2019, 131, 1077–1083. doi:10.1016/j.ijbiomac.2019.03.179.
36. Dangi, N.; Yadav, B. S.; Yadav, R. B. Pasting, rheological, thermal and gel textural properties of pearl millet starch as modified by guar gum and its acid hydrolysate. *Int. J. Biol. Macromol.* 2019, 139, 387–396. doi:10.1016/j.ijbiomac.2019.08.012.
37. Kumar, P.; Kaur, C.; Sethi, S.; Kaur Jambh, H. Effect of extruded finger millet on dough rheology and functional quality of pearl millet-based unleavened flatbread. *Cereal Chem.* 2020, 97(5), 991–998. doi:10.1002/cche.10321.
38. Li, M.; Daygon, V. D.; Solah, V.; Dhital S. Starch granule size: does it matter? *Crit. Rev. Food Sci. Nutr.* 2023, 63(19), 3683–3703. doi:10.1080/10408398.2021.1992607.
39. Conde, A. G. Starch Granule and Crystalline Structures With Relation to Enzyme Digestibility. Ph.D Dissertation, University of Arkansas, 2023.
40. Mahajan, P.; Bera, M. B.; Panesar, P. S. Structural, functional, textural characterization and *in vitro* starch digestibility of underutilized Kutki millet (*Panicum sumatrense*) flour. *J. Food Meas. Charact.* 2022, 16(6), 4800–4812.
41. Annor, G. A.; Marcone, M.; Bertoft, E.; Seetharaman, K. *In vitro* starch digestibility and expected glycemic index of kodo millet (*Paspalum scrobiculatum*) as affected by starch–protein–lipid interactions. *Cereal Chem.* 2013, 90(3), 211–217. doi:10.1094/CCHEM-06-12-0074-R.
42. Nag, N.; Sasidharan, S.; Saudagar, P.; Tripathi, T. Fundamentals of spectroscopy for biomolecular structure and dynamics. *Adv. Spectrosc. Methods Study Biomol. Struct. Dyn.* 2023, 1–35. doi:10.1016/B978-0-323-99127-8.00002-7.
43. Babu, A. S.; Mohan, R. J.; Parimalavalli, R. Effect of single and dual-modifications on stability and structural characteristics of foxtail millet starch. *Food Chem.* 2019, 271, 457–465. doi:10.1016/j.foodchem.2018.07.197.
44. Jhan, F.; Shah, A.; Gani, A.; Ahmad, M.; Noor, N. Nanoreduction of starch from underutilized millets: effect on structural, thermal, morphological and nutraceutical properties. *Int. J. Biol. Macromol.* 2020, 159, 1113–1121. doi:10.1016/j.ijbiomac.2020.05.020.

9 Advancement in Virus Research with an Adaptation of Expansion Microscopy

Sruti Bagchi Ghosh
Calcutta Institute of Pharmaceutical Technology & AHS

Arghya Bhattacharya
Bengal School of Technology

Maryam Abdul Ajees and Gagan Raju
Manipal Academy of Higher Education

9.1 INTRODUCTION

Expansion microscopy (ExM) is a sample preparation tool for biological samples that allows investigators to identify small cellular structures by expanding them via a polymer system. This technique was first proposed by Edward Boyden et al. in 2015 [1]; therefore, Ex M is a relatively new technique in the biological world. It physically expands preserved cells and tissues isotopically via a chemical process; therefore, 3D nanoscale resolution [2] imaging of samples becomes possible on a common, fast, diffraction-limited microscope. Labels are chemically equipped with anchors that enable them to bind to a web of swellable polymers (or hydrogels) that are densely [3] and evenly synthesized throughout the sample (a process called polymerization or gelation). The spacing between the polymers was estimated to be approximately a few nanometers. The polymer is densely cross-linked, so as the polymer threads swell apart from each other, the biomolecules or labels retain their spatial organization relative to one another. The polymer-embedded samples are mechanically homogenized (by high-temperature denaturation proteolysis or other processes) and then swollen by adding a solvent (water) when sodium polyacrylate polymers have been used in ExM [1–3]. One limitation of these gels is that they are not completely uniform in structure or density. This irregularity leads to small distortions in the shape of the sample when it is expanded, limiting the accuracy that can be achieved. To overcome this, a new gel called the tetragel has been developed, which forms a more predictable structure. By combining tetrahedral PEG molecules with tetrahedral sodium polyacrylates, researchers can create a lattice-like structure that is much more uniform than previously used free-radical-synthesized sodium polyacrylate hydrogels [4]. Since

 DOI: 10.1201/9781003570011-9

the process involves the synthesis of swellable polyelectrolyte hydrogels in a dense and even fashion throughout the samples, after physical magnification, molecules within a diffraction-limited region are isotropically separated in space such that they are resolvable even by conventional, ubiquitous diffraction-limited microscopes. This approach, named ExM, involves two different techniques and two different fields: the physics of swellable polyelectrolyte hydrogels and the embedding of preserved biological samples in polymer hydrogels for imaging purposes [5,6], pioneered by Peter Hausen and Christine Dreyer in the early 1980s. Flourescence imaging using ExM for examining microtubules in fixed human embryonic kidney (HEK) 293 cells labeled with the trifunctional label and imaging with confocal laser scanning microscopy before and after ExM processing was carried out. The post-ExM image was registered to the pre-ExM image via similarity transformation, resulting in a visually indistinguishable image [5,6].

9.2 EXM STEPS

1. **Cell culture**: Cell culture steps are performed in biosafety cabinets to ensure sterility. The protocol followed will depend on the sample of interest.
2. **Anchoring**: Biomolecules and/or labels are covalently equipped with molecular handles. These cross-linkers allow the polymer matrix to exert force on the biomolecules [7].
3. **Gelation**: To prevent damage to cells, the polymer chain is broken down into its constituent parts (building blocks). The sample is immersed in a monomer solution and a highly penetrating hydrogel. Once inside the cells, a chemical reaction is triggered because the monomers bind to form a network of the desired polymer [7].
4. **Homogenization**: The aim is to avoid sample distortion and ensure that the sample organization is kept intact. The sample is chemically disrupted either by enzymatic digestion or by heat/detergent treatment. The method used will depend on the sample and molecules to be visualized [7].
5. **Gel expansion**: The sample is immersed in water, which diffuses into the polyelectrolyte hydrogel through osmotic force, causing the polymer to expand. This pulls apart anchored biomolecules in an isotropic manner, creating enormous gaps between each biomolecule. The spatial organization of the expanded sample is preserved, allowing nanoscale imaging with standard fluorescence microscopes [7].
6. After expansion, further application of labels and/or amplification reactions to improve visualization may be performed, if desired.

ExM uses fluorescent dyes along with antibodies that are compatible with this process, except for cyanine dyes, which are destroyed during polymerization. Excellent alternatives to cyanine dyes are now available in spectral bands where cyanine dyes have been popular in the past. Most GFP-like (i.e., beta barrel-structured) fluorescent proteins are also preserved in this process. In contrast, infrared fluorescent proteins based on bacteriophytochromes are largely destroyed. In addition to antibodies, other protease-resistant labels, such as streptavidin, can also be applied before expansion. This has been used to visualize posttranslational modifications such as S-nitrosylation, which can

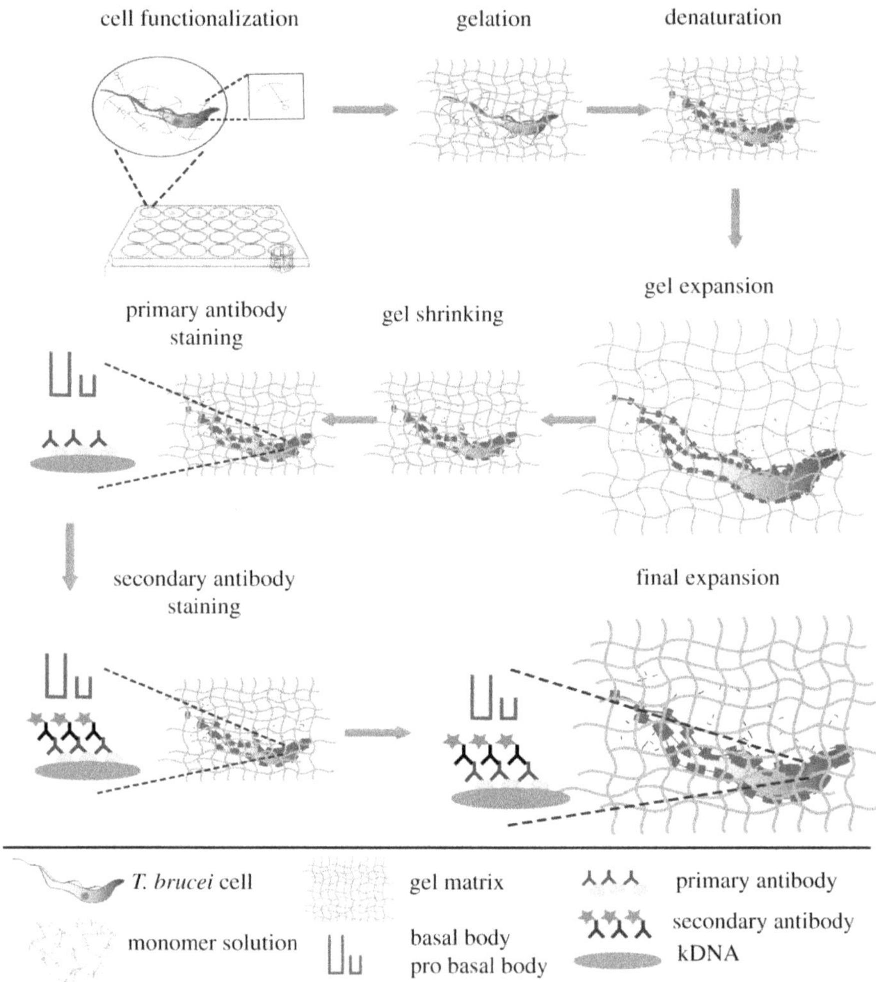

FIGURE 9.1 Detailed stepwise workflow of expansion microscopy. (The figure is reproduced with permission from reference [7].)

be marked by biotin-bearing small-molecule tags [4]. Images were transformed via a similarity transformation, resulting in visually indistinguishable images. To quantify the isotropy of ExM, the deformation vector field between the images was calculated via a nonrigid registration process. From this vector field, the root-mean-square (RMS) error of feature measurements post-ExM has been quantified (Figure 9.1).

9.3 SUBSETS OF EXM

9.3.1 Iterative ExM (iExM)

iExM [5], developed by Gao et al. (2017), involves a biological sample, which was first expanded by ExM, then a second swellable gel was formed in the space opened by the first expansion, and then the sample was expanded a second time. This double expansion process results in a linear expansion factor of ~$4.5 \times 4.5 = 20\times$ and an effective resolution of ~25 nm after two rounds of expansion (larger than the expected value of ~300 nm/20 = 15 nm due to the size of the antibodies and DNA linkers). The iExM is sufficient for resolving proteins within synapses in 3D, as well as very fine parts of neurons (such as dendritic spines in the mouse brain circuitry) [5,6].

9.3.2 Cryofixation with ExM

To combine cryofixation with ExM (Cryo-ExM), we capitalized on the well-known electron microscopy (EM) approach of cryofixation, followed by freeze substitution, which consists of embedding a vitrified sample in a resin before polymerization and subsequently processing for EM studies. This approach allows for high levels of cellular architecture preservation and far surpasses chemical fixation [3,8]. Thus, the EM resin could be replaced by the ExM hydrogel. To achieve this goal, biological samples grown on glass coverslips are first cryofixed by rapidly plunging them into liquid ethane-propane to form vitreous ice, a technique commonly used in cryo-FIB-milling studies. The vitrified coverslips were then incubated in acetone precooled with liquid nitrogen (−180°C) and placed on dry ice (−80°C) overnight. This process allows for the evaporation of dry ice and a gradual rise in sample temperature from −180°C to 0°C. During this gradual warming, acetone slowly replaces the water content of the sample, which is crucial for preserving the cell architecture [9]. Following dehydration, the samples undergo rehydration through sequential baths of ethanol mixed with increasing amounts of water. Finally, the samples are processed for ExM hydrogel embedding via the U-ExM protocol.

9.3.3 Scanning Joule ExM (SJEM)

This technique is a form of scanning probe microscopy based on atomic force microscopy that maps the temperature distribution along the surface. In this case, resolutions of 10 nm are achieved.

9.3.4 Protein Retention ExM

The first version of ExM5 (which is now called ExM 1.0) requires end users to synthesize their labeling reagents for tagging proteins of interest, which limits deployment ability in everyday biology contexts. However, it was used in one study for the visualization of Wassie et al. [9]. In contrast, in protein retention ExM (proExM9), only off-the-shelf chemicals are needed. In proExM, the succinimidyl ester of 6-((acryloyl)amino) hexanoic acid (acryloyl-X, SE; abbreviated AcX for short) is applied to a preserved sample equipped with amines on proteins with a polymerizable carbon double bond, which in turn enables the thus-equipped proteins to be anchored to the polymer during the polymerization step. The procedures below can take, depending on the protocol used and the size of the tissue to be processed, anywhere from a day or two to a week, with much of the time required being incubation times that permit chemical access of the tissue to occur. Fluorescent labels for visualization of proteins, for example, fluorescent antibodies applied to specific proteins to enable them to be visualized, can be applied either before or after expansion.

9.3.5 Expansion Pathology (ExPath)

ExPath is an extension of the preexpansion antibody administration proExM protocol that enables a variety of tissue formats common in human clinical and pathological settings to be expanded. Formalin-fixed paraffin-embedded (FFPE), hematoxylin and eosin (H&E)-stained, and fresh frozen thin-sliced samples can be preprocessed so that they can enter the proExM pipeline. ExPath precedes proExM in a series of steps—removal of paraffin using xylene and antigen retrieval in hot sodium citrate—which help make clinically prepared tissues compatible with proExM. This form of proExM has increased levels of EDTA, which enhances tissue digestion. This is particularly important because clinical samples are often heavily fixed due to prolonged exposure to formalin [1]. During this process, H&E stains are removed. For fresh frozen sections (e.g., acetone-fixed), the amount of AcX should be reduced to avoid overanchoring the samples. This adjustment is necessary because fresh-frozen sections have more free amines available to bind anchors, as there are no aldehyde fixatives occupying the amines. The ExPath protocol has been successfully applied to human skin, lung, liver, breast, prostate, pancreas, ovary, kidney, and other tissues preserved through various methods. ExPath is compatible with antibody staining, as evidenced by the labeling of numerous markers (e.g., vimentin and keratin) across diverse tissue types. Additionally, postexpansion DNA FISH has been demonstrated, including the labeling of Her2 in Her2− and Her2+ breast cancer samples from human patients. ExPath significantly reduces autofluorescence—up to an order of magnitude—by diluting or eliminating autofluorescence contributors while preserving covalently anchored information on the polymer [1]. We have shown that ExPath can resolve structures in human samples that typically require electron microscopy in pathology labs. For example, it can identify podocyte foot processes, which are effective in patients with nephrotic kidney diseases and usually necessitate electron microscopy for diagnosis. Although electron microscopy offers superior resolution compared with current ExM protocols, ExPath has a significantly shorter processing time. Furthermore, the skills and equipment required to perform ExPath are considerably less demanding than those needed for electron microscopy [1].

9.3.6 Expansion Fluorescence In Situ Hybridization (ExFISH)

ExFISH is a version of ExM, in which RNA and DNA from biological samples are retained during the ExM process [1]. In ExFISH, RNAs are covalently anchored to the hydrogel with a small molecule that we call Label X (and which is made by mixing two commercially available reagents) that binds to guanine and the hydrogel. By applying the protein-anchoring reagent (AcX) simultaneously, both RNAs and proteins can be anchored to the hydrogel for dual protein/RNA visualization. For imaging, RNAs can be labeled with FISH probes after expansion. As noted earlier, ExFISH lends itself easily to postexpansion fluorescence amplification by applying HCR so that, rather than imaging single or small numbers of fluorophores in intact tissue, each biomolecular target is coupled to dozens or perhaps hundreds of fluorophores for easier detection, even in nanoscale compartments such as dendritic spines in the intact mouse brain circuitry [10]. In the future, ExFISH may help with temporally multiplexed FISH probing, which has recently been adapted for use with intact tissues. In light sheet microscopy,

TABLE 9.1
Types of Expansion Microscopies

Types of ExM	Principle	Advantages	Disadvantages	Reference
iExM	Biological samples of interest are permeated by swellable material. After the sample is embedded in the swellable material sample can be isotropically expanded in three dimensions.	The highest expansion achieved thus revealing nanostructures in brain tissue via protein decrowding.	Preserving the chromatin ultrastructure is challenging.	[4–6]
Cryo-ExM	The process involves embedding a vitrified specimen in a resin before polymerization and subsequently processing for EM studies.	Here, water of the sample is slowly replaced by acetone, a crucial process ensuring proper cell architecture preservation.	Radiation damage is the primary limiting factor, followed by poor signal/noise ratio in images recorded with optimal electron exposure.	[3,8,9]
SJEM	This process maps the temperature distribution along with the surface.	In this case, resolutions to the tune of 10 nm are achieved.	Only the surface can be mapped.	[11,12,8]
Pro ExM	Applying the succinimidyl ester of 6-((acryloyl) amino) hexanoic acid (acryloyl-X), to a preserved specimen equips amines on proteins with a polymerizable carbon double bond, which in turn enables the thus-equipped proteins to be anchored to the polymer during the polymerization step.	Fluorescent labels for visualization of proteins—for example, fluorescent antibodies applied to specific proteins to enable them to be visualized—can be applied either before or after expansion.	The procedures below can take, depending on the protocol used and the size of the tissue to be processed, anywhere from a day or two, to a week, with much of the time needed being incubation times that allow chemical access of the tissue to occur.	[5]
ExPath	Formalin-fixed paraffin-embedded (FFPE), hematoxylin and eosin (H&E)-stained, and fresh frozen thin-sliced specimens can be preprocessed so that they can enter the proExM pipeline.	The requirement for skills and equipment to perform ExPath is much less demanding in comparison with those required for electron microscopy.	The requirement for skills and equipment to perform ExPath is much less demanding in comparison with those required for electron microscopy.	[13]
Ex FISH	RNAs are covalently anchored to the hydrogel with a small molecule that we call Label X (and which is made by mixing two commercially available reagents) that binds to guanine and to the hydrogel.	By applying the protein-anchoring reagent (AcX) simultaneously, you can anchor both RNAs and proteins to the hydrogel for dual protein/RNA visualization.	Expensive and time-consuming.	[10]

the optical pathways of excitation and detection are geometrically decoupled such that the sample is illuminated with a thin sheet of laser light, and detection is performed along an axis orthogonal to the illumination. This separation of excitation and detection light paths maximizes detection efficiency by minimizing out-of-focus fluorescence. Furthermore, the axial confinement of the excitation results in dramatically reduced photobleaching for detection, and fast cameras with high quantum efficiencies can be used; thus, the imaging speed can be increased by orders of magnitude compared with that of point-scanning techniques [10] (Table 9.1).

9.4 APPLICATIONS IN VIROLOGY RESEARCH

9.4.1 Influenza Vaccine (LAIV) under ExM-Light Sheet Microscopy

The principle of light sheet microscopy was developed more than a century ago by Siedentop and Zsigmondy [10,14] and then termed ultramicroscopy. This method was rediscovered by developmental biologists at the beginning of the 21st century, and its popularity has since increased. The high spatiotemporal resolution of light sheet microscopy has been impressively demonstrated for high-speed imaging of embryonal development, neural activity, cardiac dynamics, and physiologically representative subcellular imaging. Despite these advantages, a complication that regularly arises in light sheet microscopy concerns the mounting of the samples. The placement of samples on the microscope for imaging must be compatible with the two-objective configuration of the method, and solutions often need to be customized for the specific experimental configuration in use. Currently, commercial setups are available, and much effort is still invested in the technical development of designs that are simpler to use and that improve the versatility and utility of the method. The suitability of ExM and light sheet microscopy for imaging virus-infected samples was evaluated using human A549 cells infected with the live attenuated influenza vaccine (LAIV), a modified low-virulence variant of the influenza A virus [15,16]. This vaccine is the basis for flu vaccine formulations marketed under the names Fluenz in Europe and FluMist in the USA and Canada. ExM is particularly well suited for studying host–cell–vaccine interactions, aiding in understanding the underlying biomolecular mechanisms of the vaccine. Superresolution optical imaging is essential for analyzing the interaction between viral proteins and substructures with cell organelles, which often have dimensions comparable to or smaller than the diffraction limit of light (approximately 300 nm). However, imaging entire infected cells in multiple colors to construct a complete 3D representation via alternative superresolution techniques, such as 3D STORM and STED, is challenging [17,18] because of their prolonged acquisition times and susceptibility to photobleaching [19–23].

9.4.2 SARS CoV-2 Virus under STORM and PALM

Understanding the basic biology and molecular mechanisms behind viruses and their impact on cells is a key area of interest for researchers. High-resolution imaging plays a crucial role in this understanding. Superresolution microscopy has proven to be a successful tool in virology, demonstrating the capabilities of well-established modern techniques and widely available instruments. The question is whether these techniques can help the world understand and overcome the SARS-CoV-2 coronavirus, the cause of the COVID-19 pandemic. Jakub Chojnacki from the IrsiCaixa AIDS Research Institute in Barcelona, Spain, noted that viruses such as HIV, influenza, and SARS-CoV-2 are approximately 100nm in diameter, which is below the standard fluorescence microscopy resolution limit of approximately 240nm. As a result, studying virus structures was once the domain of EM, which is often difficult and expensive. However, in recent decades, scientists have leveraged careful control of the “on” and “off” states of fluorophores and precise localization techniques to surpass the resolution limit of fluorescence microscopy. Despite advancements, no single superresolution fluorescence microscopy technique exists in all areas, as observed by Chojnacki [24,25]. Stimulated emission depletion (STED) microscopy, for example, is typically the fastest technique for imaging structures below 100nm [18], such as viruses. STED employs one focused laser beam to excite fluorophores and a second doughnut-shaped depletion beam to quench fluorescence at the edges of the excitation spot. Scanning these beams across samples results in high-resolution images with nanometer-scale pixels, although intense light can cause photobleaching. Other prominent superresolution methods include stochastic optical reconstruction microscopy (STORM) [17] and photoactivated localization microscopy (PALM). Both techniques use lasers to randomly activate and locate small numbers of fluorophores, progressively compiling superresolution images. However, owing to the reconstruction times, these techniques are not suitable for observing very fast events and are usually applied to fixed, inactive samples when studying viruses. For investigation, virus components must be tagged with fluorescent antibodies, which Chojnacki noted was not a major difficulty when using STED. Importantly, STED allows for live-cell experiments, which is crucial because virus entry and assembly are not discrete steps but rather a series of overlapping dynamic events. STED becomes even more powerful when combined with fluorescence correlation spectroscopy (FCS), which analyzes variations in fluorescence to interpret the mobility of the studied molecules [24]. While working in Christian Eggeling’s laboratory at the University of Oxford, UK, Chojnacki and colleagues at the Institute of Research in Infectious Diseases (IRIM) in Montpellier, France, utilized this approach to study the lipid envelope surrounding some viruses in detail. This combination of STED and FCS allows for a more dynamic and detailed understanding of the virus assembly process, highlighting the advantages of studying viruses in their native, live-cell environments rather than in fixed samples.

9.4.3 Herpes Simplex Virus Type 1 (HSV-1)

Segregating Lipids under STED-ExMKOKounatidis et al. identified virus assembly platforms that were 50–80nm in size on immune cell surfaces. They reported that the virus was able to segregate certain lipids and not others during their formation at the cell surface in a live single cell. STED/FCS allows them to work on a single-virus platform at the surface of a single cell. This was the nanoview of viral life [24]. Using PALM, the IRIM team was also able to visualize HIV assembly in immune cell molecules. Muriaux’s colleague Cyril Favard said that the team then exploited a Bayesian framework method developed by Jean-Baptiste Masson at the Pasteur Institute in Paris, France. This provided a robust statistical classification of either short trajectories of single molecules or longer trajectories with missing steps [24]. This provided a robust statistical classification of either short trajectories of single molecules or longer trajectories with missing steps. In the future, correlative microscopy, combining optical methods with other techniques, such as electron microscopy or atomic force microscopy, may provide more insights. Both SARS-CoV-2 and HIV are remarkably similar in terms of particle size. They are ~35–140 nm in diameter and, as such, cannot be seen by classic optical transmission. STED and PALM can be used for structural purposes with antibody labeling on fixed samples, both with approximately 20 nm precision, but with different chromophores, she stressed. While PALM can follow single proteins in living samples, there is no equivalent for lipids, added Favard, which is why the team switched to STED-FCS. The team is studying SARS-CoV-2 mobility in solution, where it moves inside cells, and how many viruses are present within cells. Researchers have demonstrated the accuracy of this approach by using it to expand particles of herpes simplex virus type 1 (HSV-1), which have a distinctive spherical shape. After expanding the virus particles, the researchers compared the shapes to the shapes obtained by electron microscopy and reported that the distortion was lower than that seen with previous versions of ExM, allowing them to achieve an accuracy of approximately 10 nm. MIT, USA, engineers have developed a method to generate images of viral particles via light microscopy. In these images, white dots indicate clusters of proteins embedded in the viral envelope, whereas red dots represent the computationally extracted central positions of these clusters. This innovation is credited to Ruixuan Gao and the Boyden Lab. Researchers have examined how the arrangement of these proteins changes during expansion and assessed the proximity of these arrangements to a spherical shape [24]. This process allowed them to validate their method and determine how accurately they could preserve the nanostructure and relative spatial arrangements of the molecules.

9.4.4 N Protein Compartment of SARS-CoV-2

Researchers have also utilized new hydrogels to expand cells, including human kidney cells and mouse brain cells. They are now working on improving the accuracy to the point where they can image individual molecules within these cells. One limitation in achieving this degree of accuracy is the size of the antibodies used to label molecules in the cell, which are ~10–20 nm long. To image individual molecules, researchers need to develop smaller labels or add labels after the expansion is complete. They are also investigating whether other types of polymers or modified versions of the tetragel polymer could help achieve greater accuracy. If they can attain single-molecule accuracy, many new frontiers could be explored, such as Boyden [1,26]. For example, scientists could observe how different molecules interact, providing insights into cell signaling pathways, immune response activation, synaptic communication, drug–target interactions, and numerous other biological phenomena. SARS-CoV-2 is an enveloped virus that enters the respiratory tract via interactions between the receptor-binding domain on the SARS-CoV-2 spike protein and the host cell receptor angiotensin-converting enzyme 2. Upon infection, the virus induces the formation of replication organelles (ROs), such as double-membrane vesicles (DMVs), which protect viral RNA from degradation by cellular RNAs during replication [26]. After using volumetric imaging, the authors confirmed that at least one dsRNA focus is usually linked to the outer layer of the N protein compartment, which has many fused sections. A combination of expansion and light-sheet microscopy reveals the interaction between dsRNA (cyan) and the nucleocapsid protein N (magenta), demonstrating that dsRNA foci reside within the layers of the N compartments [3,25]. The scale bar represents 5 μM (considering an expansion factor of 4) (Figure 9.2).

9.5 LIMITATIONS OF EXM

Viruses are ever-present in our daily lives, ranging from Rhinovirus, which causes the common cold, to COVID-19 and HIV-1, which have triggered global health crises. These invisible hazards necessitate detailed study via powerful techniques because of their incredibly small size, rapid mutation, and adaptive capabilities. To characterize the subtle changes in their structures and identify key factors in their infection cycles [22,23], single-particle techniques with high sensitivity and resolution are needed. Superresolution microscopy (SRM) has emerged as an ideal technique for characterizing viruses. It offers high specificity, high resolution, live-cell imaging, and 3D multicolor images, making it perfectly suited for new viral studies [20]. Over the past decade, significant advancements in SRM have enabled researchers to study viral infections via various superresolution techniques [24] and approaches, addressing gaps in current viral models. This approach has been extensively reviewed, particularly in studies focusing on HIV-1 with SRM [20–25].

9.6 CONCLUSIONS

Viral studies using SRM have demonstrated how each step of the viral cycle and antiviral design can be examined in unparalleled detail, resolving models that are impossible to visualize with conventional microscopy. Analysis of viral particles has revealed the distribution of key proteins in various viruses, providing insights into their maturation process after release from infected cells. Furthermore, SRM allows for the monitoring and quantification of protein interactions from cell entry to viral assembly, identifying protein clusters and their roles in viral infection [25]. Additionally, SRM can be employed to study the molecular changes triggered by viral infections within cells, such as reorganization of the endoplasmic reticulum caused by the Zika virus, membrane modifications due to filament formation in influenza, and spatial rearrangement of the ESCRT machinery by HIV-1. SRM is also instrumental in the rational design of antivirals, combining high-throughput methods with the enhanced resolution of SRM to improve viral inactivation and vaccine production [3]. This versatile approach can accommodate different types of viruses, whether they have capsids, RNA, DNA, or virus-like particles with a single protein structure. The flexibility of the SRM is commendable. However, challenges, such as the requirement for expensive, delicate microscopes with powerful lasers, optimized sample preparation, long exposure times, and high technical expertise, exist. Recent advancements in the SRM field, including the development of new dyes, computational improvements, and new algorithms, have increased speed and sensitivity while reducing costs [3]. These advancements are expected to mitigate the disadvantages of SRM in the coming years. Despite its potential versatility and ease of use, SRM remains underutilized in virology.

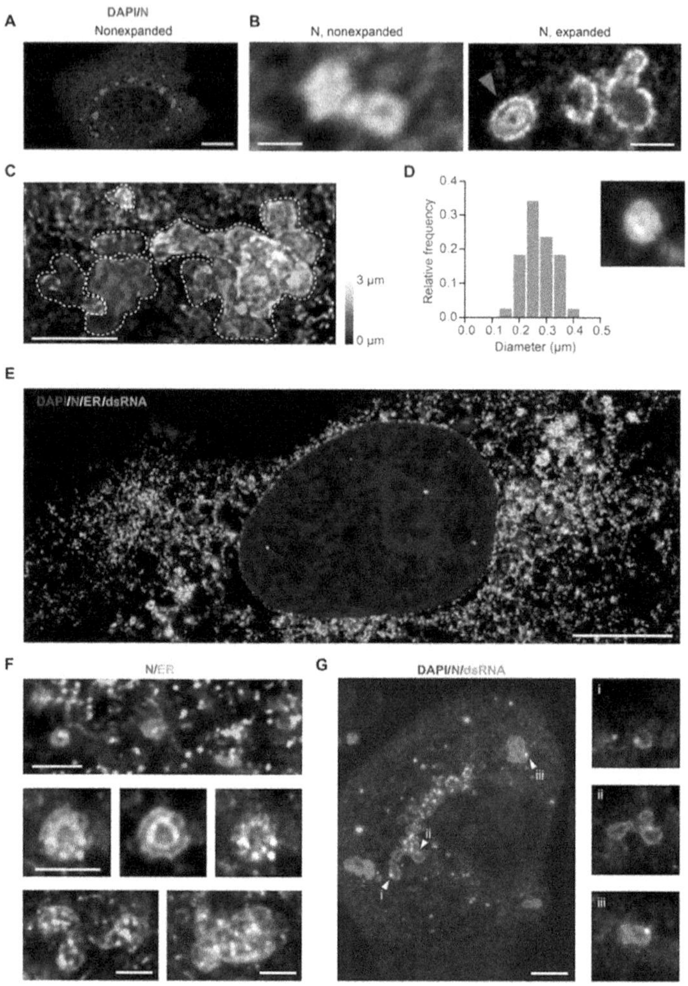

FIGURE 9.2 The SARS-CoV-2N protein is structured in layered formations intricately intertwined with the topology of the endoplasmic reticulum (ER), which contains RNA replication foci. (a) Confocal microscopy was unable to resolve the substructure of N protein puncta in the cytosol of infected Vero cells. The nuclei are shown in blue, and the N protein is shown in magenta. Scale bar: 10 µm. (b) Expansion microscopy reveals two layers of N protein within larger compartments (indicated by the red arrow). Scale bar: 1 µm. The linear expansion factor is 4.2. (c) Combining expansion microscopy with light sheet microscopy reveals the convoluted 3D nature of N protein structures. The dotted lines mark the boundaries of each convoluted compartment in the maximum intensity projection (d). The inner circular layer of the N protein double-layer compartments has an average diameter of 275 nm. Thirty-eight compartments from 10 cells at 12 h post-infection (hpi) were analyzed. The micrograph size is 1.25 µm per side. (e) A representative confocal image of an expanded, infected Vero cell stained for the nucleus (blue), ER (calnexin; orange), N protein (magenta), and dsRNA (cyan). (f) The N protein (magenta) forms layers around ER membranes (yellow). (g) A combination of expansion and light sheet microscopy shows that dsRNA foci (cyan) are located within the layers of the N protein compartments (magenta). Scale bar: 5 µm. Expansion factor: 4.2. The smaller micrographs measure 5 µm per side [26].

The integration of ExM with SRM could further enhance our understanding of viral infections and assembly, highlighting its significant yet underexploited potential in virology research.

REFERENCES

1. Chen, F, Tillberg, PW, Boyden, ES, Expansion microscopy. *Science*, 347(6221), 543–548, 2015.
2. Cohen, Y, Ramon, O, Kopelman, IJ, Mizrahi, S, Characterization of inhomogeneous polyacrylamide hydrogels. *Journal of Polymer Science Part B: Polymer Physics*, 30(9), 1055–1067, 1992.
3. Tillberg, PW, Chen F, Piatkevich KD, Zhao Y, Yu CC, Englis, Protein retention expansion microscopy of cells and tissues labeled using standard fluorescent proteins and antibodies. *Nature Biotechnology*, 34(9), 987–992, 2016.
4. Hausen, P, Dreyer, C, The use of polyacrylamide as an embedding medium for immunohistochemical studies of embryonic tissues. *Stain Technology*, 56(5), 287–293, 1981.
5. Gao, R, Yu, CC, Gao, L, Piatkevich, KD, Neve, L, Munro, JB, Upadhyayula S, Boyden, ES, A highly homogenous polymer composed of tetrahedron-like monomers for high-isotropy expansion microscopy. *Nature Nanotechnology*, 16(6), 698–707, 2021.
6. Kalichava, A, Ochsenreiter, T, Ultrastructure expansion microscopy in *Trypanosoma brucei*. *Open Biology*, 11(10), 210132, 2021.
7. Suofu, Y, Li W, Jean-Alphonse FG, Jia J, Khattar NK, Li J, Baranov SV, Leronni D, Mihalik AC, He Y, Cecon E, Dual role of mitochondria in producing melatonin and driving GPCR signaling to block cytochrome c release. *Proceedings of the National Academy of Sciences of the United States of America*, 114(38), E7997–E8006, 2017.
8. Moser, F, Pražák V, Mordhorst V, Andrade DM, Baker LA, Hagen C, Grünewald K, Kaufmann R., Cryo-SOFI enables low-dose superresolution correlative light and electron cryo-microscopy. *Proceedings of the National Academy of Sciences of the United States of America*, 116(11), 4804–4809, 2019.
9. Wassie, A, Expansion microscopy: principles and uses in biological research. *Nature Methods*, (16), 33–41, 2019.
10. Santi, P, Light sheet flourescence microscopy: a review. *Journal of Histochemistry and Cytochemistry*, 59(2), 129–138, 2011.
11. Cahoon, CK, Yu Z, Wang Y, Guo F, Unruh JR, Slaughter BD, Hawley RS, Superresolution expansion microscopy reveals the three-dimensional organization of the Drosophila synaptonemal complex. *Proceedings of the National Academy of Sciences of the United States of America*, 114(33), E6857–E6866, 2017.
12. Hofman, DP, Shtengel G, Xu CS, Campbell KR, Freeman M, Wang L, Milkie DE, Pasolli HA, Iyer N, Bogovic JA, Stabley DR, Correlative three-dimensional superresolution and block-face electron microscopy of whole vitreously frozen cells. *Science*, 367(6475), 1–12, 2020.
13. Tanaka, T, Fillmore D, Sun ST, Nishio I, Swislow G, Shah A., Phase transitions in ionic gels. *Physical Review Letters*, 45(20), 1636–1639, 1980.
14. Gambarotto, D, Zwettler FU, Le Guennec M, Schmidt-Cernohorska M, Fortun D, Borgers S, Heine J, Schloetel JG, Reuss M, Unser M, Boyden ES, Imaging cellular ultrastructure using expansion microscopy (U-Ex M). *Nature Methods*, 16, 71–74, 2019.
15. Reynaud, EG, Kržič, U, Greger, K, Stelzer, EHK, Light sheet-based fluorescence microscopy: more dimensions, more photons, and less photodamage. *Human Frontier Science Program Journal*, 2(5), 266–275, 2008.

16. Fadero, TC, Gerbich, MT, Rana, K, Suzuki, A, DiSalvo, M, Schaefer, KN, Heppert, JK, TC Boothby, B Goldstein, M Peifer, NL Allbritton, AS Gladfelter, Maddox, AS, Maddox, PS, LITE microscopy: tilted light-sheet excitation of model organisms offers high resolution and low photobleaching. *Journal of Cell Biology*, 217(5), 1869–1882, 2018.
17. Huang, B, Jones, SA, Brandenburg, B, Zhuang, X, Whole-cell 3D STORM reveals interactions between cellular structures with nanometer-scale resolution. *Nature Methods*, 5(12), 1047–1052, 2008.
18. Punge, A, Rizzoli, SO, Jahn, S, Wildanger, JD, Meyer, S, Schönle, A, Kastrup, LU, and Hell, SW, 3D reconstruction of high-resolution STED microscope images. *Microscopy Research and Technique*, 71(9), 644–650, 2008.
19. Dadonaite, B, Gilbertson, B, Knight, ML, Trifkovic, S, Rockman, S, La Bederach, A, Brown, LE, Fodor, E, Bauer, DLV, The structure of the influenza A virus genome. *Nature Microbiology*, 4(11), 1781–1789, 2019.
20. Kolpe, A, Arista-Romero, M, Schepens, B, Pujals, S, Saelens, X, Albertazzi, L, Superresolution microscopy reveals a significant impact of M2e-specific monoclonal antibodies on influenza A virus filament formation at the host cell surface. *Scientific Reports*, 9, 4450, 2019.
21. Lehmann, M, Rocha S, Mangeat B, Blanchet F, Uji-i H, Hofkens J, Piguet V, Quantitative multicolour superresolution microscopy reveals tetherin HIV-1 interaction. *PLOS Pathogens*, 7(12), 341–348, 2011.
22. Mostafa, A, Abdelwhab, EM, Mettenleiter, TC, Pleschka, S, Zoonotic potential of influenza A viruses: a comprehensive overview. *Viruses*, 10(9), 497–502, 2018.
23. Benton, DJ, Gamblin, SJ, Rosenthal, PB, Skehel, JJ, Structural transitions in influenza hemagglutinin at membrane fusion pH. *Nature*, 583(7814), 150–153, 2020.
24. Chojnacki, J, Eggeling, C, Superresolution fluorescence microscopy studies of human immunodeficiency virus. *Retrovirology*, 15(41), 1–16, 2018.
25. Laine, RF, Goodfellow, G, Young, LJ, Travers, J, Carroll, D, Dibben, O, Bright, H, Kaminski, CF, Structured illumination microscopy combined with machine learning enables the high throughput analysis and classification of virus structure. *Epidemiology and Global Health Microbiology and Infectious Diseases*, 7, 1–17, 2018.
26. Scherer, KM, Mascheroni, L, Carnell, GW, Wunderlich, LC, Makarchuk, S, Brockhoff, M, Kaminski, CF, SARS-CoV-2 nucleocapsid protein adheres to replication organelles before viral assembly at the Golgi/ERGIC and lysosome-mediated egress. *Science Advances*, 8(1), eabl4895, 2022.

10 Development of an Optimized Method for the Rapid Quantification of Exosomes

Talitha K. Kurian, Manikanth Karnati, and Budheswar Dehury
Manipal Academy of Higher Education

Subir Das
University of Zurich

10.1 INTRODUCTION

Extracellular vesicles, often referred to as EVs, are spherical secretory vesicular structures enclosed by a protein–lipid bilayer membrane that typically range from 30 to 500 nm in diameter. Most eukaryotic cell types release EVs to facilitate intercellular communication and eliminate cellular waste. In addition to the proteins present on their surfaces, EVs harbor other components, including lipids and genetic material, such as DNA, messenger RNAs (mRNAs), and microRNAs (miRNAs) [1]. Distinguished by disparities in their origin, composition, and functions, extracellular vesicles can be categorized into three primary types: microvesicles, apoptotic bodies, and exosomes. Microvesicles or MVs are structures 50–1,000 nm in size that bud directly from the plasma membrane. Owing to differences in their origins, microvesicles are enriched in a different set of proteins than are exosomes. These proteins include integrins, P-selectin, glycoprotein Ib (GPIb), and arrestin-containing protein 1 (AARDC1) [2]. MVs typically carry a greater proportion of post-translationally modified proteins, such as glycoproteins and phosphoproteins, than do exosomes [3]. When cells undergo apoptosis, they release apoptotic bodies with diameters ranging from 500 to 2,000 nm from the plasma membrane. These vesicles contain DNA-binding histones and are devoid of glycoproteins, which is a striking difference from exosomes [4]. Exosomes are small extracellular vesicles that typically measure between 30 and 100 nm in size. They are cup-shaped structures released by all mammalian cells and are present in most biological fluids (including sperm, urine, feces, breast milk, amniotic fluid, blood, saliva, cerebrospinal fluid, etc.). In 1983, the emergence of exosomes was independently documented in two papers by Harding et al. and Pan et al. These studies followed the movement of

DOI: 10.1201/9781003570011-10

labeled transferrin receptors into developing reticulocytes, revealing a remarkable process. When transferrin receptors are internalized, they repack into tiny vesicles that are approximately 50 nm in size [5,6]. These vesicles, initially presumed to be destined for lysosomal degradation, were unexpectedly released into the extracellular environment by maturing blood reticulocytes and subsequently coined "exosomes" [7].

In contrast to other forms of extracellular vesicles that originate directly from plasma membrane budding, exosomes are unique in that they are products of the endosomal pathway. The release process of exosomes begins with the development of endocytic vesicles, which are formed by the inward budding of the plasma membrane. These vesicles exhibit an inside-out membrane orientation on the outer side and contain extracellular fluid. Many endocytic vesicles coalesce and develop into endosomes, which subsequently undergo maturation to become multivesicular endosomes. As they mature, the external membrane of endosomes undergoes invagination, giving rise to small vesicles called intraluminal vesicles (ILVs), which typically have diameters ranging from 50 to 90 nm. These ILVs encapsulate cytosolic contents and exhibit an outward-facing membrane orientation. Multivesicular endosomes, also known as multivesicular bodies (MVBs), have two possible fates: they may facilitate the intracellular degradation of proteins or function as intermediates in the formation of exosomes. Within the context of the endolysosomal degradation pathway, proteins that have undergone internalization from the cell membrane or selection from the trans-Golgi network become sequestered within the ILVs of MVBs. These MVBs subsequently undergo fusion with lysosomes, which release the enzymes necessary for protein degradation. Alternatively, MVBs are directed toward the cell surface, culminating in their fusion and the subsequent release of their contents into the extracellular space, where they can form exosomes [8]. These exosomes encapsulate a range of components that mirror their specific parent cell types, including RNA, DNA, mRNA, miRNAs, proteins, lipids, and metabolites. The phospholipid bilayer membranes of exosomes also contain distinct proteins, such as tetraspanins, integrins, receptors, and transporters, which are valuable for exosome isolation and detection [9].

Once considered mere cellular waste disposal systems, exosomes are currently acknowledged for their multifaceted involvement in intercellular communication, signaling, antigen display, and the modulation of immune responses, among other biological functions. They also play pivotal roles in tumor metastasis, as exosomes originating from cancer cells carry information responsible for processes such as angiogenesis and cell proliferation. Furthermore, exosomes exhibit therapeutic promise owing to their biocompatibility, minimal immunogenicity, and capacity to traverse the blood–brain barrier, making them advantageous for precise drug delivery applications [10]. Urine-derived exosomes have the potential to function as diagnostic biomarkers for genitourinary conditions and prostate cancer. Moreover, those acquired from cerebrospinal fluid have diagnostic value for prion diseases and neurodegenerative disorders, including Parkinson's disease and Alzheimer's disease [11]. The role of exosomes as markers of diseases and prognostic indicators makes them ideal candidates for application in liquid biopsies. In this context, exosomes, which are secreted into circulating bodily fluids such as blood and urine by parent cells, can be examined to ascertain the presence or extent of various illnesses.

However, the investigation of exosomes as biomarkers for diseases demands the precise quantification of their concentrations in clinical samples. This continues to present a challenge owing to the intricacies involved in the isolation of uncontaminated exosomes from biological and cell culture fluids and the accurate detection and enumeration of these vesicles [12]. Currently, nanoparticle tracking analysis (NTA) is the gold standard technique for exosome quantification. This approach uses a camera for the capture of light scattering phenomena by suspended particles along with their Brownian motion. The NTA software subsequently utilizes these data to track particles individually and determine their size, volume, and concentration [13]. However, this method requires large volumes of exosome isolate samples and expensive instrumentation. Flow cytometry employs the use of fluorophore-conjugated antibody-coated beads to identify particles by evaluating their interaction with a laser beam as they traverse a detection cell. The instrument has a detection limit of 300 nm and therefore excludes smaller exosomes and measures concentrations inaccurately [14]. Transmission electron microscopy provides magnified images of exosomes but requires tedious manual counting and is thought to damage exosome membranes during sample preparation [15]. Dynamic light scattering (DLS) establishes a correlation between the intensity of light scattered by exosomes due to their Brownian motion and their size and distribution. However, pure homogeneous samples are required for analysis because when mixtures are analyzed, larger particles that scatter light with greater intensity are detected more than smaller particles [16]. In this study, we investigated the possible use of UV–Vis spectroscopy and fluorescence spectroscopy in the detection of exosomes. These methods are noninvasive and nondisruptive and require no manipulation of the exosomes.

10.2 EXPERIMENTAL PROCEDURES

10.2.1 Cell Culture

HepG2 hepatocytes were purchased from the American Type Culture Collection (Manassas, VA, USA) and grown in a humidified environment enriched with 5% CO_2 at 37°C in DMEM containing 10% FBS. The production of other extracellular vesicles was minimized by daily monitoring of cells for changes in the growth rate and increased proportion of dead cells. Cells seeded at a density of 2.1 × 106 were grown to 80% confluency and then maintained in serum-free DMEM for 72 h to exclude bovine exosomes. Serum-free media or conditioned media were collected from the cell cultures and processed to isolate exosomes. The conditioned media were frozen at −20°C before use.

10.2.2 Isolation by Differential Ultracentrifugation

The conditioned media was filtered through a 0.22 µm filter. The filtrate was subjected to centrifugation at 300 × g for 10 min and then at 2,000 × g for 20 min, resulting in pelleted cells and larger debris. The collected supernatant was then centrifuged at 10,000 × g for 30 min to remove the remaining cell debris and larger extracellular vesicles such as apoptotic bodies and microvesicles. The resulting supernatant was then distributed into polypropylene centrifuge tubes in 8 mL aliquots for centrifugation via a swinging bucket rotor. Exosomes were pelleted by two consecutive ultracentrifugation steps, each conducted at 100,000 × g for 70 min. Following ultracentrifugation, the supernatant was discarded, and the exosome pellet at the base of the tube was gently resuspended in PBS by vortexing. The pellets were collected in a single tube and subjected to another round of ultracentrifugation at 100,000 × g for 60 min. All the centrifugations were performed at 4°C. The resulting exosome pellet was subsequently carefully resuspended in 200 μL of PBS, safeguarded at −80°C, and stored until it was thawed for analysis.

10.2.3 Isolation by Density Gradient Ultracentrifugation

The conditioned media were processed as described above until centrifugation of the media at 10,000 × g. One milliliter of 30% sucrose was added to polypropylene centrifuge tubes, and 8 mL of the supernatant was layered carefully over the sucrose cushion. The tubes were ultracentrifuged at 100,000 × g for 70 min, after which the layer of conditioned media was carefully removed without disrupting the sucrose layer that contained the exosomes. The sucrose layer was subjected to another round of centrifugation at 100,000 × g for 70 min in 8 mL of PBS. The pellets were collected in a tube and subsequently resuspended in 1 mL of PBS. This was followed by final centrifugation at 100,000 × g for 70 min, after which the exosome pellet was resuspended in 200 μL of PBS and stored at −80°C.

10.2.4 Isolation by Precipitation

Exosomes were isolated via polyethylene glycol treatment as described in the ExtraPEG method [17]. The conditioned media were subjected to centrifugation at 300 × g for 10 min and then at 2,000 × g for 30 min at 4°C, thus effectively eliminating cell debris and large vesicles. A 15% 2× PEG mixture was subsequently prepared by combining PEG 8000, NaCl, and filtered water. An equivalent volume of the 15% polyethylene glycol (PEG) solution was gently mixed with the supernatant by inversion and kept at 4°C for 24 h. The resulting mixture was then centrifuged at 4,000 × g for 60 min at 4°C to collect the exosomes. Following this step, the centrifuge tubes were allowed to drain for 10 min to eliminate residual PEG, and the exosome pellet was suspended in 200 μL of PBS and stored at −80°C. Both the PEG solution and PBS were filtered through a 0.22 μm filter and sonicated in an ultrasonicator bath for 15 min to ensure their particle-free state.

10.2.5 Isolation by Ultrafiltration

The Amicon (Merck-Millipore) ultracentrifugation filter was equilibrated by spinning it with 15 mL of particle-free sterile PBS (autoclaved and ultrasonicated) at 4,000 × g twice. A 15 mL aliquot of conditioned media was added to the Amicon filter after it had passed through a 0.22 μm filter. The sample was spun at 4,000 × g for 20 min, and the collection tube was emptied. The exosome pellet was suspended in 14 mL of PBS and spun again at 4,000 × g for 20 min. The exosome-containing concentrated sample was collected. The process was repeated until all the conditioned media were ultrafiltered. Exosome pellets were combined, suspended in PBS, and stored at −80°C.

10.2.6 Bradford Assay

Exosome isolates were suspended in RIPA buffer (25 mM Tris-HCl, pH 7.5, 140 mM NaCl, 1% Triton X-100, 0.5% SDS, 1 mM EDTA, 1 mM PMSF, 1 mM Na_3VO_4, and 1 mM NaF). The Bio-Rad protein assay protocol was used, with BSA standards at concentrations ranging from 0.1 to 1.4 µg/mL. The absorbance at 595 nm was measured via a Varioskan Flash Multimode Microplate Reader (Thermo Scientific).

10.2.7 Dynamic Light Scattering

The size of the exosomes was measured via DLS with a Zetasizer Nano ZS-90 (Malvern). Exosome isolates were diluted with PBS to 1 mL and loaded into glass cuvettes. The 633 nm He-Ne laser was fixed at an angle of 90°, and measurements were taken at a fixed position determined by the instrument with an automatic attenuator at 25°C. The average value from three measurements was taken.

10.2.8 Spectrophotometric Measurements

Two hundred microliters of the isolated exosome sample was loaded onto a 96-well plate (Greiner Bio-One). The absorbance intensities were measured from 280 to 700 nm excitation wavelengths via a Varioskan Flash Multimode Microplate Reader (Thermo Scientific). Similarly, fluorescence spectra at excitation wavelengths of 420 and 450 nm were taken, with emission wavelengths starting at 450 and 480 nm, respectively. The step size was 2 nm for all the measurements.

10.2.9 Statistical Analysis

The results represent the mean value derived from three separate and independent readings. Graphs were generated via GraphPad Prism 5 software.

10.3 RESULTS

10.3.1 Protein Concentration

The protein concentration present in the exosome sample can be taken as a relative measure of the exosome concentration, with more exosomes being lysed with RIPA buffer to release more proteins. Figure 10.1 shows the protein concentrations measured by the Bradford assay. The results revealed that precipitation with PEG was the most effective method for exosome isolation, resulting in the highest protein concentration of 0.4408 mg/mL. The ultrafiltration and sucrose cushion methods yielded similar results, 0.2179 and 0.2141 mg/mL, respectively, and are therefore assumed to have extracted a similar number of exosomes. Ultracentrifugation resulted in the lowest protein concentration of 0.1572 mg/mL.

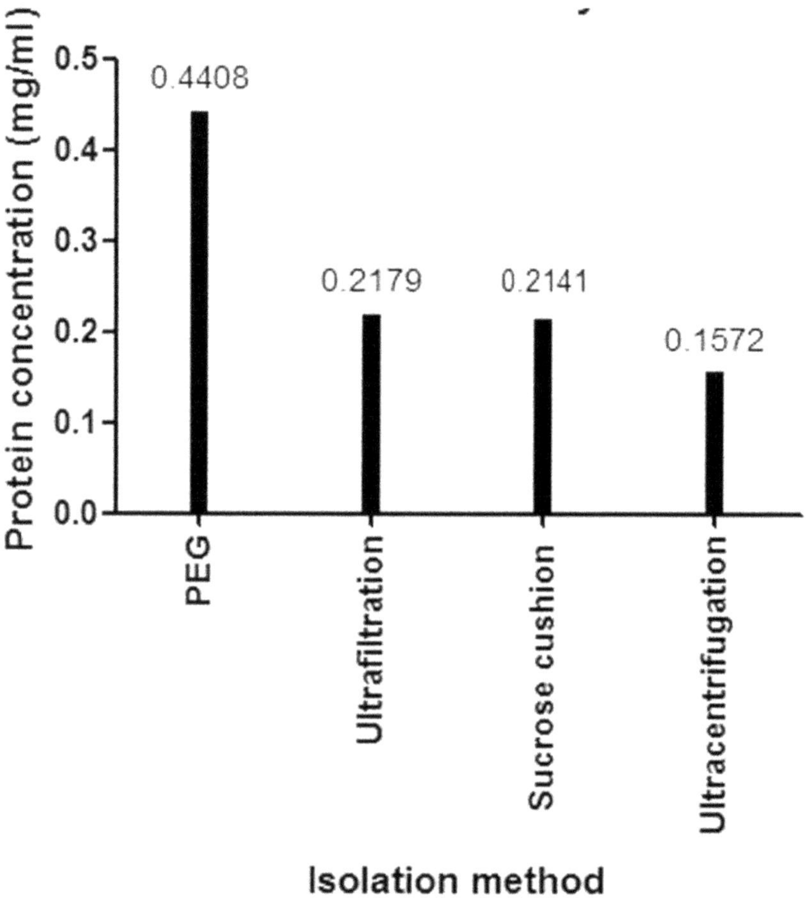

FIGURE 10.1 The protein concentration, as measured by the Bradford assay, is displayed for each exosome isolation method.

10.3.2 Spectrophotometric Measurements

Figure 10.2 displays the fluorescence spectra obtained at excitation wavelengths of 420 and 450nm. All the samples produced similar spectra, with peaks at 480 and 530nm emission wavelengths. The peak positions at both excitation wavelengths reflect the results of the protein concentration assay, with PEG-isolated samples producing the greatest intensity at the peak and ultracentrifugation-isolated samples producing the smallest intensity. Figure 10.3 displays the absorbance values at 566nm incident light, as measured via UV–Vis spectrophotometry. These values also reflect the trends of the previous measurements.

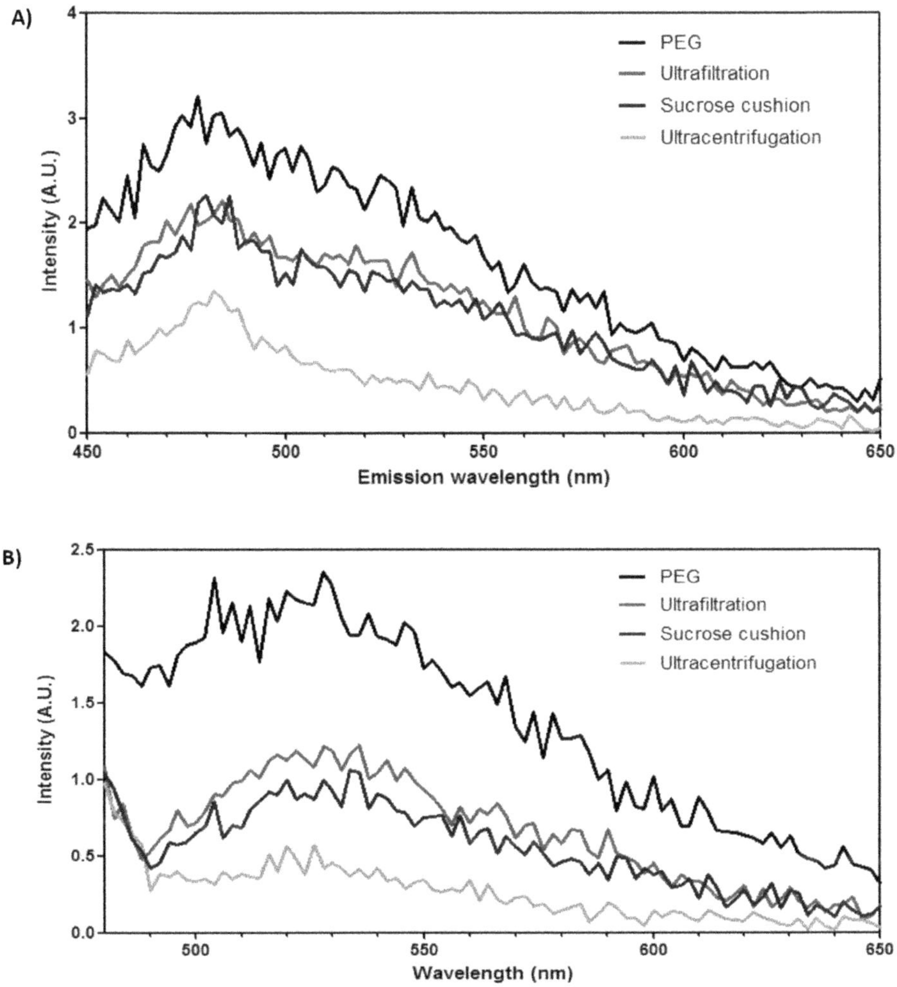

FIGURE 10.2 (a) Fluorescence spectra at an excitation wavelength of 420 nm. (b) Fluorescence spectra at an excitation wavelength of 450 nm.

10.3.3 Correlation between Protein Concentration and Spectrophotometric Readings

Figure 10.4 shows the Bradford assay results plotted against the peak intensities from the fluorescence and UV–Vis measurements. All the graphs show a linear correlation between the protein concentration and spectrophotometric readings.

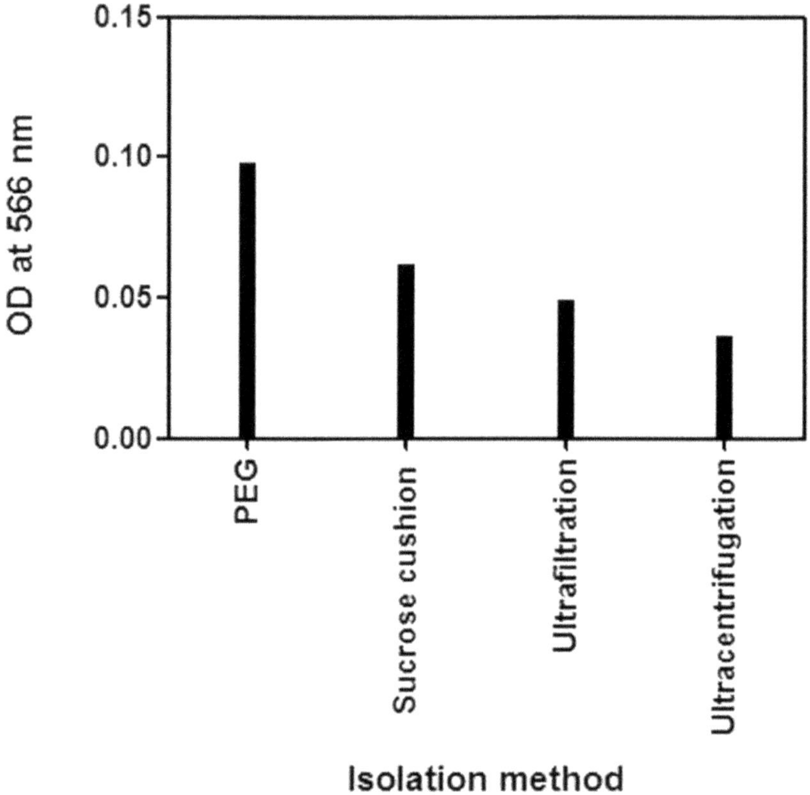

FIGURE 10.3 Absorbance values of differently isolated exosome samples at 566 nm.

10.3.4 Size Determination by DLS

The average size of the exosomes in the samples was measured via DLS. All the samples presented peaks occurring between 20 and 200 nm, which is within the range of normal exosome diameters.

10.4 CONCLUSION

One of the major obstacles in exosome research lies in obtaining pure exosome samples and accurately measuring their concentration because of their extremely small size. Even the most reliable methods of isolation, such as those considered to be the gold standard, face difficulties such as low efficiency, coprecipitation of nonexosomal molecules, and damage to the vesicular structure. Furthermore, only a limited number of techniques are available for determining the exact count of exosomes in a given sample. The equipment used in these methods is costly, requires intricate handling and upkeep, and poses challenges for integration into clinical laboratories. In recent years, however, novel methods such as microfluidic approaches and combinations of existing methods have been tested and are promising in terms of both efficiency and accuracy. The isolation of exosomes from clinical samples in a

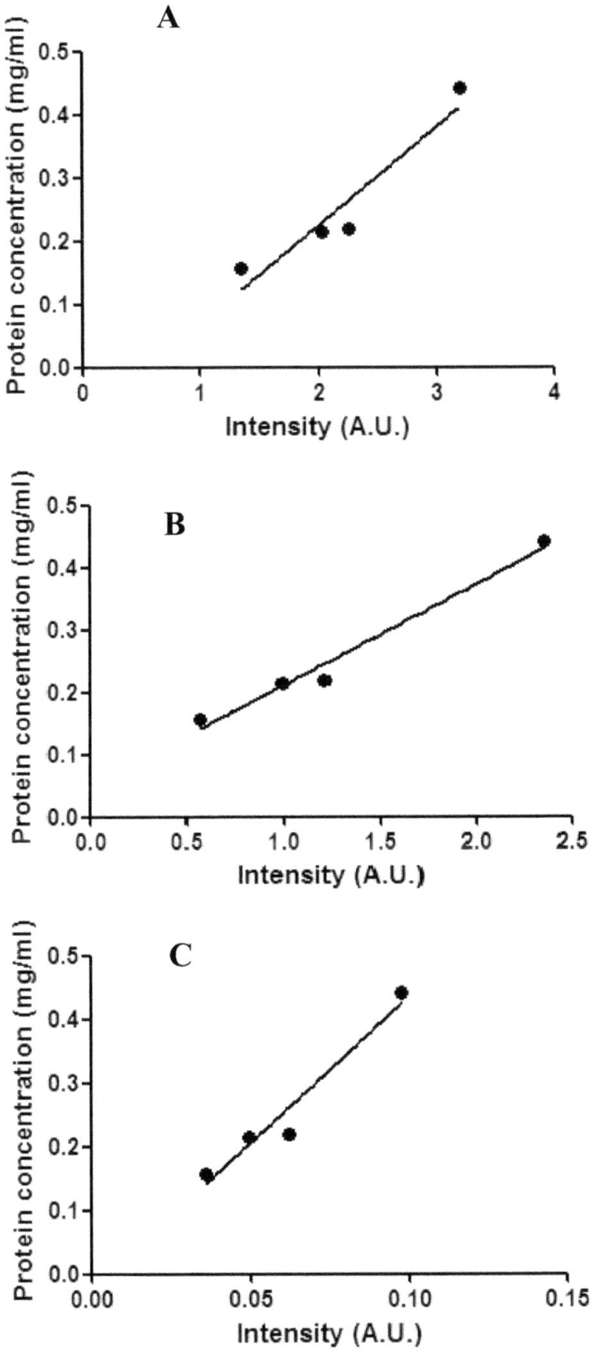

FIGURE 10.4 Bradford assay protein quantification results compared with peak intensities of different samples in (a) fluorescence spectra at 420 nm excitation, (b) fluorescence spectra at 450 nm excitation, and (c) absorption values at 566 nm. The linear correlation is clear in all the graphs.

timely, cost-effective, straightforward, or simple manner and their precise measurement would markedly increase our understanding of their potential as biomarkers of disease and therapeutic agents. In this study, we have shown that UV–Vis spectroscopy and fluorescence spectroscopy can be used to detect the presence of exosomes in a pure sample. These spectrophotometric methods can be used to quantify exosomes by plotting a calibration curve, and DLS can be used to quantify the exosomes directly [16]. The results from these measurements were linearly correlated with the protein concentration measured via the Bradford assay. This spectrophotometric method has the potential to serve as a reliable means for quantifying exosomes extracted from small sample volumes, as it does not require any modifications to the isolates. The method is cost-effective and straightforward and requires basic equipment commonly found in clinical laboratories.

ACKNOWLEDGMENTS

We also thank the Manipal School of Life Sciences, Manipal Academy of Higher Education, Manipal, Karnataka, India, for providing infrastructure and support during the course of this study.

REFERENCES

1. Ratajczak, J., Wysoczynski, M., Hayek, F., Janowska-Wieczorek, A., and Ratajczak, M. Z. (2006) Membrane-derived microvesicles: important and underappreciated mediators of cell-to-cell communication. *Leukemia*. 20, 1487.
2. Heijnen, H. F., Schiel, A. E., Fijnheer, R., Geuze, H. J., and Sixma, J. J. (1999) Activated platelets release two types of membrane vesicles: microvesicles by surface shedding and exosomes derived from exocytosis of multivesicular bodies and granules. *Blood*. 94, 3791–3799.
3. Palmisano, G., Jensen, S. S., Le Bihan, M. C., Laine, J., McGuire, J. N., Pociot, F., and Larsen, M. R. (2012) Characterization of membrane-shed microvesicles from cytokine-stimulated β-cells using proteomics strategies. *Molecular & Cellular Proteomics*. 11, 230–243.
4. Théry, C., Boussac, M., Véron, P., Ricciardi-Castagnoli, P., Raposo, G., Garin, J., and Amigorena, S. (2001) Proteomic analysis of dendritic cell-derived exosomes: a secreted subcellular compartment distinct from apoptotic vesicles. *The Journal of Immunology*. 166, 7309–7318.
5. Harding, C., Heuser, J., and Stahl, P. (1983) Receptor-mediated endocytosis of transferrin and recycling of the transferrin receptor in rat reticulocytes. *The Journal of Cell Biology*. 97, 329–339.
6. Pan, B. T., and Johnstone, R. M. (1983) Fate of the transferrin receptor during maturation of sheep reticulocytes in vitro: selective externalization of the receptor. *Cell*. 33, 967–978
7. Johnstone, R. M., Bianchini, A., and Teng, K. (1989) Reticulocyte maturation and exosome release: transferrin receptor containing exosomes shows multiple plasma membrane functions. *Blood*. 74, 1844–1851.
8. Théry, C., Zitvogel, L., and Amigorena, S. (2002) Exosomes: composition, biogenesis and function. *Nature Reviews Immunology*. 2, 569.

9. Hong, B. S., Cho, J. H., Kim, H., Choi, E. J., Rho, S., Kim, J., and Gho, Y. S. (2009) Colorectal cancer cell-derived microvesicles are enriched in cell cycle-related mRNAs that promote proliferation of endothelial cells. *BMC Genomics*. **10**, 556.
10. Andaloussi, S. E., Lakhal, S., Mäger, I., and Wood, M. J. (2013. Exosomes for targeted siRNA delivery across biological barriers. *Advanced Drug Delivery Reviews*. 65, 391–397.
11. Yang, C., and Robbins, P. D. (2011) The roles of tumor-derived exosomes in cancer pathogenesis. *Clinical and Developmental Immunology*. **2011**, 842849.
12. Van Der Pol, E., Hoekstra, A. G., Sturk, A., Otto, C., Van Leeuwen, T. G., and Nieuwland, R. (2010) Optical and non-optical methods for detection and characterization of microparticles and exosomes. *Journal of Thrombosis and Hemostasis*. 8, 2596–2607.
13. Filipe, V., Hawe, A., and Jiskoot, W. (2010) Critical evaluation of Nanoparticle Tracking Analysis (NTA) by NanoSight for the measurement of nanoparticles and protein aggregates. *Pharmaceutical Research*. 27, 796–810.
14. Momen-Heravi, F., Balaj, L., Alian, S., Tigges, J., Toxavidis, V., Ericsson, M., and Kuo, W. P. (2012) Alternative methods for characterization of extracellular vesicles. *Frontiers Physiology*. 3, 354.
15. Lässer, C., Eldh, M., and Lötvall, J. (2012). Isolation and characterization of RNA-containing exosomes. *Journal of Visualized E*xperiments. 59, e3037.
16. Palmieri, V., Lucchetti, D., Gatto, I., Maiorana, A., Marcantoni, M., Maulucci, G., and Sgambato, A. (2014) Dynamic light scattering for the characterization and counting of extracellular vesicles: a powerful noninvasive tool. *Journal of Nanoparticle Research*. 16, 2583.
17. Rider, M. A., Hurwitz, S. N., and Meckes Jr, D. G. (2016) ExtraPEG: a polyethylene glycol-based method for enrichment of extracellular vesicles. *Scientific Reports*. **6**, 23978.

11 Terahertz Spectroscopy in Biomedical Research

Olga P. Cherkasova
Russian Academy of Sciences

Yury V. Kistenev
National Research Tomsk State University

11.1 INTRODUCTION

The terahertz (THz) band refers to the frequency range from 0.1 to 10 THz [1,2]. Despite the fact that the discovery of this frequency range took place almost a hundred years ago [3], the intensive THz radiation applications began with the THz pulsed spectroscopy (TPS) devices appearing [4–6]. THz radiation has a number of features that determine the biomedical applications of THz radiation, namely:

- TГц radiation photon has energy of several orders of magnitude lower compared to the molecules' ionization threshold. Therefore, THz waves do not lead to photoionization processes in living systems and biological tissues that guarantee the safety of in vivo studies [7].
- TГц radiation is strongly absorbed by water, including tissue water. This, on the one hand, limits THz wave penetration depth into the tissue. On the other hand, it is the major source of contrast in tissue diagnostics [2].
- Due to ($1/\lambda^4$) dependency, Rayleigh scattering of THz radiation is rather small. This allows us to use the effective medium theory and complex dielectric permittivity models to describe the interaction of THz waves with tissues [8];
- TГц waves penetrate well through many dielectric materials such as dry cloth, wood, paper, plastic, etc. This makes it possible to use THz radiation for non-contact monitoring of wound healing [9];
- Biological molecules in the form of molecular crystals have characteristic absorption lines in the THz range [6,10]. This can be used for non-destructive analysis of pharmaceuticals and for distinguishing between polymorphs of chemical substances [11].

The chapter reviews modern THz spectroscopy and imaging techniques used in biophotonics, with an emphasis on THz pulsed spectroscopy. Common and emerging approaches describing the interaction of THz waves with biological fluids and tissues are discussed. The applications of THz radiation in diagnostics of benign and malignant neoplasms of different nosology and localization, diabetes, and traumatic

 DOI: 10.1201/9781003570011-11

brain injuries are presented. Also, the application of machine learning methods to the analysis of THz spectroscopy data is discussed. In conclusion, the directions of further development of THz biophotonics are discussed.

11.2 THZ TECHNOLOGY

There are two general types of THz radiation sources based on pulsed or continuous wave (CW) generation [1,6,12]. THz pulsed spectroscopy (TPS) uses femtosecond lasers [4,5,13] that provide the possibility of measuring directly refractive index n and absorption coefficient α, and hence complex permittivity of the sample in a single scan and a broad frequency range [2,14]. The scheme of the THz pulse spectrometer is shown in Figure 11.1. Here, PCA is a photo-conductive antenna, and TIA is a trans-impedance amplifier.

TPS and THz pulsed imaging (TPI) yield broader information than a THz CW-radiation device does [15,16]. The most common methods for generating and detecting THz pulses are displacement in photoconductive antennas [17–19], optical rectification in nonlinear optical crystals [20,21], generation in plasma [22], and carrier tunneling in coupled structures with double quantum wells [23].

Historically, optically pumped gas lasers [24] and backward-wave oscillators were the first sources of CW THz radiation [25]. Today, the most common CW radiation

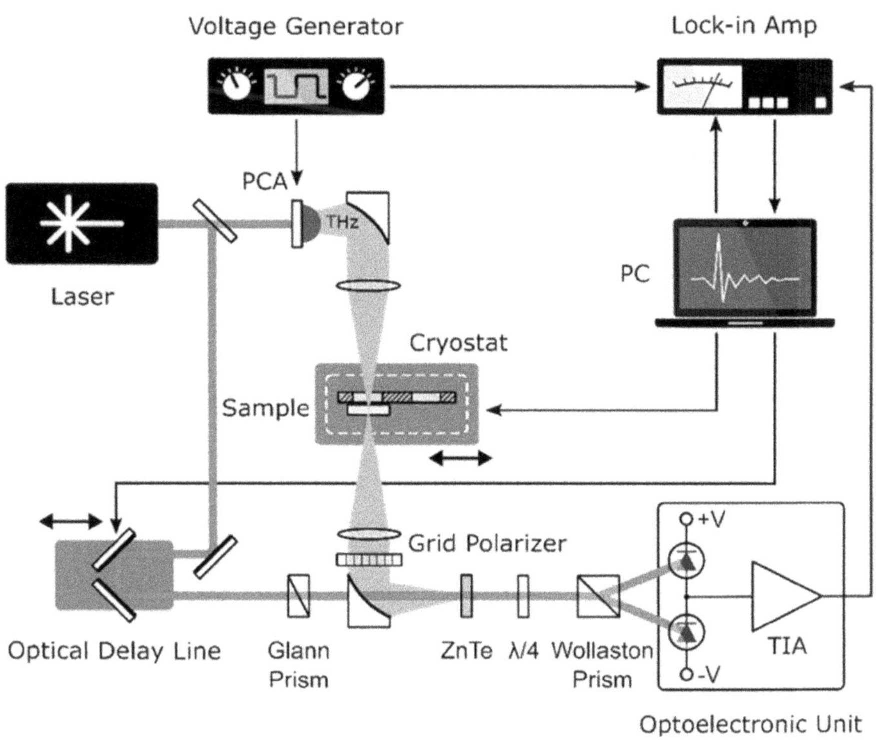

FIGURE 11.1 The scheme of the THz pulse spectrometer.

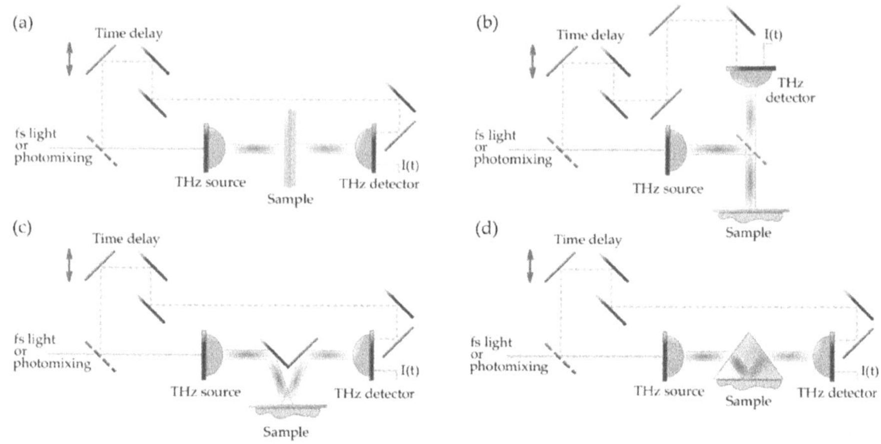

FIGURE 11.2 Schematic representation of typical measurement modes in the THz range. Transmission mode (a), normal-incidence reflection (b), generalized reflection with an incidence angle (c), and attenuated total reflection (ATR) (d). The dashed line with the entire optical part is excluded for the CW generation of THz waves, except for generation by photo mixing [29].

sources are quantum-cascade lasers [26], ultrafast high-speed transistors [27], and diodes [28].

Studies of biological samples in the THz spectral range are typically implemented in transmission, reflection, and attenuated total internal reflection (ATR) modes (Figure 11.2).

11.3 THZ SPECTROSCOPY OF BIOMOLECULES

THz spectra of pure amino acids [30], nucleotides and nucleosides [31,32], sugars [33,34], steroids [10,35], and other small organic molecules [36,37] have characteristic absorption lines. As the number of monomers increases, for example, in peptides, the narrow absorption lines disappear due to the overlap of vibrational modes of individual monomers. Resonance lines are inherent in individual amino acids and double- and triple-chain peptides [30,38], but they are smoothed in peptides of four amino acids [39] and are not observed in proteins [40,41].

The THz absorption spectra of tryptophan, some proteins, and DNA at room temperature are shown in Figure 11.3. Three absorption bands noted in the spectrum of tryptophan are responsible for torsional vibrations of the side chain and ring vibrations [42]. Absorption lines in crystalline or polycrystalline samples of biological molecules disappear at the transition to an amorphous state or dissolution of molecules [33,43]. Resonance lines are not presented in the THz spectra of aqueous solutions, biological fluids, and tissues, which is due to the changes in the molecular structure of dissolved molecules, damping of vibrational modes of these molecules, and strong absorption of THz radiation by water [2,12,44].

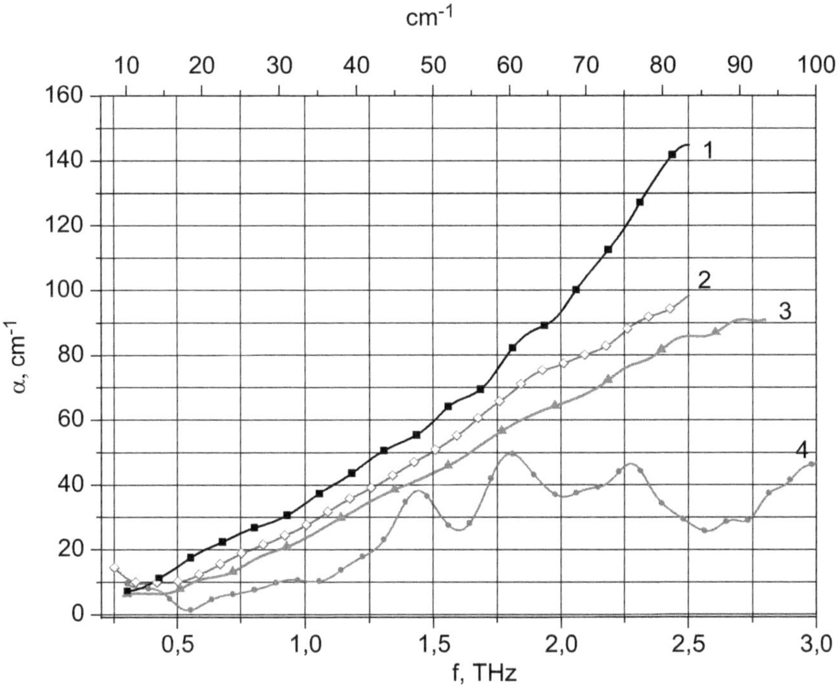

FIGURE 11.3 THz absorption spectrum of DNA from calf thymus (1), bovine serum albumin (BSA) (2), trypsin (3), and tryptophan (4).

11.4 THZ SPECTROSCOPY OF BIOFLUIDS

A local electric field of a biological molecule orients the dipoles of water molecules and forms hydrate shells. In solution, a significant part of water molecules is in the form of hydrate shells of biomolecules, changing their relaxation dynamics [45,46]. Water can be divided into free water (or bulk water that does not form strong bonds with molecules) and bound water (it surrounds molecules and interacts with them) [47,48]. Sometimes strongly bound and weakly bound water are separated, and in some cases, introduce even finer gradation of the degree of water boundness and a greater number of hydrate shells [49]. Free, strongly, and weakly bound water contribute differently to the THz response of biological solutions and tissues [2,49,50].

The reorientation of water hydrogen bonds near a biomolecule under the action of its electric field occurs in sub-picosecond time intervals [51]. In these time scales, THz pulse spectroscopy is an effective tool for studying various forms of water in biological samples, including blood and tissues [52–55]. To interpret experimental data, it is necessary to know the spectrum of bound water, which is a non-trivial task since the contribution of free water to the THz spectrum of a solution is dominant over a wide range of concentrations [2,33]. The sensitivity of THz methods to the water state (free/bound) is maximal in the range of 0.01–1.00 THz, since the characteristic frequency of the free water absorption "peak" is 20 GHz with a width of about 2–3 octaves in both directions [56].

For water, γ-relaxation dominates with a picosecond characteristic time [56,57]. The basis of γ-relaxation is the process of breaking hydrogen bonds among water molecules with their subsequent reorientation [49,50,56–58]. The response of such relaxation systems to an external electric field can be described using the Debye model. The most general formula used to analyze the dielectric response of polar solutions in the THz frequency range is as follows [57,59,60]:

$$\varepsilon = \varepsilon_\infty + \frac{\Delta\varepsilon_1}{\left(1+\left(i\omega\tau_1\right)^\alpha\right)^\beta} + \frac{\Delta\varepsilon_2}{1+i\omega\tau_2} + \frac{A_1}{\omega_{01}{}^2 - \omega^2 + i\gamma_{01}\omega} + i\sigma / \omega\varepsilon_0 \tag{11.1}$$

where ε_∞ is dielectric permittivity of water in the high-frequency range, $\omega = 2\pi f$ is circular frequency of the external electric field, τ_1 and τ_2 are relaxation times for the first ('slow') relaxation and second ('fast') relaxation processes, $\Delta\varepsilon_1$ and $\Delta\varepsilon_2$ are amplitudes of the first and the second relaxations, A_1, $\omega_{01,}$ and y_{01} are amplitude, resonance frequency, and line width of the Lorentz oscillator, δ is static conductivity, and ε_0 is dielectric constant. The results of the model (1) are shown in Figure 11.4. It is important to note that the following condition must be satisfied for static dielectric permittivity ε_s:

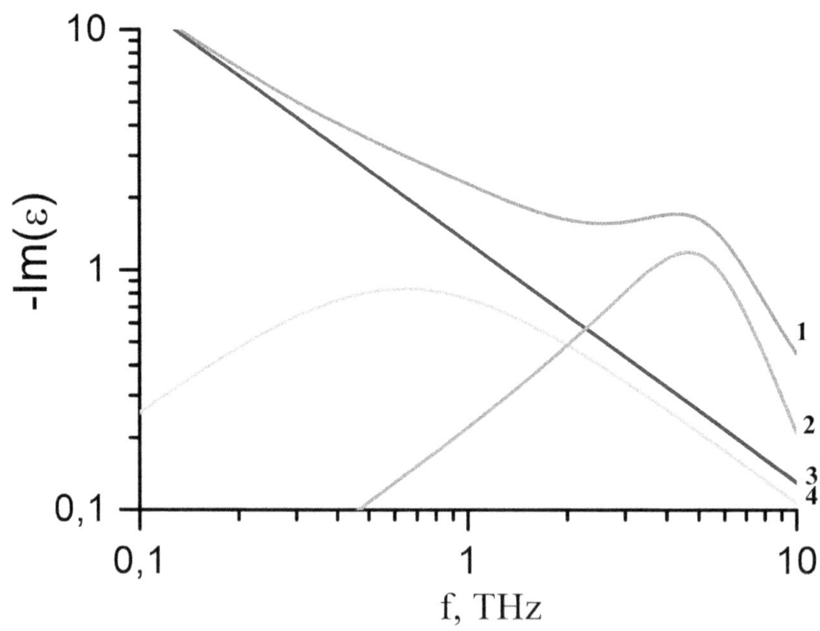

FIGURE 11.4 Dielectric loss spectrum (curve 1) for water according to the model (1): the first ("slow") relaxation process (curve 3), the second ("fast") relaxation process (curve 4), and Lorentz oscillator (curve 2) [50].

$$\varepsilon_s = \varepsilon_\infty + \Delta\varepsilon_1 + \Delta\varepsilon_2 + \frac{A_1}{\omega_{01}^2} \tag{11.2}$$

where ε_s is 78.3 for distilled water at 25°C.

At $\alpha = 1$ *and* $\beta = 1$ in (1), we have a Debye model, which is relevant for distilled water and tissues. At $\beta = 1$ *and* $\alpha < 1$, the Cole–Cole model is more adequate [61], which takes into account the symmetric broadening of the dielectric permittivity spectrum. The latter, for example, is observed when sugars and proteins dissolve in water [56,62]. When $\alpha = 1$, $\beta \neq 1$, the Davidson–Cole model is relevant, which, for example, describes the glycerol measurements in the super-cooled liquid state [63]. For salt solutions (including those in buffer), it is also necessary to take into account conductivity δ [58].

The real relaxation dynamics of the THz dielectric response of aqueous solutions, biological fluids, and tissues can be much more complex, assuming the presence of more than 2–3 characteristic relaxation times, or even a continuous spectrum of relaxation times considered in the generalized Debye model (or Fröhlich model) [64]. Nevertheless, such a complication of the THz response model and application of the generalized model in practice seems difficult.

11.5 THZ SPECTROSCOPY OF BLOOD AND BLOOD COMPONENTS

Blood is a complex liquid consisting of blood plasma or serum and the formational elements suspended in it. Blood plasma contains 90% water, about 6.6%–8.5% proteins, and other organic and mineral compounds, which are intermediate or final products of blood-borne metabolism [65]. The high water content in blood serum determines the nature of its dielectric response in the THz spectral range [2,14]. The transition of the water component from free to bound with surrounding molecules state and vice versa is the reason for the change in the THz response of blood serum [2,66]. Various constituents of blood produce a significant impact on blood optical properties, and their contribution may change during the development of pathological processes in the human body.

The level of triglycerides [67] and creatinine [68] in blood was shown to produce a significant effect on the absorption coefficient and refractive index in the 0.2–1.0 THz frequency range. The number of erythrocytes in blood correlates negatively with the absorption coefficient in this range [14,69,70]. Thrombus formation during blood coagulation also significantly affects the blood absorption in the 200–270 GHz range [71]. Many blood components, such as hemoglobin, leukocytes, proteins (including albumin and globulin), ions (including Na^+, Ca^{2+}, and K^+), lipids, and so forth, affect the blood absorption coefficient in the THz [72]. The diagram of sensitivity of physiologically relevant analytes determination using THz pulse spectroscopy methods and comparison with traditional laboratory methods are shown in Figure 11.5. It can be seen that THz technologies are still inferior to traditional analytical methods in terms of the sensitivity of biomolecule detection. However, THz pulse spectroscopy is a label-free technology and can provide information on the interaction of biomolecules in aqueous solutions.

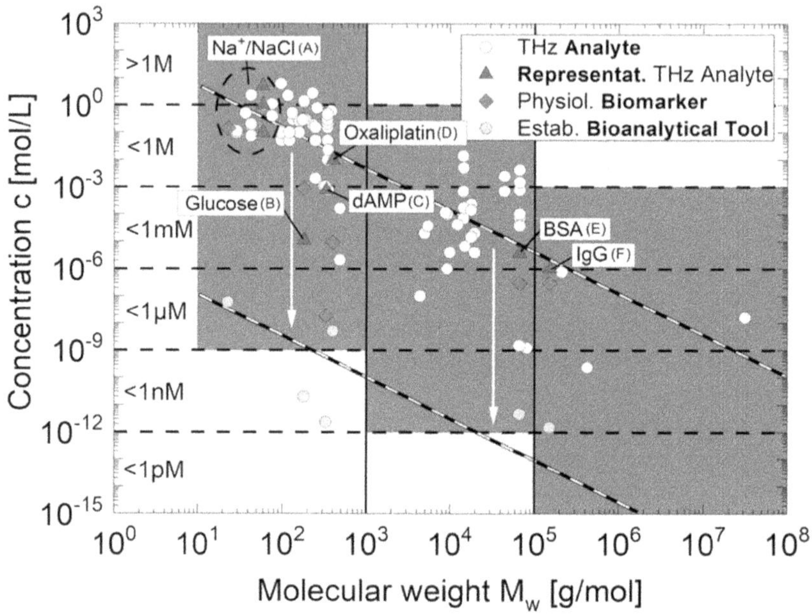

FIGURE 11.5 Diagram of sensitivity of physiologically relevant analytes determination using THz pulse spectroscopy methods (white dots) and comparison with traditional laboratory methods (black dots): the representative analytes are black triangles; the representative physiological concentration levels are black squares. The upper and the lover dashed lines represent the overall trend of decreasing detection limits with increasing molecular weights for THz and laboratory methods, accordingly [72].

The influence of blood glucose concentration on its optical properties in the THz frequency range has been studied most reliably [73–76]. In particular, when analyzing the blood of 70 patients, the absorption coefficient was demonstrated in the THz region of the spectrum and blood glucose levels to have a linear dependence [75]. The absorption index of blood plasma of rats with a severe course of experimental diabetes was shown to be more than two times lower than this value for rats with uncomplicated diabetes [77]. This suggests that the contribution in the total absorption may be given by some components found in blood plasma of patients with diabetes complications. One of such components may be glycation products [78].

A decrease in the absorption coefficient of blood plasma of mice with Ehrlich's transfected carcinoma in the THz region of the spectrum was demonstrated [79]. The authors suggested that it can be caused by changes in the blood plasma composition during carcinoma development.

The study of rat blood serum in the dynamics of experimental liver cancer development in the frequency range of 0.05–1.0 THz showed that THz spectra of samples from healthy animals and after tumor implantation have a similar shape, but differ in amplitude, which correlates with the protein content in the samples [53]. The transmission of the blood serum in the THz range was analyzed after normalization to the signal passing through the distilled water in the same cuvette, measured regularly in turn with the serum. Such an approach to normalization allows measuring tiny spectral changes more accurately.

All currently known reliable absorption spectra of blood components and solutions of biological molecules do not have narrow spectral features at frequencies below 3.0 THz [2,14,53,70]. The spectral differences of aqueous solutions of any substances are small because the shape of the measured spectrum is 90%–99% determined by water itself, which has a strong dispersion in the region of low THz frequencies. Therefore, serum or blood plasma can be represented as water with some addition of dissolved substances, and the Debye model (1) can be applied to the description of THz spectra of blood components. It has been shown that the absorption coefficient and refractive index in the 0.2–2.0 THz range of blood plasma of rats with experimental diabetes are lower compared to those of healthy animals [80]. Diabetic rats have blood glucose content 3.7 times higher compared to intact rats. Assuming that the observed changes in THz spectra are due to changes in plasma water state, it was shown that there is a 1.2-fold decrease in the ratio $\Delta\varepsilon_1/\tau_1$ in rat blood plasma when glucose concentration is increased to 24 mM [80].

It has been shown that during the dynamics of experimental U87 glioblastoma development in mice, there is a decrease in $\Delta\varepsilon_1$ and an increase in the relaxation time τ_1, which leads to a 1.6-fold decrease in the $\Delta\varepsilon_1/\tau_1$ ratio on 21 days after tumor cell injection [81]. At this term of the experiment, there is a more than 34-fold increase in tumor volume compared to 7 days after injection of U87 glioblastoma cells. It leads to a significant change in serum composition. The decrease in the $\Delta\varepsilon_1$ contribution to the dielectric constant of the samples during tumor growth dynamics is due to the fact that the fraction of free water decreases while the fraction of bound water increases. The increase in the relaxation time τ_1 in the dynamics of tumor growth indicates a "slowing down" of the reorientation of water molecules in the hydrogen bond network. Thus, changes in the composition of blood serum in the dynamics of glioma development led to changes in the parameters of complex dielectric permittivity [81].

In order to reduce the contribution of water, it was proposed to study the THz spectra of samples of lyophilized serum of patients' blood pressed into tablets. It should be noted that no spectral features were found in the THz spectrum of lyophilized samples [82,83]. It has been shown to be possible to differentiate serum from healthy individuals and diabetic patients in the range of 0.2–1.4 THz [82]. The groups are separated in the principal component space constructed on the basis of THz data.

Liquid and lyophilized blood plasma of individuals with nodular thyroid diseases, which are the most common pathology of the endocrine system, were studied by THz pulsed spectroscopy [83]. A reliable differentiation of healthy individuals and patients with thyroid nodules and differentiation of patients with benign and malignant thyroid nodules by the values of absorption coefficient and refractive index of lyophilized serum were shown. Machine learning methods were used to analyze THz experimental data. A two-stage ensemble scheme was proposed (Figure 11.6a). The first step corresponded to a sensitivity estimation. In the result, the thyroid nodule group was separated from the healthy one (Figure 11.6c). The second step corresponded to a specificity estimation. In the result, benign and malignant groups are separated (Figure 11.6d). Linear separability between healthy and thyroid nodules classes with 92% sensitivity, 85% specificity, and 88% accuracy was demonstrated using the linear Support Vector Machine (SVM).

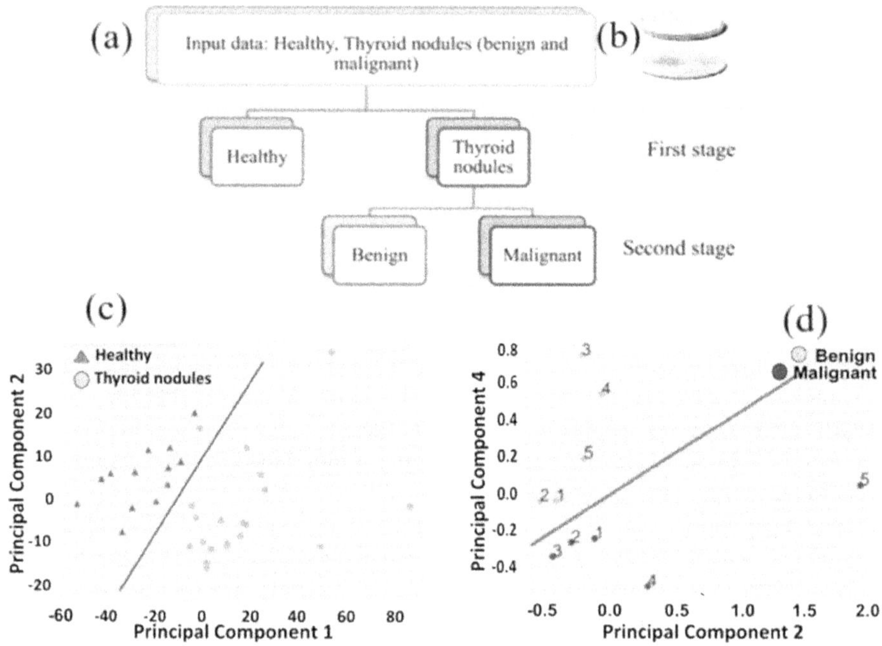

FIGURE 11.6 Differentiating malignant and benign thyroid nodules by analyzing blood plasma: a two-stage ensemble scheme (a); lyophilized blood plasma pressed in pellets (b); linear separability of the healthy and nodular thyroid diseases data by the SVM with linear kernel (c); and linear separability of the benign and malignant thyroid nodules plasma pellets transformed by principal component analysis (PCA with the Ornstein-Uhlenbeck kernel) (d) [83].

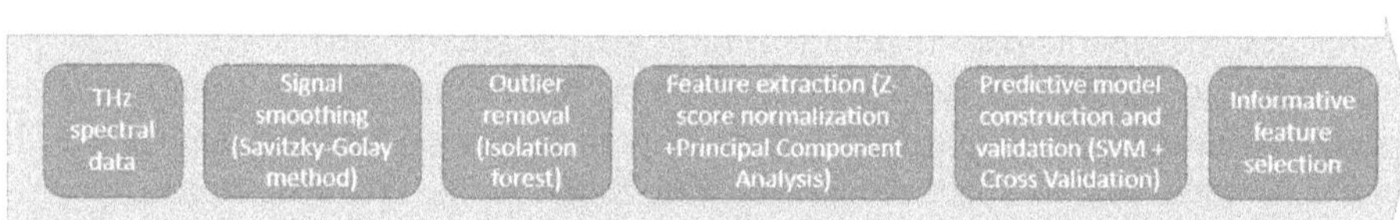

FIGURE 11.7 Proposed machine learning pipeline for THz spectral data analysis [84].

THz pulsed spectroscopy in combination with machine learning methods was used to analyze the blood serum of mice in the dynamics of experimental U87 glioblastoma [84]. Transmission spectra were measured in special cuvettes printed on a 3D printer. A machine learning pipeline was used to analyze THz spectral data (Figure 11.7).

The machine learning pipeline included THz spectra smoothing using the Savitsky–Golay filter, outlier removal using Isolation Forest, subtraction of the mean and normalizing data to the standard deviation, informative feature extraction, and dimensionality reduction by PCA. The predictive data model was created using SVM with a linear kernel. The proposed ML pipeline provided over 90% sensitivity, specificity, and accuracy of glioblastoma detection.

Application of machine learning methods to the analysis of pulsed THz blood spectroscopy data is one of the new trends and contributes to the advancement of the method in clinical practice [85–88]. These methods have been used for differential diagnosis of traumatic brain injuries [89,90] and blood plasma of patients with gliomas of different degrees of malignancy, including before and after tumor removal surgery [91].

11.6 CONCLUSION

THz methods have an essential potential in biomedical applications. THz pulsed spectroscopy provides the unique possibility to measure directly the refractive index n and absorption coefficient α, and hence complex permittivity of the sample. THz pulse spectroscopy is an effective tool for studying the reorientation of water hydrogen bonds in various forms of water in biological samples. The transition of water component from free to bound with surrounding molecules state and vice versa causes the change in the THz spectrum of blood serum. To reduce water influence, a preliminary biofluid lyophilization can be used. It opens a way to create effective methods of medical diagnostics using data from THz spectroscopy and visualization.

ACKNOWLEDGMENTS

The work was partially carried out within the State Assignment of IA&E SB RAS (FWNG-2024-0025), NRC "Kurchatov Institute" (in terms of THz technique), and was partially funded by the Ministry of Science and Higher Education of the Russian Federation grant # 075-15-2024-557 dated 04/25/2024 (data analysis by Yu.K.).

REFERENCES

1. Zhang, X.-C., Xu, J. *Introduction to THz Wave Photonics*. Springer, New York, 2010. doi:10.1007/978-1-4419-0978-7.
2. Smolyanskaya, O.A., Chernomyrdin, N.V., Konovko, A.A., Zaytsev, K.I., Ozheredov, I.A., Cherkasova, O.P., Nazarov, M.M., Guillet, J.P., Kozlov, S.A., Kistenev, Yu.V., Coutaz, J.L., Mounaix, P., Vaks, V.L., Son, J.H., Cheon, H., Wallace, V.P., Feldman, Yu., Popov, I., Yaroslavsky, A.N., Shkurinov, A.P., Tuchin, V.V. Terahertz biophotonics as a tool for studies of dielectric and spectral properties of biological tissues and liquids. *Prog. Quant. Electron.* 2018, 62, 1–77. doi:10.1016/j.pquantelec.2018.10.001.
3. Glagolewa-Arkadiewa, A. Short electromagnetic waves of wave-length up to 82 microns, *Nature*. 1924, 113, 640.
4. van Exter, M., Fattinger, C., Grischkowsky, D. Terahertz time-domain spectroscopy of water vapor, *Opt. Lett.* 1989, 14(20), 1128–1130. doi:10.1364/ol.14.001128.
5. Xu, L., Zhang, X.-C., Auston, D.H. Terahertz beam generation by femtosecond optical pulses in electrooptic materials, *Appl. Phys. Lett.* 1992, 61(15), 1784–1786. doi:10.1063/1.108426.
6. Lee, Y.S. *Principles of Terahertz Science and Technology*. Springer, New York, 2009. doi:10.1007/978-0-387-09540-0.
7. Cherkasova, O.P., Serdyukov, D.S., Ratushnyak, A.S. et al. Effects of terahertz radiation on living cells: a review, *Opt. Spectrosc.* 2020, 128, 855–866. doi:10.1134/S0030400X20060041.

8. Zaytsev, K.I., Dolganova, I.N., Chernomyrdin, N.V., Katyba, G.M., Gavdush, A.A., Cherkasova, O.P., Komandin, G.A., Shchedrina, M.A., Khodan, A.N., Ponomarev, D.S., Reshetov, I.V., Karasik, V.E., Skorobogatiy, M., Kurlov, V.N., Tuchin, V.V. The progress and perspectives of terahertz technology for diagnosis of neoplasms: a review, *J. Opt.* 2020, 22(1), 013001. doi:10.1088/2040–8986/ab4dc3.
9. Shirkavand, A., Tuchin, V.V., Jahangiri, F., Mohajerani, E. A review on terahertz non-destructive applications for wound and diabetic foot screening, *Opt. Quant. Electron.* 2022, 54, 467. doi:10.1007/s11082-022-03828-z.
10. Smirnova, I.N., Sapozhnikov, D.A., Kargovsky, A.V. et al. Lowest-lying vibrational signatures in corticosteroids studied by terahertz time-domain and Raman spectroscopies, *Vib. Spectrosc.* 2012, 62, 238–247. doi:10.1016/j.vibspec.2012.05.013.
11. Taday, P.F., Pepper, M., Arnone, D.D. Selected applications of terahertz pulses in medicine and industry, *Appl. Sci.* 2022, 12, 6169. doi:10.3390/app12126169.
12. Chen, X., Lindley-Hatcher, H., Stantchev, R.I., Wang, J., Li, K., Serrano, A.H., Taylor, Z.D., Castro-Camus, E., Pickwell-MacPherson, E. Terahertz (THz) biophotonics technology: instrumentation, techniques, and biomedical applications, *J. Chem. Phys. Rev.* 2022, 3(1), 011311. doi:10.1063/5.0068979.
13. Yu, L., Hao, L., Meiqiong, T., Jiaoqi, H., Wei, L., Jinying, D., Xueping, C., Weiling, F., Yang, Z. The medical application of terahertz technology in non-invasive detection of cells and tissues: opportunities and challenges, *RSC Adv.* 2019, 9(17), 9354–9363. doi:10.1039/c8ra10605c.
14. Angeluts, A.A., Balakin, A.V., Evdokimov, M.G., Esaulkov, M.N., Nazarov, M.M., Ozheredov, I.A., Sapozhnikov, D.A., Solyankin, P.M., Cherkasova, O.P., Shkurinov, A.P. Characteristic responses of biological and nanoscale systems in the terahertz frequency range, *Quant. Electron.* 2014, 44(7), 614–632. doi:10.1070/QE2014v044n07ABEH015565.
15. Nikitkina, A.I., Bikmulina, P.Y., Gafarov, E.R., Kosheleva, N.V., Efremov, Yu. M., Bezrukov, E.A., Butnaru, D.V., Dolganova, I.N., Chernomyrdin, N.V., Cherkasova, O.P., Gavdush, A.A., Timashev, P.S. Terahertz radiation and the skin: a review, *J. Biomed. Opt.* 2021, 26(4), 043005. doi:10.1117/1.JBO.26.4.043005.
16. D'arco, A., Di Fabrizio, M., Dolci, V. et al. THz pulsed imaging in biomedical applications, *Condens. Matter.* 2020, 5(2), 25. doi:10.3390/condmat5020025.
17. Tani, M., Herrmann, M., Sakai, K. Generation and detection of terahertz pulsed radiation with photoconductive antennas and its application to imaging, *Meas. Sci. Technol.* 2002, 13(11), 1739. doi:10.1088/0957-0233/13/11/310.
18. Yachmenev, A.E., Lavrukhin, D.V., Glinskiy, I.A. et al. Metallic and dielectric metasurfaces in photoconductive terahertz devices: a review, *Opt. Eng.* 2019, 59(6), 061608. doi:10.1117/1.OE.59.6.061608.
19. Ponomarev, D.S., Gorodetsky, A., Yachmenev, A.E. et al. Enhanced terahertz emission from strain-induced InGaAs/InAlAs superlattices, *J. Appl. Phys.* 2019, 125(15), 151605. doi:10.1063/1.5079697.
20. Zhang, X.C., Ma, X.F., Jin, Y., Lu, T.M., Boden, E.P., Phelp, P.D., Stewart, K.R., Yakymyshyn, C.P. Terahertz optical rectification from a nonlinear organic crystal, *Appl. Phys. Lett.* 1992, 61(26), 3080–3082. doi:10.1063/1.107968.
21. Schneider, A., Neis, M., Stillhart, M. et al. Generation of terahertz pulses through optical rectification in organic DAST crystals: theory and experiment, *J. Opt. Soc. Am. B* 2006, 23(9), 1822–1835. doi:10.1364/JOSAB.23.001822.
22. Shumyatsky, P., Alfano, R.R. Terahertz sources, *J. Biomed. Opt.* 2011, 16(3), 033001. doi:10.1117/1.3554742.
23. Sun, G., Xu, G.B., Ding, Y.J.J., Zhao, H.P., Liu, G.Y., Zhang, J. Efficient terahertz generation within InGaN/GaN multiple quantum wells, *IEEE J. Sel. Top. Quant. Electron.* 2011, 17(1), 48–53. doi:10.1109/JSTQE.2010.2049343.

24. Dodel G. On the history of far-infrared (FIR) gas lasers: thirty-five years of research and application, *Infrared Phys. Technol.* 1999, 40(3), 127–139. doi:10.1016/S1350-4495(99)00005-5.
25. Komandin, G.A., Chuchupal, S.V., Lebedev, S.P. et al., BWO generators for terahertz dielectric measurements, *IEEE Trans. Terahertz Sci. Technol.* 2013, 3(4), 440–444. doi:10.1109/TTHZ.2013.2255914.
26. Williams, B. Terahertz quantum-cascade lasers, *Nat. Photon.* 2007, 1(9), 517–525. doi:10.1038/nphoton.2007.166.
27. Yun, J., Oh, S.J., Song, K., Yoon, D., Son, H.Y., Choi, Y., Huh, Y., Rieh, J. Terahertz reflection-mode biological imaging based on InP HBT source and detector, *IEEE Trans. Terahertz Sci. Technol.* 2017, 7(3), 274–283. doi:10.1109/TTHZ.2017.2673549.
28. Kasagi, K., Suzuki, S., Asada, M. Large-scale array of resonant-tunneling-diode terahertz oscillators for high output power at 1 THz, *J. Appl. Phys.* 2019, 125, 151601. doi:10.1063/1.5051007.
29. Cherkasova, O., Peng, Y., Konnikova, M., Kistenev, Y., Shi, C., Vrazhnov, D., Shevelev, O., Zavjalov, E., Kuznetsov, S., Shkurinov, A. Diagnosis of glioma molecular markers by terahertz technologies, *Photonics.* 2021, 8, 22. doi:10.3390/photonics8010022.
30. Wang, B., Qin, X., Meng, K., Zhu, L., Li, Z. Classification of amino acids using hybrid terahertz spectrum and an efficient channel attention convolutional neural network, *Nanomaterials.* 2022, 12, 2114. doi:10.3390/nano12122114.
31. Fischer, B.M., Walther, M., Uhd Jepsen, P. Far-infrared vibrational modes of DNA components studied by terahertz time-domain spectroscopy, *Phys. Med. Biol.* 2002, 47(21), 3807–3814. doi:10.1088/0031–9155/47/21/319.
32. King, M.D., Korter, T.M. Noncovalent interactions between modified cytosine and guanine DNA base pair mimics investigated by terahertz spectroscopy and solid-state density functional theory, *J. Phys. Chem.* A. 2011, 115, 14391–14396. doi:10.1021/jp208883t.
33. Cherkasova, O.P., Nazarov, M.M., Konnikova, M., Shkurinov, A.P. THz spectroscopy of bound water in glucose: direct measurements from crystalline to dissolved state, *J. Infrared Millim. Terahertz Waves.* 2020, 41(9), 1057–1068. doi:10.1007/s10762-020-00684-4.
34. Takahashi, M., Ishikawa, Y. Terahertz vibrations of crystalline α-D-glucose and the spectral change in mutual transitions between the anhydride and monohydrate, *Chem. Phys. Lett.* 2015, 642, 29–34. doi:10.1016/j.cplett.2015.11.003.
35. Cherkasova, O., Nazarov, M., Sapozhnikov, D. et al. Vibrational spectra of corticosteroid hormones in the terahertz range, *Proc. SPIE.* 2010, 7376, 73760P. doi:10.1117/12.871047.
36. Fu, X., Liu, Y., Chen, Q., Fu, Y., Cui, T.J. Applications of terahertz spectroscopy in the detection and recognition of substances, *Front. Phys.* 2022, 10, 86953. doi:10.3389/fphy.2022.869537.
37. Ajito, K., Kim, J.-Y., Ueno, Y., Song, H.-J., Ueda, K., Limwikrant, W., Yamamoto, K., Moribe, K. Nondestructive multicomponent terahertz chemical imaging of medicine in tablets, *J. Electrochem. Soc.* 2014, 161(9), B171–B175. doi:10.1149/2.0201409jes.
38. Kutteruf, M.R., Brown, C.M., Iwaki, L.K., Campbell, M.B., Korter, T.M., Heilweil, E.J. Terahertz spectroscopy of short-chain polypeptides, *Chem. Phys. Le*tt. 2003, 375, 337–343. doi:10.1016/S0009-2614(03)00856-X.
39. Neu, J., Stone, E.A., Spies, J.A., Storch, G., Hatano, A.S., Mercado, B.Q., Miller, S.J., Schmuttenmaer, C.A. Terahertz spectroscopy of tetrameric peptides, *J. Phys. Chem. Le*tt. 2019, 10, 2624–2628. doi:10.1021/acs.jpclett.9b01091.
40. Mancini, T., Mosetti, R., Marcelli, A., Petrarca, M., Lupi, S., D'Arco, A. Terahertz spectroscopic analysis in protein dynamics: current status, *Radiation.* 2022, 2, 100–123. doi:10.3390/radiation2010008.

41. Cherkasova, O.P., Nazarov, M.M., Shkurinov, A.P., Fedorov, V.I. Terahertz spectroscopy of biological molecules, *Radiophys. Quantum Electron.* 2009, 52(7), 518–523. doi:10.1007/s11141-009-9152-9.
42. Yu, B., Zeng, F., Yang, Y. et al. Torsional vibrational modes of tryptophan studied by terahertz time-domain spectroscopy, *Biophys. J.* 2004, 86 (3), 1649–1654. doi:10.1016/S0006-3495(04)74233-2.
43. Cherkasova, O., Konnikova, M., Kistenev, Y., Vaks, V., Coutaz, J.-L., Shkurinov, A. Terahertz spectroscopy of biological molecules in solid, liquid, and gaseous states In: V.P. Gupta (Ed.), *Molecular and Laser Spectroscopy.* Elsevier, Amsterdam, 2022, pp. 433–478. doi:10.1016/B978-0-323-91249-5.00008-9.
44. Nazarov, M.M., Shkurinov, A.P., Kuleshov, E.A., Tuchin, V.V. Terahertz time-domain spectroscopy of biological tissues, *Quant. Electron.* 2008, 38(7), 647–654. doi:10.1070/QE2008v038n07ABEH013851.
45. Torii, T., Chiba, H., Tanabe, T., Oyama, Y. Measurements of glucose concentration in aqueous solutions using reflected THz radiation for applications to a novel sub-THz radiation non-invasive blood sugar measurement method, *Digit. Health.* 2017, 3, 1–5.
46. Duponchel, L., Simon, L., Basak, H. et al. Terahertz microfluidic sensor for in situ exploration of hydration shell of molecules, *Chemom. Intell. Lab. Syst.* 2013, 123, 28–35.
47. Nandi, N., Bhattacharyya, K., Bagchi, B. Dielectric relaxation and solvation dynamics of water in complex chemical and biological systems, *Chem. Rev.* 2000, 100, 2013.
48. Comez, L., Paolantoni, M., Sassi, P., Corezzi, S., Morresi, A., Fioretto, D. Molecular properties of aqueous solutions: a focus on the collective dynamics of hydration water, *Soft Matter.* 2016, 12, 5501.
49. Charkhesht, A., Regmi, C.K., Mitchell-Koch, K.R., Cheng, S., Vinh, N.Q. High-precision megahertz-to terahertz dielectric spectroscopy of protein collective motions and hydration dynamics, *J. Phys. Chem. B.* 2018, 122, 6341.
50. Nazarov, M.M., Cherkasova, O.P., Shkurinov, A.P. Study of the dielectric function of aqueous solutions of glucose and albumin by THz time-domain spectroscopy, *Quant. Electron.* 2016, 46, 488.
51. Born, B., Havenith, M. Terahertz dance of proteins and sugars with water, *J. Infrared Millim. Terahertz Waves.* 2009, 30, 1245.
52. Nazarov, M.M., Cherkasova, O.P., Shkurinov, A.P. A comprehensive study of albumin solutions in the extended terahertz frequency range, *J. Infrared Millim. Terahertz Waves.* 2018, 39(9), 840–853. doi:10.1007/s10762-018-0513-3.
53. Nazarov, M.M., Cherkasova, O.P., Lazareva, E.N., Bucharskaya, A.B., Navolokin, N.A., Tuchin, V.V., Shkurinov, A.P. A complex study of the peculiarities of blood serum absorption of rats with experimental liver cancer, *Opt. Spectrosc.* 2019, 126(6), 721–729.
54. Zaytsev, K.I., Kudrin, K.G., Karasik, V.E., Reshetov, I.V., Yurchenko, S.O. In vivo terahertz spectroscopy of pigmentary skin nevi: pilot study of non-invasive early diagnosis of dysplasia, *Appl. Phys. Lett.* 2015, 106(5), 053702.
55. Heugen, U., Schwaab, G., Bründermann, E., Heyden, M., Yu, X., Leitner, D., Havenith, M. Solute induced retardation of water dynamics probed directly by terahertz spectroscopy, *Proc. Natl. Acad. Sci. USA* 2006, 103, 12301.
56. Wolf, M., Gulich, R., Lunkenheimer, P., Loidl, A. *Biochim. Biophys.* Acta*: Proteins Proteomic.* 2012, 1824(5), 723.
57. Raicu, V., Feldman, Y. (Eds.), *Dielectric Relaxation in Biological Systems: Physical Principles, Methods, and Applications.* Oxford University Press, Oxford, 2015, p. 464.
58. Penkov, N., Shvirst, N., Yashin, V. et al. *J. Phys. Chem. B.* 2015, 119(39), 12664.
59. Nagai, M., Yada, H., Arikawa, T., Tanaka, K. Terahertz time-domain attenuated total reflection spectroscopy in water and biological solution, *J. Infrared Millim. Terahertz Waves.* 2006, 27(4), 505.

60. Yada, H., Nagai, M., Tanaka, K. *Chem. Phys. Lett.* 2008, 464, 166.
61. Cole, K.S., Cole, R.H. Dispersion and absorption in dielectrics I, Alternating current characteristics. *J. Chem. Phys.* 1941, 9, 341–351.
62. Shiraga, K., Suzuki, T., Kondo, N. et al. Quantitative characterization of hydration state and destructuring effect of monosaccharides and disaccharides on water hydrogen bond network, *Carbohydr. Res.* 2015, 406, 46.
63. Davidson, D.W., Cole, R.H. Dielectric relaxation in glycerol, propylene glycol, and n-propanol, *J. Chem. Phys.* 1951, 19, 1484–1490.
64. Frohlich, H. *Theory of Dielectrics: Dielectric Constant and Dielectric Loss.* Clarendon Press, Oxford, UK, 1958, 192 pp.
65. Speicher, M.R., Pantel, K. Tumor signatures in the blood, *Nat. Biotechnol.* 2014, 32(5), 441–443.
66. Nazarov, M., Shkurinov, A., Tuchin, V.V., Zhang, X.-C. (Eds.), *Handbook of Photonics for Biomedical Science.* CRC Press, Boca Raton, FL, 2010.
67. Tseng, T.-F., You, B., Gao, H.-C., Wang, T.-D., Sun, C.-K. Pilot clinical study to investigate the human whole blood spectrum characteristics in the sub-THz region, *Opt. Express.* 2015, 23(7), 9440–9451. doi:10.1364/OE.23.009440.
68. Gusev, S.I., Balbekin, N.S., Sedykh, E.A., Grineva, E.N., Khodzitsky, M.K. Influence of creatinine and triglycerides concentrations on blood optical properties of diabetics in THz frequency range. *J. Phys. Conf. Ser.* 2016, 735(1), 012088. doi: 10.1088/1742-6596/735/1/012087
69. Jeong, K., Huh, Y.-M., Kim, S.-H., Park, Y., Son, J.-H., Oh, S.J., Suh, J.-S. Characterization of blood using terahertz waves, *J. Biomed. Opt.* 2013, 18(10), 107008, doi:10.1117/1.JBO.18.10.107008.
70. Reid, C., Reese, G., Gibson, A., Wallace, V. Terahertz time domain spectroscopy of human blood, *IEEE J. Biomed. Health Inform.* 2013, 17(4), 774–778. doi:10.1109/JBHI.2013.2255306.
71. Sun, C.K., Chen, H.Y., Tseng, T.F., You, B., Wei, M.L., Lu, J.Y., Chang, Y.L., Tseng, W.L., Wang, T.D. High sensitivity of t-ray for thrombus sensing, *Sci. Rep.* 2018, 8(1), 3948. doi:10.1038/s41598-018-22060-y.
72. Weisenstein, C., Wigger, A.K., Richter, M. et al. THz detection of biomolecules in aqueous environments—status and perspectives for analysis under physiological conditions and clinical use, *J. Infrared Millim. Terahertz Waves.* 2021, 42, 607–646. doi:10.1007/s10762-021-00792-9.
73. Gusev, S.I., Demchenko, P.S., Cherkasova, O.P., Fedorov, V.I., Khodzitsky, M.K. Influence of glucose concentration on blood optical properties in THz frequency range. *Chin. Opt.* 2018, 11(2), 182.
74. Gusev, S.I., Nazarov, R. Kh., Demchenko, P.S., Tianmiao, Z., Khodzitsky, M.K. Terahertz time-domain spectroscopy in the assessment of diabetic complications. In: A. Dunaev and V. Tuchin (Eds.), *Biomedical Photonics for Diabetes Research,* 1st ed. Boca Raton, CRC Press, 2022, p. 25.
75. Chen, H., Chen, X., Ma, S., Wu, X., Yang, W., Zhang, W., Lie, X. Quantify glucose level in freshly diabetic's blood by terahertz time-domain spectroscopy. *J. Infrared Millim. Terahertz Waves.* 2018, 39, 399.
76. Cherkasova, O.P., Nazarov, M.M., Shkurinov, A.P. tudy of blood and its components by terahertz pulsed spectroscopy. In: *EPJ Web of Conferences.* 2018, vol. 195, p. 10003. doi:10.1051/epjconf/201819510003.
77. Cherkasova, O.P., Nazarov, M.M., Smirnova, I.N., Angeluts, A.A., Shkurinov, A.P. Application of time-domain THz spectroscopy for studying blood plasma of rats with experimental diabetes. *Physics of Wave Phenomena.* 2014, 22(3), 185.

78. Shirshin, E., Cherkasova, O., Tikhonova, T., Berlovskaya, E., Priezzhev, A., Fadeev, V. *J. Biomed. Opt.* 2015, 20(5), 051033. doi: 10.1117/1.JBO.20.5.051033.
79. Smolyanskaya, O.A., Kravtsenyuk, O.V, Panchenko, A.V., Odlyanitskiy, E.L., Guillet, J.P., Cherkasova, O.P., Khodzitsky, M.K. Study of blood plasma optical properties in mice grafted with Ehrlich carcinoma in the frequency range 0.1–1.0 THz, *Quant. Electron.* 2017, 47, 1031. doi:10.1070/QEL16383.
80. Cherkasova, O.P., Nazarov, M.M., Angeluts, A.A. et al. Analysis of blood plasma at terahertz frequencies, *Opt. Spectrosc.* 2016, 120, 50–57. doi:10.1134/S0030400X16010069.
81. Cherkasova, O.P., Konnikova, M.R., Nazarov, M.M., Vrazhnov, D.A., Kistenev, Yu.V., Shkurinov, A.P. Terahertz spectroscopy of mouse blood serum in the dynamics of experimental glioblastoma, *J. Biomed. Photon. Eng.* 2023, 9(3), 030308.
82. Lykina, A.A., Nazarov, M.M., Konnikova, M.R., Mustafin, I.A., Vaks, V.L., Anfertev, V.A., Domracheva, E.G., Chernyaeva, M.B., Kistenev, Yu. V., Vrazhnov, D.A., Prischepa, V.V., Kononova, Yu. A., Korolev, D.V., Cherkasova, O.P., Shkurinov, A.P., Babenko, A.Y., Smolyanskaya, O.A. Terahertz spectroscopy of diabetic and non-diabetic human blood plasma pellets, *J. Biomed. Opt.* 2021, 26(4), 043006. doi:10.1117/1.JBO.26.4.043006.
83. Konnikova, M.R., Cherkasova, O.P., Nazarov, M.M., Vrazhnov, D.A., Kistenev, Yu.V., Titov, S.E., Kopeikina, E.V., Shevchenko, S.P., Shkurinov, A.P. Malignant and benign thyroid nodule differentiation through the analysis of blood plasma with terahertz spectroscopy, *Biomed. Opt. Express.* 2021, 12(2), 1020–1035. doi:10.1364/BOE.412715.
84. Vrazhnov, D., Knyazkova, A., Konnikova, M., Shevelev, O., Razumov, I., Zavjalov, E., Kistenev, Y., Shkurinov, A., Cherkasova, O. Analysis of mouse blood serum in the dynamics of u87 glioblastoma by terahertz spectroscopy and machine learning, *Appl. Sci.* 2022, 12, 10533.
85. Kistenev, Y., Borisov, A., Vrazhnov, D. *Medical Applications of Laser Molecular Imaging and Machine Learning.* SPIE Press, Bellingham, 2021.
86. Koul, S.K., Kaurav, P. Machine learning and biomedical sub-terahertz/terahertz technology. In: A. Dunaev and V. Tuchin (Eds.), *Sub-Terahertz Sensing Technology for Biomedical Applications.* Springer, Singapore, 2022, pp. 123–145.
87. Park, H., Son, J.-H. Machine learning techniques for thz imaging and time-domain spectroscopy, *Sensors.* 2021, 21, 1186.
88. Jiang, Y., Li, G., Ge, H., Wang, F., Li, L., Chen, X., Lu, M., Zhang, Y. Machine learning and application in terahertz technology: a review on achievements and future challenges. *IEEE Access.* 2022, 10, 53761–53776.
89. Vrazhnov, D.A., Ovchinnikova, D.A., Kabanova, T.V., Paulish, A.G., Kistenev, Y.V., Nikolaev, N.A., Cherkasova, O.P. Terahertz time-domain spectroscopy of blood serum for differentiation of glioblastoma and traumatic brain injury. *Appl. Sci.* 2024, 14, 2872. doi:10.3390/app14072872.
90. Wang, Y., Wang, G., Xu, D. et al. Terahertz spectroscopic diagnosis of early blast-induced traumatic brain injury in rats. *Biomed. Opt. Express.* 2020, 11, 4085–4098.
91. Cherkasova, O., Vrazhnov, D., Knyazkova, A., Konnikova, M., Stupak, E., Glotov, V., Stupak, V., Nikolaev, N., Paulish, A., Peng, Y., Kistenev, Yu., Shkurinov, A. Terahertz time-domain spectroscopy of glioma patient blood plasma: diagnosis and treatment. *Appl. Sci.* 2023, 13, 5434. doi:10.3390/app13095434.

12 Purity Analysis of Adulterated Vegetable Oils by Optical Spectroscopy

Gagan Raju and Shyju Bhaskar
Manipal Academy of Higher Education

Rajib Biswas
Tezpur University

Budheswar Dehury
Manipal Academy of Higher Education

12.1 INTRODUCTION

Vegetable oils are edible oils extracted from plant sources, primarily from seeds such as sunflower, mustard, and palm. They serve many purposes and are present in nearly all processed foods. Additionally, they find applications in various industrial sectors, including cosmetics, soap manufacturing, fuel production, candle making, and lubricants. The issue of adulteration in food and cooking products, particularly vegetable oils, is a global concern. Adulteration typically involves mixing vegetable oils with lower-grade oils or other oils that closely resemble them in terms of their physical appearance. For example, olive oil may be adulterated with lower-quality olive oil or oils that bear a similar physical resemblance, such as sunflower oil. The adulterants introduced into vegetable oils can have two major consequences: they can compromise the authenticity and purity of the oil, or they can pose potential health risks to consumers. An example of an adulterant is argemone oil, derived from the seeds of Argemone mexicana (Mexican prickly poppy), which is often used to adulterate mustard oil [1]. Argemone seeds resemble mustard seeds, potentially leading to adulteration even before the oil extraction process. When used as an adulterant in vegetable oils, it can result in a clinical condition known as epidemic dropsy, a type of edema. Additionally, argemone oil can cause various health issues, such as dyspnea, diarrhea, and glaucoma [2,3]. Vegetable oils are also sometimes adulterated with artificial nonnatural petroleum derivative oils such as mineral oils, which are unsafe for consumption. Assessing the quality and purity of oils is a significant concern because of the expense, time consumption, and sample preparation required for

DOI: 10.1201/9781003570011-12

traditional techniques such as chemical methods and chromatography. To overcome these challenges, optical spectroscopy techniques such as fluorescence and Raman spectroscopy are used. These methods are less time-consuming, cost-effective, and do not necessitate extensive sample preparation. By using spectroscopy methods, vegetable oils can be analyzed with a minimum sample quantity [4,5].

12.2 MATERIALS AND METHODS

12.2.1 Sample Collection

Different types and brands of vegetable oils, such as olive oil, coconut oil, sunflower oil, palm oil, and mustard oil, were collected from markets and mills. The adulteration concentration ranges from 0% to 100% with an increment of 10% for olive oil, sunflower oil, olive oil, and palm oil. All the oils were opened on the same day. The readings were taken periodically.

12.2.2 Fluorescence Spectroscopy

The collected vegetable oils were subjected to fluorescence spectroscopy in both pure and adulterated forms via a Corning Flat Bottom 96-well plate and a Varioskan flash. The oil samples were pipetted into a 96-well plate, and analysis was carried out for different types of oil collected from the mill and the local market. The readings were taken from an emission of 280–800 nm. The results were plotted via the software Origin Pro [6].

12.2.3 Raman Spectroscopy

All the samples were also subjected to Raman spectroscopy via the Ava Raman system. The integration time used was 30 s, and the results were plotted via Origin Pro [7,8].

12.2.4 Fourier Transform Infrared (FT-IR) Spectroscopy

FTIR spectra were recorded on a Bruker alpha FTIR spectrometer (Germany) at 375–7,500 cm^{-1} with KBr Optics for different samples of vegetable oils. The vegetable oil samples were mixed with KBr and mounted onto a KBr pellet holder. A spectrum of vegetable oils was generated [9].

12.3 RESULTS AND DISCUSSION

12.3.1 Fluorescence Spectroscopy

Figure 12.1 shows the fluorescence spectra of different samples of different types of pure vegetable oils. Figure 12.2 shows the fluorescence spectra of the collected samples of adulterated vegetable oils. Fluorescence spectroscopy can be used to estimate and quantify the concentrations of different fluorophores (such as chlorophyll

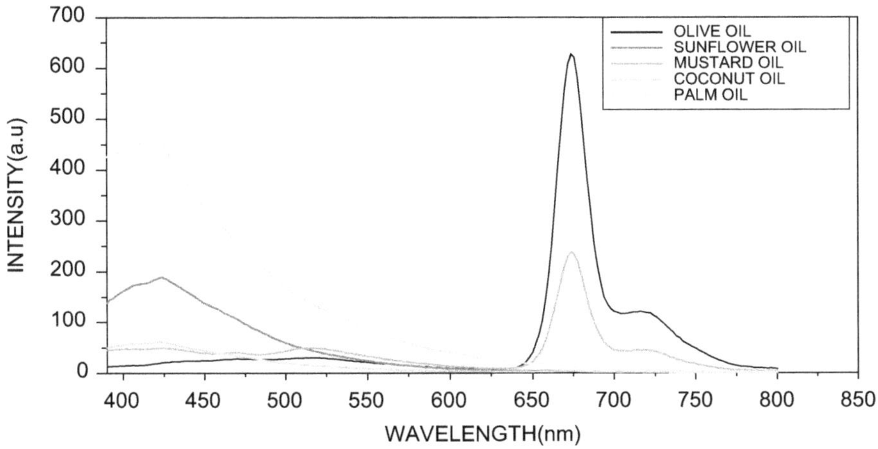

FIGURE 12.1 Fluorescence spectra of different types of vegetable oils at an emission wavelength of 360 nm.

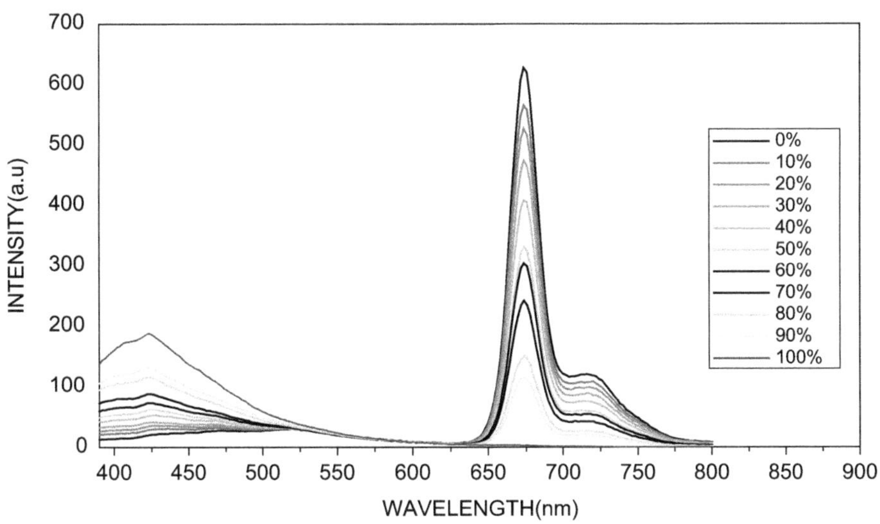

FIGURE 12.2 Fluorescence spectra of olive oil adulterated with sunflower oil at an emission wavelength of 360 nm.

and vitamin E) present in vegetable oils. Here, chlorophyll, which is present in the region of approximately 650–730 nm, and vitamin E, which is present in the region of approximately 500–600 nm, were detected. Olive oil has the highest amount of chlorophyll, and palm oil, coconut oil, and sunflower oil have few chlorophyll pigments [10–12].

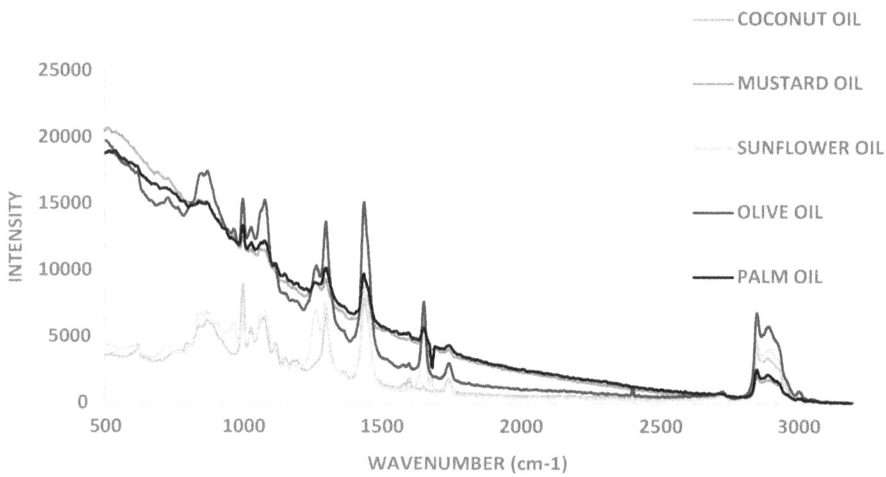

FIGURE 12.3 Raman spectra of different types of vegetable oil.

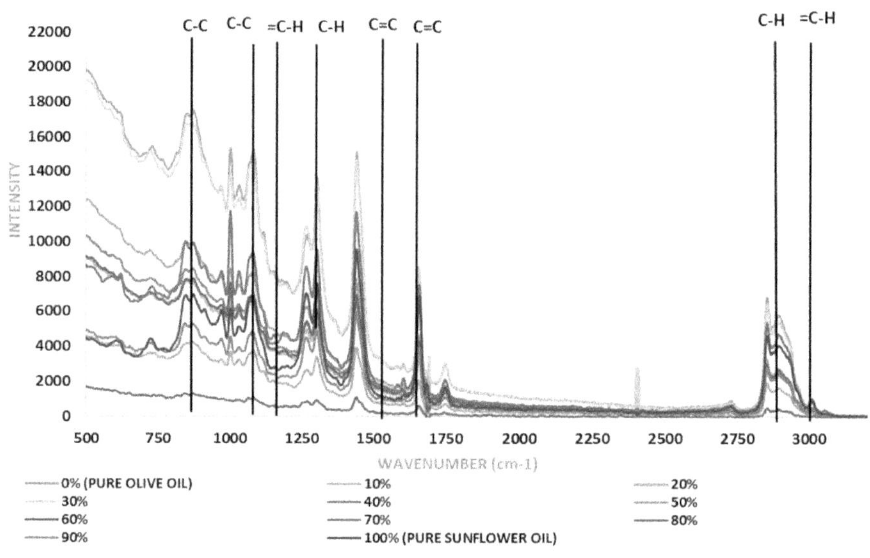

FIGURE 12.4 Raman spectra of olive oil adulterated with sunflower oil at different concentrations.

12.3.2 Raman Spectroscopy

Figure 12.3 shows the Raman spectra of different types of pure vegetable oils. Figure 12.4 shows the Raman spectra of the adulterated vegetable oils. In Raman spectroscopy, the spectra provide information about the chemical nature of vegetable oil samples. This property is unique to each oil. Each peak corresponds to a particular chemical bond in the spectra, such as the wavenumber zone at 3,015 cm^{-1}, which represents the =C-H bond; 2,970–2,860 cm^{-1}, which represents the C-H bond; 1,750 cm^{-1}, which represents the C=O bond; and 1,100–800 cm^{-1}, which represents the C-C bond (Table 12.1).

12.3.3 Fourier Transform Infrared (FT-IR) Spectroscopy

Figure 12.5 shows the FT-IR spectra of the vegetable oils. FT-IR was used to interpret the chemical composition of the oils. Each peak corresponds to different bonds in the spectra; for example, the peak at 3,470 cm^{-1} represents O-H bonds, the peak at 2,925 cm^{-1} represents CH_2 bonds, and the peak at 1,465 cm^{-1} represents CH_2 bonds (Table 12.2) [13, 14,15].

12.4 CONCLUSION

In this study, different optical spectroscopy techniques, such as fluorescence spectroscopy, Raman spectroscopy, and FT-IR, were used. We can conclude that the purity of different types of vegetable oils can be analyzed via optical spectroscopy methods. By using fluorescence spectroscopy, the chlorophyll and vitamin E profiles of vegetable oils can be studied. The fluorescence spectra provide insight into the excitation of fluorophores in vegetable oil. Different emission wavelengths have been used in fluorescence spectroscopy. Thus, vegetable oils can be characterized on the basis of the chlorophyll profile. Raman spectroscopy provides information regarding the chemical composition of the oil (for example, the types of bonds present in the oils), which can be used to interpret the fatty acids in the oil [16,17].

TABLE 12.1
Raman Shifts and Vibrational Bonds Present in Vegetable Oil

Band/cm	Bonds
3,470	vO-H
3,008	vC-H
2,925	vCH_2
2,854	vCH_2
1,748	vC=O
1,651	vC=C
1,465	δCH_2
1,373	δCH_3
1,163	vC-O
721	CH_2

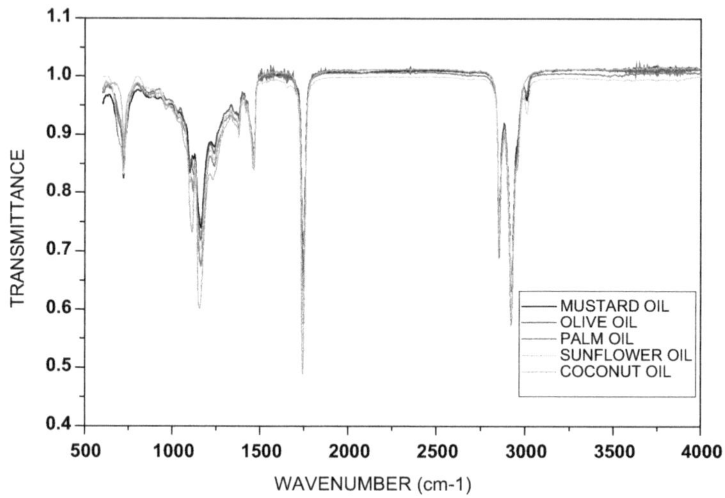

FIGURE 12.5 FT-IR spectra of different types of vegetable oil.

TABLE 12.2
FT-IR Wavenumbers and Corresponding Bonds

Raman Shift (cm^{-1})	Change Observed	Vibrational Bond
670	Increased	νC-C (aliphatic)
760	Increased	νC-C (aliphatic)
1,219	Increased	C-O-C asymmetric stretching
1,740	Increased	ν C-O
1,085	Decreased	ν C-C
1,305	Decreased	In-phase methylene twisting
1,268	Decreased	(δ=C-H) deformation in cis double bond
1,443	Decreased	δCH
1,654	Decreased	C=C cis double bond

ACKNOWLEDGMENTS

We thank the ASEAN-India Science & Technology Development Fund (AISTDF), Anusandhan National Research Foundation (ANRF), Government of India, India (Project Number- ANRF/F/84/2025-2026), and the Indian Council of Medical Research (ICMR) (Project Number-Dev/SG-00170/2024) Government of India, India, for financial support. The authors thank the Manipal School of Life Sciences, Manipal Academy of Higher Education, Manipal, Karnataka, India, for providing the infrastructure.

REFERENCES

1. Babu, C. K., Khanna, S. K., & Das, M. (2007). Adulteration of mustard cooking oil with argemone oil: do Indian food regulatory policies and antioxidant therapy both need revisitation? *Antioxidants & Redox Signaling*, 9(4), 515–525.
2. Arlorio, M., Coisson, J. D., Bordiga, M., Travaglia, F., Garino, C., Zuidmeer, L., Martelli, A. (2010). Olive oil adulterated with hazelnut oils: simulation to identify possible risks to allergic consumers. *Food Additives and Contaminants—Part A Chemistry, Analysis, Control, Exposure and Risk Assessment*, 27(1), 11–18.
3. Seppänen-Laakso, T., Laakso, I., & Hiltunen, R. (2002). Analysis of fatty acids by gas chromatography, and its relevance to research on health and nutrition. *Analytica Chimica Acta*, 465(1–2), 39–62.
4. Xu, L., Xu, X., Xiong, H., Chen, L., & Li, Y. (2014). Rapid detection of vegetable cooking oils adulterated with inedible used oil using fluorescence quenching method with aqueous CTAB-coated quantum dots. *Sensors and Actuators, B: Chemical*, 203, 697–704.
5. McReynolds, N., Auñón Garcia, J. M., Guengerich, Z., Smith, T. K., & Dholakia, K. (2016). Optical spectroscopic analysis for the discrimination of extra-virgin olive oil. *Applied Spectroscopy*, 70(11), 1872–1882
6. Kongbonga, Y. G. M., Ghalila, H., Onana, M. B., Majdi, Y., Lakhdar, Z. B., Mezlini, H., & Sevestre-Ghalila, S. (2011). Characterization of vegetable oils by fluorescence spectroscopy. *Food and Nutrition Sciences*, 2(7), 692–699.
7. Jiménez-Sanchidrián, C., & Ruiz, J. R. (2016). Use of Raman spectroscopy for analyzing edible vegetable oils. *Applied Spectroscopy Reviews*, 51(5), 417–430.
8. Huang, F., Li, Y., Guo, H., Xu, J., Chen, Z., Zhang, J., & Wang, Y. (2016). Identification of waste cooking oil and vegetable oil via Raman spectroscopy. *Journal of Raman Spectroscopy*, 47(7), 860–864.
9. Rusak, D. A., Brown, L. M., & Martin, S. D. (2003). Classification of vegetable oils by principal component analysis of FTIR spectra. *Journal of Chemical Education*, 80(5), 541.
10. Da Silva, C. E. T., Filardi, V. L, Pepe, I. M., Chaves, M. A., & Santos, C. M. S. (2015). Classification of food vegetable oils by fluorimetry and artificial neural networks. *Food Control*, 47, 86–91.
11. Krause, G. H., & Weis, E. (1991). Chlorophyll fluorescence and photosynthesis: the basics. *Annual Review of Plant Physiology and Plant Molecular Biology*, 42(1), 313–349.
12. Guimet, F., Ferre, J., Boque, R., Vidal, M., & Garcia, J. (2005). Excitation-emission fluorescence spectroscopy combined with three-way methods of analysis as a complementary technique for olive oil characterization. *Journal of Agricultural and Food Chemistry*, 53(24), 9319–9328.
13. Baeten, V., Hourant, P., Morales, M. T., & Aparicio, R. (1998). Oil and fat classification by FT-Raman spectroscopy. *Journal of Agricultural and Food Chemistry*, 46(7), 2638–2646.
14. Alexa, E., Dragomirescu, A., Pop, G., Jianu, C., & Dragoş, D. (2009). The use of FT-IR spectroscopy in the identification of vegetable oils adulteration. *Journal of Food, Agriculture and Environment*, 7(2), 20–24.
15. Allam, M. A., & Hamed, S. F. (2007). Application of FTIR spectroscopy in the assessment of olive oil adulteration. *Journal of Applied Sciences Research*, 18(February), 102–108.
16. Rusak, D. A., Brown, L. M., & Martin, S. D. (2003). Classification of vegetable oils by principal component analysis of FTIR spectra. *Journal of Chemical Education*, 80(5), 541.

13 A Study of Computer-Assisted Diagnosis of Alzheimer's Disease by Deep Learning Models

Santanu Roy
Dayananda Sagar College of Engineering

13.1 INTRODUCTION

Alzheimer's Disease (AD) is a severe and life-limiting neurodegenerative disorder [1] that predominantly affects older individuals. AD is the most prevalent cause of dementia, accounting for approximately 60%–80% of all dementia cases [2]. As a neurodegenerative form of dementia, AD typically begins with mild cognitive impairment (MCI) and progressively deteriorates over time. It damages neurons, leading to memory loss, impaired cognitive abilities, and difficulties in carrying out routine tasks [3,4]. Thus, AD is characterized as a progressive, multifactorial neurological disorder. Individuals diagnosed with MCI are at a significantly higher risk of developing AD compared to the general population [5]. The clinical symptoms of AD often emerge only after several years of pathological changes in the brain, with the disease process believed to commence two decades or more before symptoms become apparent [6]. According to Alzheimer's Disease International (ADI), over 70 million individuals globally are currently living with dementia. This figure is projected to rise to 152 million by 2050, indicating that a new case of dementia develops every three seconds [2,7]. The prevalence of AD is age-dependent, with increasing incidence observed in older populations. According to the World Health Organization's (WHO) 2019 ranking of the top ten global causes of mortality, AD is identified as the seventh leading cause of death worldwide. Projections suggest that by 2050, approximately one in every 85 individuals globally will be living with AD [2].

Early manifestations of AD include short-term memory loss, language impairments, difficulties in logical reasoning, and eventually a decline in the ability to manage daily activities [4]. AD is primarily caused by the pathological accumulation of abnormal proteins, notably beta-amyloid plaques and tau neurofibrillary tangles, within the brain [8]. These toxic deposits interfere with normal cellular processes, leading to the progressive degeneration and death of neurons [8]. As neural connections deteriorate

 DOI: 0.1201/ 9781003570011-13

and synaptic communication between brain cells becomes increasingly impaired, individuals experience a gradual decline in cognitive functions, including memory, reasoning, and behavior. Neuro-anatomical alterations in the hippocampus [9] can be detected by assessing its volume and morphology, along with the characteristics of gray matter, using advanced neuro-imaging modalities such as computed tomography (CT) [10], positron emission tomography (PET) [11], and magnetic resonance imaging (MRI) [12]. Among these imaging techniques, MRI is the most commonly utilized due to its non-invasive nature, widespread accessibility, and minimal exposure to ionizing radiation [13]. Furthermore, evaluating changes in the cerebrospinal fluid (CSF) system [14] contributes to determining the stage of Alzheimer's disease. As the disorder advances, there is a progressive enlargement of the CSF spaces, accompanied by a marked reduction in the volume of both the cerebral cortex and hippocampus [9,15]. These subtle morphological changes in the cerebral cortex and hippocampus are often challenging for clinicians to detect through visual inspection alone [9]. Thus, manual extraction and interpretation of Alzheimer's-specific biomarkers from MRI scans are highly intricate and tedious tasks for radiologists. Visual fatigue is a common issue, especially after reviewing a large number (100) of such MRI slices. Hence, there is a critical demand for an automated computer-aided diagnosis (CAD) system [16] to facilitate the early and accurate detection of AD from MRI data.

The primary obstacle confronting Alzheimer's researchers is the absence of an effective and dependable treatment for AD to date. Nevertheless, current therapeutic approaches can alleviate symptoms or decelerate their progression of AD [4]. Therefore, the early diagnosis of AD, particularly during its prodromal stage, is of critical importance [16]. CAD systems are employed to ensure accurate and timely detection of AD, thereby helping to reduce the substantial caregiving costs associated with AD, which are projected to rise significantly. In the context of early AD detection from imbalanced MRI images, numerous researchers exploit mainly four different approaches by using deep learning models: (1) Utilize the conventional convolutional neural network (CNN) model and use handcrafted features along with the original data, which helps to reduce the solution space of the model, thus achieving a higher efficacy by this method. (2) Employ CNN models and use extensive image processing techniques and oversampling to increase the data sample especially in minor classes, (3) Utilize CNN models along with attention module or problem specific module, like channel attention, spatial attention, and so forth, which automatically handles the class imbalance issue to a certain extent, and (4) Utilize Vision Transformer (ViT) models, lightweight ViTs, and hybrid CNN-ViT models to tackle the class imbalance issue from MRI datasets.

13.2 EXISTING DEEP LEARNING MODELS FOR ALZHEIMER'S DETECTION

In this study, we begin by explaining the fundamental principles of CNN, pre-trained CNN models, and ViT models. Subsequently, we provide a summary of various CNN and ViT models, specifically proposed for the classification of different categories of dementia. This is followed by a comprehensive analysis of the implementation and performance of some of these existing models.

13.2.1 Brief Description of a Basic CNN Model and Numerous Pre-Trained CNN Models

The concept of CNNs emerged in the early 2010s, primarily for addressing computer vision tasks. A typical CNN architecture [17] is composed of three fundamental components: (1) Convolutional layers, (2) pooling layers, and (3) fully connected (FC) layers or dense layers, which are subsequently followed by an output layer, as illustrated in Figure 13.1. Convolutional layers automatically extract the features in two dimensions (or 1D in 1D-CNN), by utilizing the convolution operation. In these convolutional layers, all filter coefficients serve as trainable parameters. In contrast, pooling layers contain no learnable parameters. Pooling operations [17] are primarily employed to progressively reduce the spatial dimensions (typically by half) after each convolutional layer, thereby drastically reducing the overall number of trainable parameters from the model. Additionally, the pooling layer introduces the concept of a multi-scale hierarchical representation [18], which enables the network to learn distinctive and multiscale features across successive convolutional blocks. Thereafter, the flatten layer converts all these 2D features into 1D features, which are followed by dense layers or FC layers. Typically, the Softmax activation function [19] is employed in the final dense (output) layer for multi-class classification tasks, whereas the Sigmoid activation function is utilized for binary classification scenarios. In contrast, the Rectified Linear Unit activation function [20] is commonly adopted in all the intermediate layers of CNN architectures. Furthermore, Batch Normalization layers [21] are commonly integrated after each convolutional layer. These layers serve to standardize the feature activations by centering them around zero and constraining their range approximately within [−0.5, 0.5]. This normalization layer simplifies the loss function and thus accelerates the convergence rate during the training phase. As a consequence, the CNN model can alleviate the overfitting problem to some extent.

Numerous standard universal CNN model architectures are readily available within the Keras library, including AlexNet [22], VGG-16 [23], VGG-19, Inception-V2 [24], Xception [25], MobileNet-V3 [26], ResNet-50 [27], and ResNet101. These models

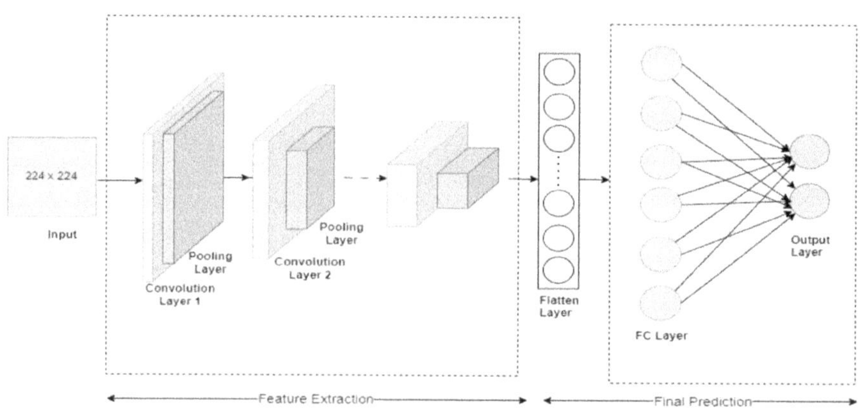

FIGURE 13.1 Fundamental block diagram of the convolutional neural network (CNN) model.

have been extensively implemented on the large-scale ImageNet dataset [28], which comprises over 3 billion images across 1,000 distinct classes. enabling the application of transfer learning. In this context, model training is initiated from these pre-initialized weights rather than commencing from random initialization. Consequently, the optimization process is guided toward the global minimum of the loss function from the very first epoch, ensuring accelerated convergence. Hence, those pre-trained CNN models can perform efficiently in terms of accuracy, precision, recall, and F1 score, even on small and imbalanced datasets.

Among these pre-trained CNN models, VGG-16 [23] initially garnered significant attention from the research community due to its architectural simplicity and uniform design. The model consistently applies max-pooling operations after each convolutional block, employing a pool size of 2×2 with a stride of =2, effectively reducing the spatial dimensions of the input tensor by half. One of the properties of VGG-16 is that the spatial dimensions of the feature maps remain constant within a single convolutional block; however, a reduction in spatial resolution occurs between consecutive convolutional blocks. In an effort to further improve performance, He et al. [27] attempted to increase the network depth beyond that of VGG-16. However, they observed a counterintuitive phenomenon, wherein both training and test errors increased as the number of layers grew. Their analysis revealed that this degradation in performance was not attributable to overfitting, but rather to the vanishing gradient problem, which hinders effective learning in very deep CNN architectures. To overcome this limitation, the authors introduced an identity-based skip connection, which is a direct link between the input of a convolutional layer and the output of a subsequent layer. This architectural innovation, known as the residual structure, enables improved gradient flow during backpropagation and facilitates the training of deeper networks. The introduction of the residual structure in 2016 marked a significant breakthrough in the field of computer vision. Consequently, the ResNet family of architectures, such as ResNet-34, ResNet-50, ResNet-101, and ResNet-152, has been widely adopted by researchers for a wide range of computer vision tasks.

Another major breakthrough in the domain of computer vision was the discovery of the Inception-v3 [24] architecture by the Google Research team. The Inception-v3 model demonstrated the capability to extract features at multiple scales, which was a pivotal innovation. This multi-scale feature extraction mechanism proved to be highly effective in various object detection and image classification tasks, where target objects often appear at varying spatial resolutions. Particularly in medical image analysis, such as tumor detection, pathological features may exhibit substantial variability in size and shape across different samples. Therefore, for the first time, this architectural design offered a practical solution to the challenges posed by scale variation in real-world scenarios. An illustration of a single Inception unit block is provided in Figure 13.2. As shown, convolutional layers with different kernel sizes (e.g., 3×3 and 5×5) are utilized to facilitate multi-scale feature extraction. Notably, 1×1 convolutional layers are employed prior to the 3×3 and 5×5 convolutional layers to reduce (or regulate) the spectral dimensionality, thereby significantly minimizing the number of trainable parameters within the CNN architecture. Additionally, 1×1 convolution layers are used in other parallel branches of the block, although for distinct purposes. In one such branch, the 1×1 convolution does

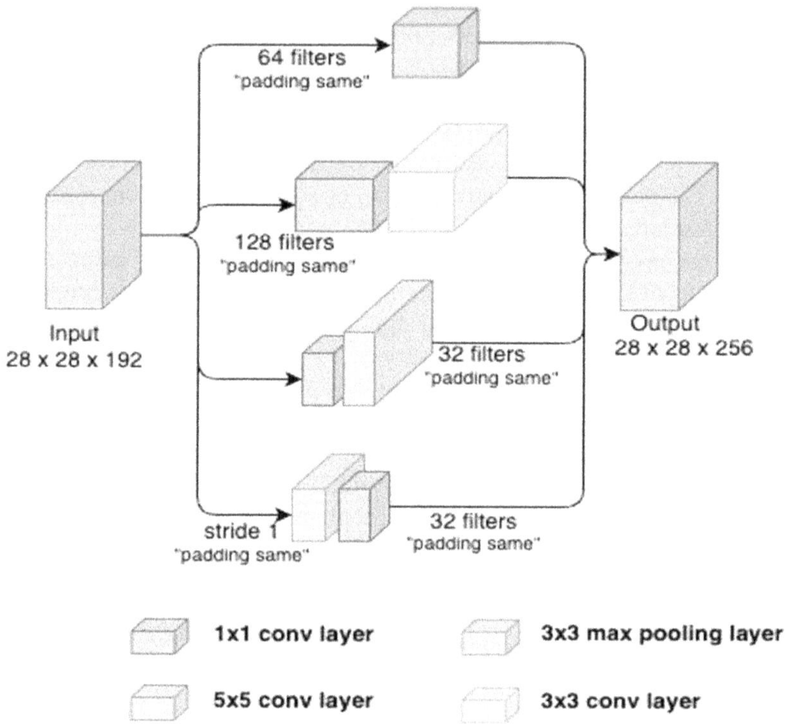

FIGURE 13.2 Fundamental unit block diagram of the Inception-V3 model.

not serve to extract spatial features; instead, it scales the channel-wise activations, effectively performing an implicit form of feature augmentation. In the fourth parallel branch, a max-pooling operation with a stride of 1 is followed by a 1×1 convolutional layer, which further enriches the feature space by introducing statistically diverse representations relative to the original features. Collectively, these parallel connections contribute to the generation of a rich and diverse set of features throughout the network, thus enhancing the generalization ability of the model.

In recent years (2015–2023), numerous researchers have invented a wide range of modifications and advancements to the existing Inception-V3 and ResNet CNN architectures. For instance, Xception [25] can be considered an evolved variant of Inception-V3, wherein similar Inception-style modules are repeated; however, it introduces depth-wise separable convolution (DWSC) layers for the first time. These DWSC layers offer a substantial reduction in computational complexity compared to traditional convolutional layers. Another notable modification is MobileNet-V2 [26], which builds upon the ResNet architecture and incorporates Global Average Pooling (GAP) [29] in place of the conventional flattening operation. This design choice significantly decreases the number of trainable parameters, particularly within the fully connected layers that follow the convolutional backbone. Subsequently, the authors extensively incorporated DWSC layers [25] throughout the MobileNet architecture, thereby rendering the model significantly more lightweight compared to conventional

CNN architectures. This design enables efficient deployment on resource-constrained devices such as smartphones. Furthermore, several other advancements have been explored, such as the use of dilated convolutions [30], group convolutions [31], Shuffle-Net [32], and ConvNeXt [33], all aimed at reducing the computational burden of deep neural networks while also maintaining or improving the model performance.

Despite the notable success of CNN-based deep learning models, a key limitation lies in their tendency to overly rely on specific features inherent to the training dataset [34]. As a consequence, these models often exhibit limited generalization capability, particularly when there exist statistical discrepancies between the training and testing (or unseen) datasets. Such variations are commonly observed when MRI scans are acquired from diverse patient populations. In real-world clinical settings, if such a CNN-based deep learning model is deployed in hospitals where clinicians would prefer to directly input their own untested hospital-specific data into the deep learning model for Alzheimer's disease diagnosis, there exists a potential risk of degraded model performance on unseen testing data. Hence, CNN-based models may not be considered entirely reliable or preferable diagnostic tools by clinicians, due to their limitations in generalization ability.

Another limitation of CNN models is their tendency to disregard intermediate or low-level features extracted in the earlier layers [35], becoming predominantly dependent on high-level (global) features obtained from the deeper layers. This over-reliance on later-stage representations may hinder the model's ability to capture fine-grained or localized information critical for certain tasks. To address this limitation, several researchers have proposed the incorporation of direct skip connections from the outputs of earlier convolutional layers to the output of the last convolutional layer, through a concatenation layer, which is followed by the GAP layer [29]. Such approaches enable the model to effectively capture both local and global features from medical images, thereby enhancing its representational capacity and diagnostic performance.

13.2.2 Basic Principle of ViTs

The concept of the transformer model comes from the notion of self-attention [36], which enables the model to capture long-range dependencies within data sequences. Transformer models have become very popular for the natural language processing (NLP) problem [37]; however, recently, they have also drawn the attention of researchers in the computer vision field [38,39]. Numerous researchers wanted to mimic the principles of transformers [38,40] for the computer vision problem, where the dataset is an image or video. Such a model architecture is named as ViT. Here, the input images are first partitioned into a sequence of fixed-size patches that serve a role analogous to word tokens in NLP applications. Each image patch is then flattened into a one-dimensional vector and subjected to a linear projection. This projection incorporates positional encoding to retain spatial information and encode the relative position of each patch within the original image. These 1D vectors are then fed into the original block of the transformer, which consists of two main modules: (1) a self-attention module and (2) a multilayer perceptron. The self-attention module is the central component of a transformer.

The self-attention mechanism enables the model to focus on various regions (or patches) of an image independently. Through this mechanism, the model systematically captures contextual dependencies among various patches. This is accomplished by computing three learnable matrices, which are query (Q), key (K), and value (V), for each patch [37].

$$\text{Attention}(Q, K, V) = \text{Soft max}\left(\frac{QK^T}{d_k^{1/2}}\right)V \tag{13.1}$$

Here, in Equation (13.1), the entire set of query vectors is collectively represented in the matrix Q, while the corresponding key and value vectors are similarly aggregated into matrices K and V, respectively. Subsequently, a scaled dot-product operation is performed between the query matrix Q and the transpose of the key matrix K, denoted as K^T, to facilitate the attention computation. In essence, for a given patch, the self-attention mechanism calculates a similarity score between its corresponding query vector and the key vectors of all other patches in the image, according to Equation (13.1). These similarity scores are then normalized by the factor $d_k^{1/2}$, typically using a softmax function, to determine attention weights (Figure 13.3).

This allows each patch representation to be contextually enriched with information from all other patches, thereby modeling the relationships among these patches to extract meaningful contextual information. This enables the model to enhance its generalization ability for the image classification task. The other component, MLP, works very similar way as it does in the conventional ANN or CNN model [17]. Instead of relying solely on MLPs and self-attention mechanisms, the fundamental building block of the ViT also integrates normalization layers and skip connections, as illustrated in Figure 13.2. The normalization layer standardizes the weighted patch embeddings, thus improving the converging rate of the model. While the skip connections facilitate improved gradient flow and augment a greater variety of features, together these two elements enhance the overall efficacy and generalization ability of the model.

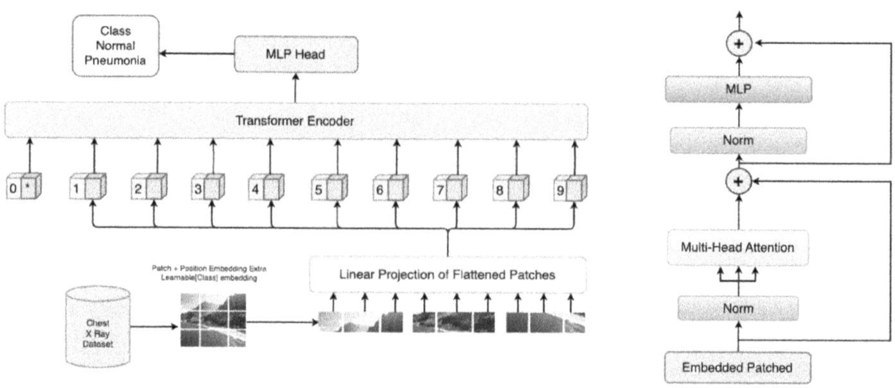

FIGURE 13.3 Fundamental block diagram of the vision transformer (ViT) model.

ViTs have demonstrated the ability to overcome several limitations associated with CNNs. Unlike CNNs, which may become overly dependent on dataset-specific features, ViTs process input images by partitioning them into smaller patches and modeling the relationships among these patches to extract meaningful contextual information. This patch-based attention mechanism enhances the model's capacity for generalization, particularly in scenarios involving significant statistical deviation between training and testing datasets. Consequently, ViT models have achieved notable success in a wide range of computer vision tasks and have garnered growing attention within the medical imaging domain as well.

Despite a remarkable success in the computer vision field, ViT models are not compatible with small and imbalanced datasets, which are prevalent for medical datasets. ViTs are highly prone to overfitting [41], especially when trained on limited or imbalanced datasets, primarily due to their computationally intensive, heavy architecture. To mitigate this limitation, recent research has focused on designing lightweight ViT architectures. Furthermore, unlike CNNs, ViTs inherently lack a multi-scale hierarchical structure [42]. Such hierarchical architectures are crucial in capturing features at various spatial resolutions, thereby enabling the extraction of a diverse set of patterns essential for object classification tasks. To address this, Swin Transformers [43] have been introduced, which incorporate principles from both CNNs and ViTs to establish a hierarchical representation within the transformer framework.

13.2.3 Existing Methods or Frameworks Proposed for AD Detection

A wide range of studies [44–77] have extensively utilized diverse CNN and ViT frameworks to develop automated CAD systems for AD detection using MRI data. These are explained below:

E2AD2C: A 2D-CNN framework: Helaly et al. [44] introduced a 2D CNN-based framework named E2AD2C, which is applied to an MRI dataset, shown in Figure 13.4. This dataset consists of 300 patients divided into four classes: AD, EMCI, LMCI, and NC. Their E2AD2C architecture comprises three convolutional layers, two fully connected layers, and a final output layer. The model architecture, influenced by the classical VGG-16 structure, was designed with a reduced number of trainable parameters, that is, only 8.2 lakhs. The authors also applied several pre-processing strategies, such as over-sampling, under-sampling, data augmentation, MRI filtering, and normalization, prior to feeding the data into this classifier. Although the authors reported superior performance in their study, their approach is heavily dependent on a set of image processing techniques utilized during the initial pre-processing stage of their pipeline. However, this strong reliance on specific pre-processing methods renders their framework somewhat impractical for broader applications. This is because such a fixed sequence of image processing steps may be effective for a particular MRI dataset, but is unlikely to generalize well to other MRI datasets that exhibit varying statistical properties.

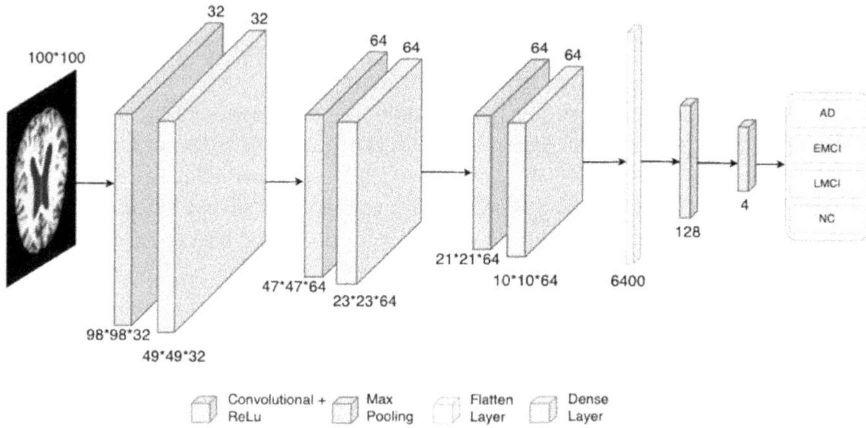

FIGURE 13.4 Block diagram of E2AD2C. (The first three blocks represent the Convolutional ReLu and Max pooling layers, followed by the flattened layer containing 6400 features. The flattened layer is followed by the dense layers).

Handcrafted feature extraction by Achilleos et al.: Achilleos et al. [45] introduced a manual feature extraction approach, wherein Haralick texture features [46] were computed to characterize hippocampal atrophy—an anatomical region critically implicated in the early detection of AD from MRI scans. These texture descriptors were subsequently integrated with hippocampal volumetric measures, and the combined feature set was evaluated using a ten-fold cross-validation strategy with a Decision Tree classifier for a four-class classification problem. Their experimental findings demonstrated that, by employing Random Forest classifiers augmented with predefined "Argumentation Rules," an accuracy of approximately 90%–91% could be attained. Nonetheless, their framework exhibited suboptimal sensitivity. Furthermore, the proposed methodology lacks full automation, and its performance is substantially influenced by the manually configured Argumentation Rules. The manual extraction of such diagnostically significant features is computationally intensive and laborious. Consequently, a considerable body of recent research has shifted toward CNNs for automated feature extraction from MRI data. This automated paradigm not only reduces the computational burden but also enhances the robustness and overall efficacy of the model.

Weighted Categorical Cross Entropy (WCCE)-based lightweight CNN by Masud et al.: Another potential direction of approaching these imbalanced MRI datasets is to deploy WCCE [47], which assigns weights for every class, which is inversely proportional to the number of images in that class. The conventional WCCE is presented by the following equation.

$$L_{\text{WCCE}}(y,p) = -\frac{1}{N}\sum_{i=1}^{N}\sum_{c=1}^{K}\alpha_c * \log(p_{i,c}) \tag{13.2}$$

In Equation (13.2), $p_{i,c}$ denotes the predicted probability that the ith sample belongs to class c, N represents the total number of samples, and α_c refers to the class-specific weighting factor. In the standard CCE formulation, this weighting factor α_c is uniformly assigned across all classes, i.e., $\alpha_c = 1$. In contrast, the WCCE introduces a non-uniform weighting scheme, where α_c is assigned inversely proportional to the frequency of class c in the dataset. For instance, consider a binary classification scenario comprising 80 samples of class "Apple" and 20 samples of class "Orange". To address the resulting class imbalance, the model must place greater emphasis on the minor "Orange" class. Accordingly, the weights can be set in a 1:4 ratio for Apple and Orange classes, respectively, reflecting the inverse of their instance ratios (i.e., Apple: Orange = 4:1). By incorporating such class-specific weighting, the neural network allocates proportionally greater attention to minority classes during training, thereby enhancing the model's robustness against class imbalance and improving its generalization, especially in the minor classes.

Masud et al. [48] employed such a WCCE loss function incorporated in a lightweight CNN architecture to classify four distinct categories of dementia from MRI scans. Their network architecture, as depicted in Figure 13.5, consists of only four convolutional layers and a single fully connected layer, with two MaxPooling layers interspersed between the convolutional layers. Due to the minimal number of trainable parameters, the model is prone to underfitting. Furthermore, the WCCE loss function demonstrates limited effectiveness in scenarios involving severe class imbalance. In this context, Tyagi et al. [49] were among the first to observe that the conventional application of WCCE is suboptimal for addressing highly imbalanced problems. Although WCCE may improve classification performance for minor classes, it often introduces performance instability for the major class. To address this limitation, Roy and Khurana [50] proposed a novel attention-guided loss function, wherein the WCCE weights are not only inversely proportional to the class frequencies but also are adjusted based on an additional statistical inference, namely, "the inter-class variance" derived from the dataset. While this approach provides a more informed weighting scheme, the computation of such statistical inference is resource-intensive and increases the overall computational complexity.

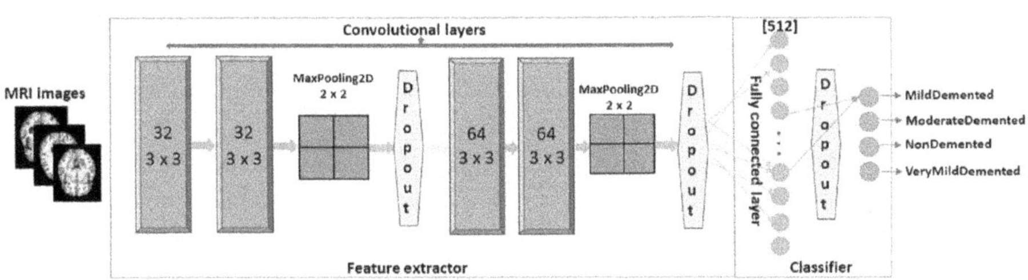

FIGURE 13.5 Block diagram of lightweight CNN by Masud et al. [48].

Multi-scale convolutional neural network (MSC-Net) and channel attention: Another significant approach to mitigating the class imbalance problem involves the incorporation of channel-wise or spatial attention mechanisms into CNN architectures. Hu et al. [51] introduced a Squeeze-and-Excitation Network (SE-Net), also known as "channel attention", is a significant breakthrough in the field of computer vision. By integrating the SE block within a CNN, the network is now able to explicitly model and emphasize inter-channel dependencies. This, in turn, enables the model to prioritize channels that extract more significant features, thereby improving the overall performance of the CNN model. The convolutional block attention module (CBAM), introduced by Woo et al. [52], is another significant advancement that incorporates a combination of channel and spatial attention in the CNN architecture. Their spatial attention module (SAM) starts from a concatenation of Max and Average pooling (stride 1), and thereafter, the feature map is passed through a single 7×7 convolutional filter, with a sigmoid activation function incorporated in it. Therefore, this combination now focuses only on one channel and thereby extracts spatial (multi-scale) global features, which enables their model to enhance the generalization capability on the large-scale ImageNet dataset. There has been a surge in the development of such attention-based modules in recent studies [53–54].

Liu et al. [55] proposed a novel multi-scale convolutional neural network (MSC-Net), incorporating dilated convolutional filters [56] with varying dilation rates (e.g., 1, 2, 3, and 4), and subsequently connected them in parallel within each block. Instead of processing adjacent pixels, those dilated convolutions introduce gaps (dilations) between the pixels in the filter; this effectively increases the receptive field of the filter without increasing its size. For example, a 3×3 dilated convolutional filter is analogous to a 5×5 convolutional filter. Hence, the parallel concatenation of the dilated convolutional layers facilitates the extraction of diverse features across multiple spatial scales, while incurring minimal computational overhead. The authors emphasized that multi-scale feature extraction plays a critical role in the diagnosis of AD from MRI datasets, as neuro-pathological markers may manifest at varying spatial resolutions and anatomical scales. Additionally, they have employed an SE-Net module, which works as a channel attention module. By integrating this SE-Net block into the multi-scale CNN model, the network can give more attention to the channels that extract more significant features, thereby improving the overall performance of the CNN model. Many more research studies [56–60] in a similar direction have been proposed for AD detection, that is, either by employing channel attention or spatial attention on top of CNN models.

AD-lite net by Roy et al.: S. Roy et al. [61] have recently proposed a very lightweight CNN model for Alzheimer's detection from MRI images' datasets. Their proposed model was designed for a specific purpose of alleviating the class imbalance problem; the model is illustrated in Figure 13.5. The authors deliberately selected three diverse MRI datasets to rigorously evaluate model generalization ability (Figure 13.6).

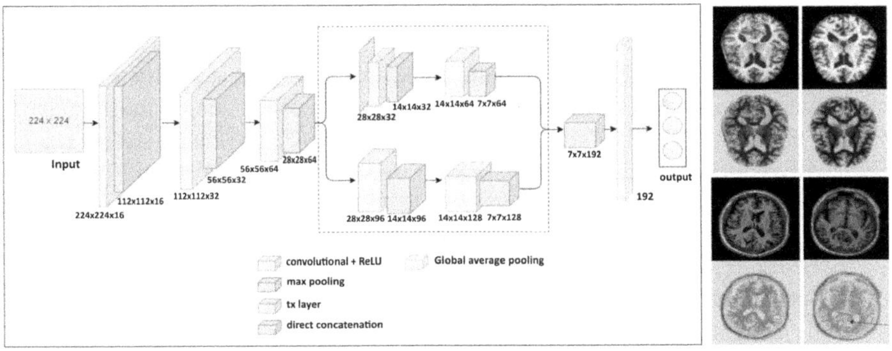

(a) Entire architecture of AD-Lite Net (b) Complementary features

FIGURE 13.6 Block diagram of AD-lite net.

Among them, the AD-dataset exhibited a slight class imbalance. The ADNI dataset, although relatively balanced, comprised a limited number of samples; whereas the OASIS dataset was considerably larger in scale, but it was severely class-imbalanced. The primary objective of their project was to develop a universal CNN model, trained from scratch, that could perform consistently and effectively across these diverse and challenging MRI datasets.

The AD-Lite Net model consists of two parallel connections. (1) One backbone or base CNN model, which consists of multiple numbers of convolutional layers and DWSC layers. These DWSC layers facilitated efficient feature extraction while significantly reducing computational complexity. The integration of DWSC layers along with a GAP layer substantially reduces the total number of trainable parameters in the AD-Lite Net architecture, constraining it to approximately 320,000. (2) A second parallel concatenation was developed, originating from a custom-designed transformation layer, referred to as the "custom tx layer", which maps the learned features into a distinct vector space. This transformation is intended to extract a complementary set of features, as illustrated in Figure 13.5b.

This "custom tx layer" can be represented by the following equation.

$$I_{tx}\left(f\right)_{x\times y} = m * \left(255\ - I\left(f\right)_{x\times y}\right) \tag{13.3}$$

where $I_{tx}(f)$ is the output of that transformation layer, $I(f)$ is the input tensor coming to the transformation layer, $x \times y$ is the spatial dimension of the feature map, and m here is a trainable parameter; the model automatically augments some complementary feature maps such that it will be imperative with respect to the final classification task.

The authors suggested that these complementary representations play a crucial role in capturing additional discriminative patterns pertinent to the diagnosis of AD. Therefore, with this parallel concatenation, the base model converges significantly faster and smoother than the previous one, thus enabling the model's generalization ability. The authors reported that their proposed AD-Lite Net architecture achieved higher efficacy across all three heterogeneous MRI datasets, attaining classification accuracy in the range of 98%–99%. Furthermore, they formally introduced a *lemma* stating that: *"Any parallel concatenation module integrated atop a CNN architecture, provided it extracts distinct and salient features, inherently facilitates automatic feature augmentation, thereby mitigating class imbalance to a certain extent.*" This theoretical insight justifies the inclusion of parallel concatenation in the AD-Lite Net model. One more research [62] was actually influenced by this research, but applied in different applications.

Lightweight ViT: Recently, Heo et al. [63] introduced the pooling-based Vision Transformer (PiT), a lightweight variant of the conventional ViT architecture. For the first time, the authors incorporated pooling layers within the ViT framework, achieving two primary objectives: (1) The reduction in the number of tokens (or patches), which substantially decreases the number of trainable parameters and helps to mitigate overfitting, particularly relevant for small or imbalanced datasets; and (2) The integration of a multi-scale hierarchical structure, which is crucial for effectively handling computer vision tasks. These attributes make PiT a very lightweight Vision Transformer, having only 46 lakhs parameters. Thus, PiT becomes well-suited for applications involving small-scale medical datasets. However, recent findings by Roy et al. [61] have shown that, despite PiT's suitability for limited data, its performance deteriorates significantly in the presence of severe class imbalance. This suggests that the self-attention mechanism alone is insufficient to effectively address class imbalance challenges. In a similar fashion, Zhu et al. [64] introduced a sparse self-attention (SSA) mechanism aimed at the early detection of Alzheimer's disease using MRI data. The proposed SSA block is capable of significantly reducing the number of feature representations from N to $logN$, thereby maintaining global contextual information while minimizing computational overhead. Several other studies [65–68] have also explored analogous strategies focused on reducing the computational complexity of ViT-based architectures for Alzheimer's disease classification from MRI scans.

Multi-scale spatial self-attention network (Mss-Net): Ye et al. [69] proposed a multi-scale hybrid CNN-ViT model, named "MSS-Net", which was designed to effectively capture both global and local (fine-grained) contextual features from brain MRI images. Initially, they employed a slice selection method as a pre-processing task, applied to volumetric (3D) MRI scans. Each 3D MRI image was decomposed into a series of 2D slices. Subsequently, the entropy of each 2D slice was computed, and the slice exhibiting the highest entropy was selected for further analysis. Their proposed architecture incorporates two distinct attention mechanisms. The first is a novel SAM, referred to as "SA-ROI", which integrates max-pooling and soft-pooling operations through a parallel concatenation to extract distinctive spatial features. In addition, the authors developed an SSA convolution (SSC) module. In this SSC, both ViT encoder blocks and convolutional layers are employed in an alternative way within its attention structure. This SSC module is further enhanced by a multiscale feature fusion mechanism. Their proposed SSC not only strengthens (multi-scale) spatial feature representations but also enables the model to provide significant attention to the relevant or suspicious regions during training. As a result, their proposed framework demonstrates strong efficacy when evaluated on complex and challenging MRI datasets.

Novel SAM-net by Roy et al.: Another trained-from-scratch CNN model, "SAM-Net", was recently proposed by Roy et al. [70]. The model architecture of SAM-net is illustrated in Figure 13.7. Their architecture comprises a lightweight backbone CNN, designed for computational efficiency. A novel SAM was integrated into the framework and connected in parallel with the base CNN, enabling the model to enhance spatial feature representations through attention-driven refinement. The authors were the first to observe that incorporating conventional attention mechanisms such as CBAM or SE-Net into the deeper layers of a CNN architecture may result in significant information loss. To mitigate this issue, they introduced a novel SAM, which employs 3×3 dilated convolutional filters with a dilation rate of 2, as opposed to the standard 7×7 convolutional filters. These dilated filters, combined with 'same' padding, effectively preserve the spatial resolution of feature maps while enabling the extraction of multiscale global features with reduced computational overhead. Moreover, unlike conventional SAMs, such as those derived from CBAM [52], which predominantly compress information across channels into a single aggregated map, the proposed attention mechanism preserves and utilizes multiple channel-wise representations. This strategy effectively mitigates the risk of substantial inter-channel information loss at deeper layers. Furthermore, due to employing multiple filters, a larger set of neurons

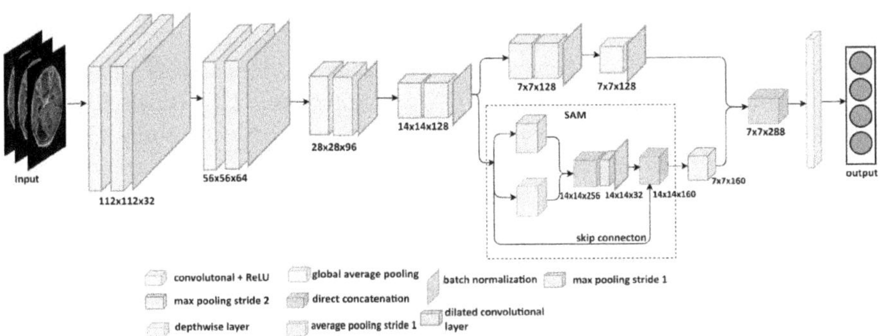

FIGURE 13.7 Block diagram of SAM-net.

associated with the SAM block actively contributes to the classifier's final decision. Hence, it enhances the impact of the SAM in their proposed CNN model.

Multi-patch attention network former (MAD-former): Ye et al. [71] proposed a novel hybrid CNN-ViT architecture, referred to as "MAD-former," designed to efficiently perform multi-scale feature extraction from MRI datasets. The proposed framework is primarily composed of three integral components: (1) a 3D brain feature extraction network (3D-BFEN), which employs five sequential 3D convolutional layers to capture discriminative volumetric features from MRI scans, and (2) a multi-scale attention structure, incorporating both multi-patch embedding and multi-headed self-attention mechanisms. Finally, this is followed by (3) an interpretable attention selection module (IASM). The design of MSAS facilitates the extraction of local patch features at two distinct spatial resolutions, specifically 8×8 and 4×4. The architecture of their ViT encoder closely aligns with that of the standard ViT; however, the integration of 3D-BFEN significantly reduces the input patch dimensions, thereby markedly decreasing the computational overhead of the ViT module. Furthermore, the utilization of a multi-scale patch self-attention mechanism enhances the efficacy of the attention operation compared to conventional ViT encoders. On the other hand, the IASM mechanism lies in identifying salient attention patterns by measuring the sparsity of the query (Q) matrix. This selective attention is subsequently leveraged to guide the decision-making process of the MAD-former model and to support interpretability analysis. Consequently, the IASM also enables the model to concentrate on the most informative image patches while simultaneously preserving a substantially reduced computational overhead. The authors demonstrated that their proposed MAD-former outperforms several existing state-of-the-art models by substantial margins in terms of classification accuracy. Similarly, numerous researchers [72–77] have employed the CNN-ViT-based model in order to classify several grades of AD.

13.3 RESULTS AND ANALYSIS

This section consists of three subsections: (1) Section 13.3.1 outlines the training configurations of all the implemented CNN architectures. (2) Section 13.3.2 provides a concise overview of the evaluation metrics employed in this study. (3) Finally, Section 13.3.3 presents a comparative analysis of the performance of various existing CNN and ViT models applied to an imbalanced OASIS MRI dataset.

This OASIS dataset is publicly accessible online [78]. It exhibits a pronounced class imbalance. It comprises four diagnostic categories, Mild Dementia, Moderate Dementia, Non-Dementia, and Very Mild Dementia, with corresponding image counts of 5,002, 488, 67,200, and 13,725, respectively. Such a substantial imbalance among class distributions poses a significant challenge, potentially undermining the performance of conventional CNN-based models due to their limited capacity to generalize effectively across underrepresented classes.

13.3.1 Training Specifications

The training of the existing CNN and ViT models was conducted in accordance with the following specifications:

1. All the CNN and ViT models were developed using the Keras Sequential API. The experiments were conducted on the Google Colab Pro platform, which provides access to a Tesla P100 GPU.
2. Each dataset was initially divided into training and testing sets using an 80%–20% stratified split to maintain proportional class representation, thereby addressing the challenges posed by class imbalance. Subsequently, the training portion underwent a secondary stratified 80%–20% split to derive validation data.
3. Before being passed to the classifier model, all images were uniformly rescaled to dimensions of 224×224.
4. Adams-optimizer is preferred for running all the existing models, with a learning rate of 0.00095.
5. A batch size of 64 was consistently utilized across all experiments for training the existing CNN and ViT models.
6. All the CNN and ViT models were trained on the "OASIS" dataset for seven epochs, with a learning rate decay of 5% initiated after the fourth epoch to enhance training stability and convergence.
7. For training ViT-based models, we have run the model for at least 15 epochs.
8. For training other existing CNN or hybrid CNN-ViT models (novel framework) proposed for AD detection, exactly the same learning rate is chosen as mentioned in their paper, with seven epochs.
9. No manual feature extraction is done in all the experiments.
10. No pre-processing, like data augmentation, is done in any of the experiments.

13.3.2 Quality Metrics

$$\text{Precision} = \frac{\text{TP}}{\text{TP} + \text{FP}} \tag{13.4}$$

$$\text{Recall} = \frac{\text{TP}}{\text{TP} + \text{FN}} \tag{13.5}$$

$$\text{Accuracy} = \frac{\text{TP} + \text{TN}}{\text{TP} + \text{FN} + \text{TN} + \text{FP}} \tag{13.6}$$

$$\text{F1score} = \frac{2 * \text{Precision} * \text{Recall}}{\text{Precision} + \text{Recall}} \tag{13.7}$$

The standard "accuracy" metric [79] is employed to evaluate the overall correctness of the model's predictions. In addition to accuracy, other evaluation metrics such as precision, recall, and the F1-score are also assessed to ensure that the model's performance is not adversely influenced by any class imbalance issues. The mathematical formulations for these evaluation metrics are presented below. In Equations (13.4)–(13.7), TP denotes True Positives, referring to instances that are actually positive and are correctly identified as such. TN represents True Negatives, that is, instances that are truly negative and are accurately classified as negative. FP corresponds to False Positives, where instances belonging to the negative class are incorrectly predicted as positive. Similarly, FN indicates False Negatives, referring to positive instances that are mistakenly predicted as negative.

Additionally, an extra metric, "secs/epoch", is introduced in Table 13.1 to represent the mean time (in seconds) the model requires per epoch during training. This serves as an indicative measure of the model's time complexity. Due to the unavailability of a dedicated GPU, it was not feasible to compute the standard FLOPs in this study. Furthermore, the total number of trainable parameters has been reported in Table 13.1 to provide readers with an estimate of the model's computational complexity.

13.3.3 Performance Comparison and Analysis

We have implemented several pre-trained CNN architectures, including VGG16, Xception, DenseNet-121, and MobileNet, all of which were fully fine-tuned on the ImageNet dataset. The corresponding results are presented in Table 13.1. In addition to these, three existing CNN-based models were also implemented: (1) 2D-M2IC [6], (2) MSC-Net [22], and (3) AD-Lite Net. These models were originally proposed for AD detection and were trained from scratch. Furthermore, we have evaluated and compared the performance of these CNN models with selected ViT variants, namely, (1) the standard ViT and (2) PiT, a lightweight ViT architecture.

From Table 13.1, it is evident that the MobileNet model demonstrated superior performance compared to all other pre-trained CNN models when evaluated on the imbalanced MRI dataset. Notably, the VGG-16 model yielded significantly lower accuracy. This underperformance can be attributed to VGG-16's limited ability to handle class imbalance, primarily due to its inadequate feature extraction capabilities across both spatial and spectral domains. Furthermore, with approximately 147 million trainable parameters, VGG-16 is highly prone to overfitting.

In contrast, models such as MobileNet-V2, DenseNet-121, and Xception exhibited substantially improved performance over VGG-16. Among them, Xception, a modified architecture derived from Inception-V3, was the first to integrate DWSC layers, as discussed in Section 13.2.2. Despite having a relatively large number of trainable parameters, Xception generalized effectively on both the training and testing sets of this imbalanced MRI dataset, achieving an impressive accuracy of 95.1%, possibly due to its exceptional feature extraction capabilities across both spatial and spectral domains. MobileNet-V2, on the other hand, incorporates both DWSC layers and GAP, resulting in a significantly lightweight architecture. With only 3.2 million trainable parameters, it stands out for its computational efficiency without compromising performance. Hence, the MobileNet-V2 model has attained an accuracy of 97% from this severely imbalanced OASIS dataset. DenseNet-121, another widely recognized

TABLE 13.1
Comparisons of Testing Results for Several Existing Models on an Imbalanced MRI Dataset "OASIS" (The best results are presented in bold)

Models	Accuracy	Precision	Recall	F1-Score	Secs/ Epochs	No of Params (in Lakhs)
VGG-16	0.251	0.235	0.129	0.167	87	147
Xception	0.951	0.952	0.950	0.951	112	208
MobileNet-V2	0.973	0.974	0.973	0.973	30	32
DenseNet-121	0.962	0.964	0.961	0.962	76	71
ViT	0.601	0.643	0.521	0.572	176	858
Pooling-based ViT	0.285	0.410	0.125	0.273	30	46
E2AD2C	0.937	0.926	0.939	0.937	**3**	8.2
MSC-Net+SE-net	0.877	0.623	0.989	0.764	113	114
AD-lite net	**0.996**	**0.996**	**0.996**	**0.996**	12	**2.3**

CNN architecture, also leverages DWSC layers within its dense blocks. As a result, the total parameter count is moderate, around 7.1 million, which facilitates robust generalization and leads to commendable accuracy and F1 score (96.2%) on the imbalanced OASIS dataset. Overall, all the pre-trained CNN models (except VGG-16) performed effectively on this imbalanced dataset, primarily because the training did not commence from randomly initialized weights. Instead, transfer learning from the ImageNet dataset enabled the models to converge toward a global minimum from the very first epoch, thereby facilitating more stable and efficient learning.

From Table 13.1, it can be noticed that both the ViT models, ViT and PiT, did not perform well. This indicates that their attention mechanism could not handle the class imbalance problem in an efficient way. Moreover, this can be observed from Table 13.1 that the ViT pre-trained model has a total of 85.8 million trainable parameters. Due to such a higher number of trainable parameters, the model often goes to overfitting, especially for a small or skewed dataset. On the other hand, PiT is a lighter version of ViT, which has only 46 lakhs parameters. Despite having significantly fewer parameters, pre-trained PiT underperforms on the OASIS dataset, achieving only 28.5% accuracy. This is the evidence that traditional ViT cannot alleviate the severe class-imbalance problem.

Besides that, some of the train-from-scratch models were also implemented, and their results are reported in Table 13.1. Both E2AD2C and AD-Lite Net demonstrated efficient performance on the challenging and imbalanced OASIS dataset, primarily owing to their lightweight architectures. In contrast, MSCNet integrated with SE-Net attention did not exhibit satisfactory results, as evidenced by the substantial disparity between accuracy and precision, suggesting that the channel attention mechanism was ineffective in addressing the class imbalance issue inherent in the dataset. On the other hand, it can be observed from Table 13.1 that AD-Lite Net outperformed all other existing models. The authors attribute this superior performance to the integration of a novel parallel concatenation strategy, which effectively enhances the model's generalization capability. Consequently, the model exhibited robust performance across three heterogeneous MRI datasets characterized by distinct statistical distributions, according to the authors. In our results and analysis, it also achieved superior results, registering the highest accuracy, precision, recall, and F1-score on the imbalanced "OASIS MRI dataset".

Furthermore, the training and validation accuracy versus epoch plots for all existing models are illustrated in Figures 13.8 and 13.9. These figures present two distinct types of visual analyses. Specifically, Figure 13.8 provides a model-wise depiction, offering insights into the individual learning behavior of each model along with their corresponding validation performance. In contrast, Figure 13.9a and b collectively display the training and validation curves of all models in a consolidated manner, facilitating a comprehensive comparison of their respective performances. As observed in the graphical representations of Figure 13.8, the validation curves of MSC-Net exhibit significant fluctuations and diverge notably from its training curve. This inconsistency indicates that MSC-Net failed to attain optimal performance on the "OASIS" dataset. In contrast, Figure 13.9a illustrates that the MobileNet model demonstrates the fastest convergence to peak accuracy among all evaluated models. Moreover, as depicted in Figure 13.9b, it is apparent that only the validation curves of AD-Lite Net and MobileNet exhibit minimal fluctuations in comparison to those of MSC-Net and DenseNet. These observations substantiate the reason for the superior efficacy of AD-Lite Net and MobileNet over the other existing models.

FIGURE 13.8 Accuracy versus epochs for numerous pre-trained models (separately) from left to right in (a) AD-lite net, (b) MSC-Net+SE-net, and (c) mobile net-V2.

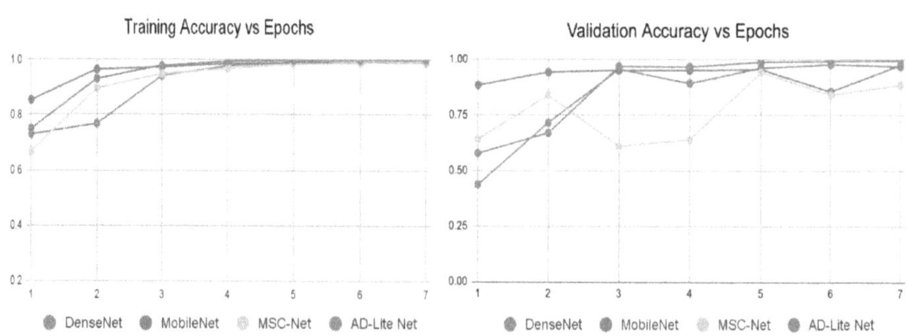

FIGURE 13.9 Training and validation accuracy graph for numerous existing models together.

13.4 CONCLUDING REMARKS

This chapter presents a concise overview of several pre-trained CNN models, as well as various existing CNN and hybrid CNN-ViT frameworks developed for AD detection using CXR images. Initially, each of these models and frameworks was described in detail, including their architectural components. Subsequently, some of those models were implemented and evaluated on the severely class-imbalanced and highly challenging OASIS MRI dataset, which is publicly accessible via the Kaggle platform. The experimental findings demonstrated that pre-trained models exhibited superior performance compared to novel CNN architectures trained from scratch, except AD-Lite Net. Owing to its lightweight architecture with a substantially reduced number of parameters, ADLite Net exhibited minimal overfitting. Furthermore, by enhancing the model's generalization capability through the incorporation of a parallel concatenation layer, it achieved the highest accuracy and F1 score on the challenging OASIS dataset. Notably, both MobileNet and AD-Lite Net consistently outperformed the other CNN-based frameworks on this OASIS dataset, effectively addressing the inherent challenges associated with class imbalance and data complexity.

REFERENCES

1. Waziry, R., and Williams, O. A. Alzheimer disease: a life-limiting disease with a burden of a terminal illness. *Neurology: Clinical Practice* 13(6), e200208 (2023).
2. Alzheimer's Association. 2016 Alzheimer's disease facts and figures. *Alzheimer's and Dementia* 12(4), 459–509 (2016).
3. Reitz, C., Brayne, C., and Mayeux, R. Epidemiology of Alzheimer disease. *Nature Reviews Neurology* 7(3), 137–152 (2011).
4. James, B. D., and Bennett, D. A. Causes and patterns of dementia: an update in the era of redefining Alzheimer's disease. *Annual Review of Public Health* 40(1), 65–84 (2019).
5. Cullen N. C., Leuzy A., Palmqvist S., Janelidze S., Stomrud E., Pesini P., Sarasa L, Allué JA, Proctor N. K., Zetterberg H., Dage JL, Blennow K, Mattsson-Carlgren N., Hansson O. Individualized prognosis of cognitive decline and dementia in mild cognitive impairment based on plasma biomarker combinations. *Nature Aging* 1(1), 114–123 (2021). doi: 10.1038/s43587-020-00003-5. Epub 2020 Nov 30. PMID: 37117993.
6. Self, W. K., and Holtzman, D. M. Emerging diagnostics and therapeutics forAlzheimer disease. *Nature Medicine* 29(9), 2187–2199 (2023).
7. Kim, A. Y., Al Jerdi, S., MacDonald, R., and Triggle, C. R. Alzheimer's disease and its treatment–yesterday, today, and tomorrow. *Frontiers in Pharmacology* 15, 1399121 (2024).
8. Anitha, K., Singh, M. K., Kohat, K., Chenchula, S., Padmavathi, R., Amerneni, L. S., Vardhan, V., Chavan, M. R., and Bhatt, S. Recent insights into the neurobiology of Alzheimer's disease andadvanced treatment strategies. *Molecular Neurobiology* 62(2), 2314–2332 (2025).
9. Barrantes, F. J. Cognitive synaptopathy: synaptic and dendritic spine dysfunction inage-related cognitive disorders. *Frontiers in Aging Neuroscience* 16, 1476909 (2024).
10. Burns, A., Jacoby, R., Philpot, M., and Levy, R. Computerised tomography in Alzheimer's disease: methods of scan analysis, comparison with normal controls, and clinical/radiological associations. *The British Journal of Psychiatry* 159(5), 609–614 (1991).
11. Nordberg, A. PET imaging of amyloid in Alzheimer's disease. *The Lancet Neurology* 3(9), 519–527 (2004).

12. Li, S. J., Li, Z., Wu, G., Zhang, M. J., Franczak, M., and Antuono, P. G. Alzheimer disease: evaluation of a functional MR imaging index as a marker. *Radiology* 225(1), 253–259 (2002).
13. Chen, W., Song, X., and Zhang, Y. Alzheimer's disease neuroimaging initiative. assessment of the Virchow-Robin spaces in alzheimer disease, mild cognitive impairment, and normal aging, using high-field MR imaging. *American Journal of Neuroradiology* 32(8), 1490–1495 (2011).
14. Sweetman, B., and Linninger, A. A. (2011). Cerebrospinal fluid flow dynamics in the central nervous system. *Annals of Biomedical Engineering* 39, 484–496.
15. Bjorkli, C., Sandvig, A., and Sandvig, I. Bridging the gap between fluid biomarkers for Alzheimer's disease, model systems, and patients. *Frontiers in Aging Neuroscience* 12, 272 (2020).
16. Illan, I. A., Górriz, J. M., López, M. M., Ramírez, J., Salas-Gonzalez, D., Segovia, F., Chaves, R. and Puntonet, C. G. Computer aided diagnosis of Alzheimer's disease using componentbased SVM. *Applied Soft Computing* 11(2), 2376–2382 (2011).
17. Xu, Y., Xiao, T., Zhang, J., Yang, K., and Zhang, Z. Scale-invariant convolutional neural networks. arXiv preprintarXiv. 1411.6369 (2014).
18. Bronstein, M. M., Bruna, J., LeCun, Y., Szlam, A., Vandergheynst, P. Geometric deep learning: going beyond euclidean data. *IEEE Signal Processing Magazine* 34(4), 18–42 (2017).
19. Wang, B., Luo, X., Li, Z., Zhu, W., Shi, Z., and Osher, S. Deep neural nets with interpolating function as output activation. *Advances in Neural Information Processing Systems (NeurIPS)*, 31, 751–761 (2018).
20. Agostinelli, F., Hoffman, M., Sadowski, P., and Baldi, P. Learning activation functions to improve deep neural networks. arXiv preprint arXiv:1412.6830 (2014).
21. Ioffe, S., and Szegedy, C. Batch normalization: accelerating deep network trainingby reducing internal covariate shift. In: *International Conference on Machine Learning (ICML)*, Lille, France, pp. 448–456, 2015.
22. Krizhevsky, A., Sutskever, I., and Hinton, G. F. Imagenet classification with deep convolutional neural networks. *Advances in Neural Information Processing Systems (NeurIPS)*, 25, 1097–1105 (2012).
23. Simonyan, K., and Zisserman, A. Very deep convolutional networks forlarge-scale image recognition. arXiv preprint arXiv:1409.1556 (2014).
24. Szegedy, C., Liu, W., Jia, Y., Sermanet, P., Reed, S., Anguelov, D., Erhan, D., Vanhoucke, V., and Rabinovich, A. Going deeper with convolutions. In: *Proceedings of the IEEE Conference on Computer Vision and Pattern Recognition (CVPR)*, Boston, MA, USA, 2015.
25. Chollet, F. Xception: deep learning with depthwise separable convolutions. In: *Proceedings of the IEEE Conference on Computer Vision and Pattern Recognition (CVPR)*, Honolulu, HI, 2017.
26. Sandler, M., Howard, A., Zhu, M., Zhmoginov, A., and Chen, L. C. Mobilenetv2: inverted residuals and linear bottlenecks. In: *Proceedings of the IEEE Conference on Computer Vision and Pattern Recognition (CVPR)*, Salt Lake City, Utah, 2018.
27. He, K., Zhang, X., Ren, S., and Sun, J. Deep residual learning for image recognition. In: *Proceedings of the IEEE Conference on Computer Vision and Pattern Recognition (CVPR)*, Las Vegas, Nevada, USA, 2016.
28. Deng, J., Dong, W., Socher, R., Li, L.J., Li, K., and Fei-Fei, L. Imagenet: a large-scale hierarchical image database. In: *2009 IEEE Conference on Computer Vision and Pattern Recognition (CVPR)*, Miami, Florida, USA, 2009.
29. Lin, M., Chen, Q., and Yan, S. Network in network. arXiv preprint arXiv:1312.4400 (2013).

30. Li, Y., Zhang, X., and Chen, D. Csrnet: dilated convolutional neural networks forunderstanding the highly congested scenes. In: *Proceedings of the IEEE Conference on Computer Vision and Pattern Recognition (CVPR)*, Salt Lake City, Utah, USA, pp. 1091–1100, 2018.
31. Zhang, T., Qi, G. J., Xiao, B., and Wang, J. Interleaved group convolutions. In: *Proceedings of the IEEE International Conference on Computer Vision (ICCV)*, Venice, Italy, pp. 4373–4382, 2017.
32. Zhang, Q. L., and Yang, Y. B. Sa-net: shuffle attention for deep convolutional neural networks. In: *ICASSP* 2021-2021 *IEEE International Conference on Acoustics, Speech and Signal Processing (ICASSP) IEEE*, Virtual, pp. 2235–2239, 2021.
33. Woo, S., Debnath, S., Hu, R., Chen, X., Liu, Z., Kweon, I. S., and Xie, S. Convnextv2: co-designing and scaling convnets with masked autoencoders. In: *Proceedings of the IEEE/CVF Conference on Computer Vision and Pattern Recognition (CVPR)*, Vancouver, Canada, pp. 16133–16142, 2023.
34. Shin, H. C., Roth, H. R., Gao, M., Lu, L., Xu, Z., Nogues, I., Yao, J., Mollura, D., and Summers, R. M. Deep convolutional neural networks for computer-aided detection: CNN architectures, dataset characteristics and transfer learning. *IEEE Transactions on Medical Imaging* 35(5), 1285–1298 (2016).
35. Li, H., Chen, J., Lu, H., and Chi, Z. CNN for saliency detection with low-level feature integration. *Neurocomputing* 226, 212–220 (2017).
36. Xie, Z., Lin, Y., Yao, Z., Zhang, Z., Dai, Q., Cao, Y., Hu, H. Self-supervised learningwith swin transformers. arXiv preprint arXiv:2105.04553 (2021).
37. Vaswani A., Shazeer N., Parmar N., Uszkoreit J., Jones L., Gomez A. N., Kaiser Ł., Polosukhin I. Attention is all you need. *Advances in Neural Information Processing Systems (NeurIPS)* 30, 5998–6008 (2017).
38. Dosovitskiy, A. An image is worth 16x16 words: transformers for image recognition at scale. arXiv preprint arXiv:2010.11929 (2020).
39. Yin, H., Vahdat, A., Alvarez, J. M., Mallya, A., Kautz, J., and Molchanov, P. A-vit: adaptive tokens for efficient vision transformer. In: *Proceedings of the IEEE/CVF Conference on Computer Vision and Pattern Recognition (CVPR)*, New Orleans, LA, USA, pp. 10809–10818, 2022.
40. Yuan, L. et al. Tokens-to-token vit: training vision transformers from scratch on imagenet. In: *Proceedings of the IEEE/CVF International Conference on Computer Vision (CVPR)*, Virtual, pp. 558–567, 2021.
41. Shah, S. A., Taj, I., Usman, S. M., Hassan Shah, S. N., Imran, A. S., and Khalid, S. A hybrid approach of vision transformers and CNNs for detection of ulcerative colitis. *Scientific Reports* 14(1), 24771 (2024).
42. Liu, Z., Mao, H., Wu, C. Y., Feichtenhofer, C., Darrell, T., and Xie, S. A convnet for the 2020s. In: *Proceedings of the IEEE/CVF Conference on Computer Vision and Pattern Recognition (CVPR)*, Virtual, pp. 11976–11986, 2022.
43. Liu Z., Lin Y., Cao Y., Hu H., Wei Y., Zhang Z., Lin S., Guo B. Swin transformer: hierarchical vision transformer using shifted windows. In: *Proceedings of the IEEE/CVF International Conference on Computer Vision (ICCV),* Virtual, pp. 10012–10022, 2021.
44. Helaly, H. A., Badawy, M., and Haikal, A. Y. Deep learning approach for early detection of Alzheimer's disease. *Cognitive Computation* 14(5), 1711–1727 (2022).
45. Achilleos, K. G., Leandrou, S., Prentzas, N., Kyriacou, P. A., Kakas, A. C., and Pattichis, C. S. Extracting explainable assessments of Alzheimer's disease via machine learning on brain MRI imaging data. In: *2020 IEEE 20th International Conference on Bioinformatics and Bioengineering (BIBE)*, IEEE, Virtual, pp. 1036–1041, 2020.
46. Haralick, R. M., Shanmugam, K., and Dinstein, I. Texturalfeatures for image classification. *IEEE Transactions on Systems, Man, and Cybernetics* 6, 610–621 (1973).

47. Roy S., Tyagi M., Bansal V., Jain V. SVD-CLAHE boosting and balanced loss function for COVID19 detection from an imbalanced Chest X-Ray dataset. *Computers in Biology and Medicine* 150, 106092 (2022).
48. Masud, M., Almars, A. M., Rokaya, M. B., Meshref, H., Gad, I., and Atlam, E. S. A novel light-weight convolutional neural network model to predict Alzheimer's disease applying weighted loss function. *Journal of Disability Research* 3(4), 20240042, (2024).
49. Tyagi M., Roy S., Bansal V. Custom weighted balanced loss function for Covid 19 detection from an imbalanced CXR dataset. In: *26th International Conference on Pattern Recognition (ICPR),* IEEE, Virtual, pp. 2707–2713, 2022.
50. Roy, S., and Khurana, R. Mobile freeze-net with attention-based loss function for Covid-19 detection from an imbalanced CXR dataset. In: *Proceedings of the 38th ACM/SIGAPP Symposium on Applied Computing,* Tallinn, Estonia, pp. 611–613, 2023.
51. Hu, J., Shen, L., and Sun, G. Squeeze-and-excitation networks. In: *Proceedings of the IEEE Conference on Computer Vision and Pattern Recognition (CVPR),* Salt Lake City, Utah, USA, pp. 7132–7141, 2018.
52. Woo, S., Park, J., Lee, J. Y., and Kweon, I. S. CBAM: convolutional block attention-module. In: *Proceedings of the European Conference on Computer Vision (ECCV),* Munich, Germany, pp. 3–19, 2018.
53. Wang, Q., Wu, B., Zhu, P., Li, P., Zuo, W., and Hu, Q. ECA-net: efficient channel attention for deep convolutional neural networks. In: *Proceedings of the IEEE/CVF Conference on Computer Vision and Pattern Recognition (CVPR),* Seattle, WA, USA, pp. 11534–11542, 2020.
54. Qin, Z., Zhang, P., Wu, F., and Li, X. Fcanet: frequency channel attention networks. In: *Proceedings of the IEEE/CVF International Conference on Computer Vision (ICCV),* Montreal, Canada, pp. 83–792, 2021.
55. Liu, Z., Lu, H., Pan, X., Xu, M., Lan, R., and Luo, X. Diagnosis of Alzheimer's disease via an attention-based multi-scale convolutional neural network. *KnowledgeBased Systems* 238, 107942 (2022).
56. Zhu W., Sun L., Huang J., Han L., Zhang D. Dual attention multi-instance deep learning for Alzheimer's disease diagnosis with structural MRI. *IEEE Transactions on Medical Imaging* 40(9), 2354–2366 (2021).
57. Zhang Q., Long Y., Cai H., Chen Y. W. Lightweight neural network for Alzheimer's disease classification using multi-slice sMRI. *Magnetic Resonance Imaging* 107, 164–170 (2024).
58. Zhu J., Tan Y., Lin R., Miao J., Fan X., Zhu Y., Liang P., Gong J., He H. Efficient self-attention mechanism and structural distilling model for Alzheimer's disease diagnosis. *Computers in Biology and Medicine* 147, 105737 (2022).
59. Al-Otaibi, S., Mujahid, M., Khan, A. R., Nobanee, H., Alyami, J., and Saba, T. Dual attention convolutional autoencoder for diagnosis of alzheimer's disorder in patients using neuroimaging and MRI features. *IEEE Access* 12, 58722–58739 (2024).
60. Liu, J., Xu, Y., Liu, Y., Luo, H., Huang, W., and Yao, L. Attention-guided 3DCNN with lesion feature selection for early Alzheimer's disease prediction using longitudinal sMRI. *IEEE Journal of Biomedical and Health Informatics* (2025). https://doi.org/10.1109/JBHI.2025.0000000.
61. Roy, S., Gupta, A., Tiwari, S., and Sahu, P. AD-Lite net: a lightweight and concatenated CNN model for Alzheimer's detection from MRI images. In: *International Conference on Pattern Recognition (ICPR),* Springer, Cham, pp. 1–16, 2025.
62. Roy, S., Suresh, A., Sahu, P., A. Roy and Gupta, T. R. Novel pooling-based VGG-lite for pneumonia and Covid-19 detection from imbalanced chest X-ray datasets, *IEEE Transactions on Emerging Topics in Computational Intelligence*, 2025, doi: 10.1109/TETCI.2025.3577509.

63. Heo, B., Yun, S., Han, D., Chun, S., Choe, J., Oh, S.J. Rethinking spatial dimensions of vision transformers. In: *Proceedings of the IEEE/CVF International Conference on Computer Vision (ICCV),* Montreal, Canada,. pp. 11936–11945, 2021.
64. Zhu J., Tan Y., Lin R., Miao J., Fan X., Zhu Y., Liang P., Gong J., He H. Efficient self-attention mechanism and structural distilling model for Alzheimer's disease diagnosis. *Computers in Biology and Medicine* 147, 105737 (2022).
65. Lu, S. Y., Zhang, Y. D., and Yao, Y. D. An efficient vision transformer for Alzheimer's disease classification using magnetic resonance images. *Biomedical Signal Processing and Control* 101, 107263 (2025).
66. Zhu J., Tan Y., Lin R., Miao J., Fan X., Zhu Y., Liang P., Gong J., He H. Efficient self-attention mechanism and structural distilling model for Alzheimer's disease diagnosis. *Computers in Biology and Medicine* 147, 105737 (2022).
67. Kadri, R., Bouaziz, B., Tmar, M., and Gargouri, F. Efficient multimodel method based on transformers and CoAtNet for Alzheimer's diagnosis. *Digital Signal Processing* 143, 104229 (2023).
68. Yu Q., Ma Q., Da L., Li J., Wang M., Xu A., Li Z., Li W. Alzheimer's disease neuro-imaging initiative: a transformer-based unified multimodal framework for Alzheimer's disease assessment. *Computers in Biology and Medicine* 180, 108979 (2024).
69. Ye, J., Pan, D., Zeng, A., Zhang, Y., Chen, Q., and Liu, Y. MssNet: an efficient spatial attention model for early recognition of Alzheimer's disease. *IEEE Transactions on Emerging Topics in Computational Intelligence* 9, 1454–1468 (2025).
70. Roy, S., Gupta, A., Tiwari, S., Sharma, H., Kapoor, M., and Singh, S. A novel spatial attention module (SAM) for Alzheimer's detection from MRI images. In: *Proceedings of the Fifteenth Indian Conference on Computer Vision Graphics and Image Processing ICVGIP*, India, pp. 1–10, 2024.
71. Ye, J., Zeng, A., Pan, D., Zhang, Y., Zhao, J., Chen, Q., and Liu, Y. MAD-former: a traceable interpretability model for Alzheimer's disease recognition based on multi-patch attention. *IEEE Journal of Biomedical and Health Informatics* 28(7), 1320–1332 (2024).
72. Almalki, H., Khadidos, A. O., Alhebaishi, N., and Senan, E. M. Early detection of Alzheimer's disease progression stages using hybrid of CNN and transformer encoder models. *Scientific Reports* 15(1), 1–22 (2025).
73. Bai, J., Zhang, Z., Yin, Y., Jin, W., Ali, T. A. A., Xiong, Y., and Xiao, Z. LGGNeXt: a next generation CNN and transformer hybrid model for the diagnosis of Alzheimer's disease using 2D structural MRI. *IEEE Journal of Biomedical and Health Informatics* 29, 2808–2818 (2024).
74. Majee, A., Gupta, A., Raha, S., and Das, S. Enhancing MRI-based classification of Alzheimer's disease with explainable 3D hybrid compact convolutional transformers. In: *2024 International Joint Conference on Neural Networks (IJCNN)*, IEEE, Beijing, China, pp. 1–8, 2024.
75. Hu, Z., Wang, Z., Jin, Y., and Hou, W. VGG-TS winformer: transformer-based deep learning model for early Alzheimer's disease prediction. *Computer Methods and Programs in Biomedicine* 229, 107291 (2023).
76. Mohan, R., Arunachalam, R., Verma, N., and Bernatin, T. ViT-FusCNN: vision transformer-based weighted fusion of convolution networks for Alzheimer's disease detection from MRI images with enhanced mother optimization. *Signal, Image and Video Processing* 19(7), 1–15 (2025).
77. Xin, J., Wang, A., Guo, R., Liu, W., and Tang, X. CNN and swin-transformer based efficient model for Alzheimer's disease diagnosis with sMRI. *Biomedical Signal Processing and Control* 86, 105189 (2023).
78. https://www.kaggle.com/datasets/ninadaithal/imagesoasis.
79. Huang, J., and Ling, C. X. Using AUC and accuracy in evaluating learning algorithms. *IEEE Transactions on Knowledge and Data Engineering* 17(3), 299–310 (2005).

Index

For Product Safety Concerns and Information please contact our EU representative GPSR@taylorandfrancis.com
Taylor & Francis Verlag GmbH, Kaufingerstraße 24, 80331 München, Germany

www.ingramcontent.com/pod-product-compliance
Ingram Content Group UK Ltd.
Pitfield, Milton Keynes, MK11 3LW, UK
UKHW020159200926
475702UK00004B/530
* 9 7 8 1 0 3 2 9 3 7 6 8 7 *